C O M A P ' S

Mathematics: Modeling Our World

DEVELOPED BY

COMAP, Inc.

57 Bedford Street, Suite 210
Lexington, Massachusetts 02420

PROJECT LEADERSHIP

Solomon Garfunkel
COMAP, INC., LEXINGTON, MA

Landy Godbold
THE WESTMINSTER SCHOOLS, ATLANTA, GA

Henry Pollak
TEACHERS COLLEGE, COLUMBIA UNIVERSITY, NY, NY

W. H. Freeman and Company
New York

www.whfreeman.com

The Consortium for Mathematics and Its Applications (COMAP)

57 Bedford Street, Suite 210
Lexington, MA 02420

Published and distributed by

W. H. Freeman and Company

41 Madison Avenue, New York, NY 10010

This book was prepared with the support of NSF Grant ESI-9255252. However, any opinions,
findings, conclusions, and/or recommendations herein are those of the authors
and do not necessarily reflect the views of the NSF.

ISBN 0-7167-4153-9

Printed in the United States of America.

4 5 6 7 VH 02 03 01

EDITOR: Landy Godbold
AUTHORS: Allan Bellman, WATKINS MILL HIGH SCHOOL, GAITHERSBURG, MD; John Burnette, KINKAID SCHOOL, HOUSTON, TX; Horace Butler, GREENVILLE HIGH SCHOOL, GREENVILLE, SC; Claudia Carter, MISSISSIPPI SCHOOL FOR MATH AND SCIENCE, COLUMBUS, MS; Nancy Crisler, PATTONVILLE SCHOOL DISTRICT, ST. ANN, MO; Marsha Davis, EASTERN CONNECTICUT STATE UNIVERSITY, WILLIMANTIC, CT; Gary Froelich, COMAP, INC., LEXINGTON, MA; Landy Godbold, THE WESTMINSTER SCHOOLS, ATLANTA, GA; Bruce Grip, ETIWANDA HIGH SCHOOL, ETIWANDA, CA; Rick Jennings, EISENHOWER HIGH SCHOOL, YAKIMA, WA; Paul Kehle, INDIANA UNIVERSITY, BLOOMINGTON, IN; Darien Lauten, OYSTER RIVER HIGH SCHOOL, DURHAM, NH; Sheila McGrail, CHARLOTTE COUNTRY DAY SCHOOL, CHARLOTTE, NC; Geraldine Oliveto, THOMAS JEFFERSON HIGH SCHOOL FOR SCIENCE AND TECHNOLOGY, ALEXANDRIA, VA; Henry Pollak, TEACHERS COLLEGE, COLUMBIA UNIVERSITY, NY, NY, J.J. Price, PURDUE UNIVERSITY, WEST LAFAYETTE, IN; Joan Reinthaler, SIDWELL FRIENDS SCHOOL, WASHINGTON, D.C.; James Swift, ALBERNI SCHOOL DISTRICT, BRITISH COLUMBIA, CANADA; Brandon Thacker, BOUNTIFUL HIGH SCHOOL, BOUNTIFUL, UT; Paul Thomas, MINDQ, FORMERLY OF THOMAS JEFFERSON HIGH SCHOOL FOR SCIENCE AND TECHNOLOGY, ALEXANDRIA, VA

Dear Student,

Mathematics: Modeling Our World is a different kind of math book than you may have used, for a different kind of math course than you may have taken. In addition to presenting mathematics for you to learn, we have tried to present mathematics for you to use. We have attempted in this text to demonstrate mathematical concepts in the context of how they are actually used day to day. The word "modeling" is the key. Real problems do not come at the end of chapters in a math book. Real problems don't look like math problems. Real problems ask questions such as: How do we create computer animations? Where should we locate a fire station? How do we effectively control an animal population? Real problems are messy.

Mathematical modeling is the process of looking at a problem, finding a mathematical core, working within that core, and coming back to see what mathematics tells you about the problem with which you started. You will not know in advance what mathematics to apply. The mathematics you settle on may be a mix of several ideas in geometry, algebra, and data analysis. You may need to use computers or graphing calculators. Because we bring to bear many different mathematical ideas as well as technologies, we call our approach "integrated."

Another very important and very real feature of this course is that frequently you will be working in groups. Many problems will be solved more efficiently by people working in teams. In today's world, this is very much what work looks like. You will also see that the units in this book are arranged by context and application rather than by math topic. We have done this to reemphasize our primary goal: presenting you with mathematical ideas the way you will see them as you go on in school and out into the work force. There is hardly a career that you can think of in which mathematics will not play an important part and understanding mathematics will not matter to you.

Most of all, we hope you have fun. Mathematics is important. Mathematics may be the most useful subject you will learn. Using mathematics to solve truly interesting problems about how our world works can and should be an enjoyable and rewarding experience.

Solomon Garfunkel
CO-PRINCIPAL INVESTIGATOR

Landy Godbold
CO-PRINCIPAL INVESTIGATOR

Henry Pollak
CO-PRINCIPAL INVESTIGATOR

Mathematics: Modeling Our World

UNIT 3
Landsat 150

UNIT 4
Prediction 278

UNIT 1

Pick a Winner: Decision Making in a Democracy

Mathematics plays an important role in many aspects of your life. That's why you have studied it since you were a child. In this unit, the percentages, decimals, and calculations with which you are already familiar will be used in ways that are probably new to you. You will see new mathematical representations that help you better understand problems and find better solutions. You will learn more about the way mathematics is used in the world.

Although mathematics is used to solve problems, it does not always give perfect solutions to real-world problems. When solutions to problems are not perfect, mathematics can be used to improve them. In this unit, you will do just that—you will use mathematics to understand why solutions are not perfect, and you will use mathematics to improve on those solutions. You will do so by examining a problem that is very important to our society.

DECISION MAKING IN A DEMOCRACY

You live in a democracy. As a United States citizen, when you reach the age of 18 you will have the right to vote. In your lifetime you will vote for candidates for school boards, governor, senator, president, and many others. All of the other citizens of this country have the same right. However, they do not all feel the same way about the candidates and issues on which they vote. In order for the country, states, and cities to function, the votes of all citizens must be combined to produce a single decision on who will lead the nation, your state, and your city. Therefore, the central question of this unit is: How does a group make a democratic decision?

David Barber

Voting has been chosen as the topic of the first unit in this textbook because it is an experience shared by everyone. Although this unit is about voting, it is also about mathematics and the way mathematics is used to solve real-world problems. Mathematics is used to determine the winner whenever people vote. The mathematics is often simple, but it does an important job by selecting the people who run our governments, the products that are in our stores, and the programs that are on our televisions.

LESSON ONE

Democratic Elections in the United States

KEY CONCEPTS

Elections

Election flaw

Percentages

Preference diagram

Majority

Plurality

David Barber

PREPARATION READING

Is the System Flawed?

*E*lections are used to choose public officials from the president to members of school boards. In the case of school boards, several members are chosen. There is only one president. The simplest elections are those that choose a single winner. When you use mathematics to solve real-world problems like deciding the winner of an election, it's easiest to start simple. So, the first kind of election you will consider in this unit is the election with only one winner.

Most single-winner elections in the United States are conducted by a simple procedure. The voters mark the candidate of their choice on a ballot. Sometimes they actually use a pencil, but more commonly they punch a hole in a card or pull a lever on a voting machine. The ballots are counted, and the winner is the candidate who receives the most votes. It seems like common sense. Could there be anything wrong with so simple a system?

Library of Congress/Corbis

A 1940 voting machine.

In order to use mathematics to improve the election process, you must first understand why many people feel that the process is flawed. The purpose of this lesson is to help you understand the election flaw that plagues the simple system used to select many office holders in the United States.

ACTIVITY

ELECTIONS IN THE UNITED STATES

1

In this activity, you will consider the way public officials are elected in the United States. You will also see that representing election data in different ways can help you better understand elections.

1. Because the number of voters in an election can be fairly large, percentages are often used to represent the vote totals. Here, for example, are the vote totals for the 1992 presidential election.

   Bill Clinton   43,682,624

   George Bush   38,117,331

   Ross Perot   19,217,213

   Represent each candidate's vote total as a percentage of the total votes cast for these three candidates. It is customary to round these percentages to the nearest tenth.

2. One reason mathematicians use different representations is that some representations show certain things more clearly than other representations do. Does the percentage representation make it easier to see anything more clearly about the 1992 presidential election? In other words, why do you think many people prefer percentages to vote totals?

3. Steven Brams is a political scientist and mathematician at New York University. He is considered one of the world's leading authorities on the mathematical aspects of elections. In his writings he describes an extremely close race for one of New York's seats in the United States Senate. Here are the results of that 1970 election.

   Charles Goodell   1,434,472

   Richard Ottinger   2,171,232

   James Buckley   2,288,190

   Represent each candidate's vote total as a percentage.

4. Based on polls and other knowledge of the election, Brams believes that those who voted for Goodell preferred Ottinger over Buckley, that those who voted for Ottinger preferred Goodell over Buckley, and that those who voted for Buckley were about evenly split in their preference for Goodell or Ottinger. (These are simplifications since, for example, a few people who voted for Goodell preferred Buckley over Ottinger.)

Sound confusing? Mathematicians think so, too, and sometimes use **preference diagrams** to represent how voters ranked the candidates. The highest ranking candidate is written at the top of the diagram, and the lowest ranking candidate is written at the bottom. Either the total number of voters or the percentage of voters is written below the diagram. **Figure 1.1** shows preference diagrams for Brams' study of the 1970 New York Senate race.

On a copy of Transparency T1.1, write the correct percentage under each preference diagram. (Reread the information at the beginning of this item and apply it to your answer to Item 3 on page 5.)

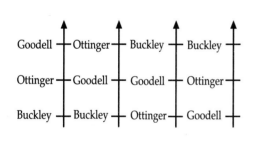

Figure 1.1.
Preference diagrams for the 1970 New York Senate race.

5. Brams found the percentage of voters who ranked each candidate first, second, and third. He noticed something about the winner that many people consider a flaw in the way elections are conducted.

a) Describe what you think the flaw is.

b) When the **plurality** method is used, this flaw can happen without anyone knowing. Why?

1. Why are alternate ways of representing data important?

2. Why is the election method used in many elections considered flawed by many people?

INDIVIDUAL WORK 1

The Plurality Method

*I*n the following items you will consider several sets of data from real elections. You will represent the data in different ways in order to determine whether the elections produce the flaw you saw in the *Elections in the United States* activity.

1. In the United States, we decide many of our elections by a method called plurality. "Plurality" means that the winner is the candidate with the most votes. But getting the most votes does not always mean getting a **majority**. To get a majority of the votes, the winner must get over half of the total votes cast. The 1992 presidential election is an example of an election in which the winner did not get a majority. Here again are the results of that election.

Bill Clinton 43,682,624   **George Bush** 38,117,331   **Ross Perot** 19,217,213

Based on polls conducted during and before the election, political scientists like Steven Brams believe that Perot's supporters were about evenly split in their preferences for Clinton and Bush. Suppose that about 2/3 of Clinton voters preferred Perot to Bush and about 2/3 of Bush's supporters preferred Perot to Clinton. Draw preference diagrams to show this information. Use percentages.

2. Based on your answer to Item 1, does the 1992 presidential election exhibit the same flaw you identified in the 1970 New York Senate election? Explain your answer.

3. Go back and reconsider your results for the 1970 New York Senate election and the 1992 presidential election. Suppose that the candidate who received the fewest votes had not been in the election. Tell who you think would have won in each of the two cases. (In this and all future questions about presidential elections, you should use the vote totals to determine the winner. The president of the United Sates is actually chosen by the electoral college, and it's possible, although unlikely, to win the popular vote and lose in the electoral college.)

4. Another closely contested presidential election was the 1968 race between Richard Nixon, Hubert Humphrey, and George Wallace. Here are the results of that election.

Richard Nixon 31,785,480 **Hubert Humphrey** 31,275,166 **George Wallace** 9,906,473

Represent the vote totals of each candidate as a percentage.

5. Suppose that 2/3 of Wallace supporters preferred Humphrey to Nixon, about 2/3 of Nixon's supporters preferred Humphrey to Wallace, and about 2/3 of Humphrey's supporters preferred Nixon to Wallace. Draw preference diagrams to show this information. Use percentages.

6. Based on your answer to Item 5, does the 1968 presidential election exhibit the same flaw you identified in the 1970 New York Senate election? Explain your answer.

7. Suppose George Wallace had not been in the election. Who do you think would have won?

8. In terms of election results, one of the most interesting presidents is Grover Cleveland. He won the presidency in 1884 by 50.1% to 49.9% over James Blaine. Four years later, he received 50.4% of the vote to 49.6% for Benjamin Harrison, but lost the election by 168–223 in the electoral college. (Historians consider the 1888 presidential election one of the most corrupt in U. S. history.) In 1892, Grover Cleveland gave it a third try, but this time there was a strong third candidate in the race. Here are the results of the 1892 presidential election.

Grover Cleveland is the only president in U. S. history elected to terms that were not consecutive. He is one of only two people to have won the popular vote, but lost in the electoral college.

Grover Cleveland	5,555,426
Benjamin Harrison	5,182,690
James B. Weaver	1,029,846

Suppose 2/3 of the Weaver voters preferred Harrison to Cleveland, and that about a fourth of the supporters of Harrison and Cleveland preferred Weaver to the other candidate. Draw preference diagrams to show this information. Use percentages.

9. Based on your answer to Item 8, does the 1892 presidential election exhibit the same flaw you identified in the 1970 New York Senate election? Explain your answer.

10. In this lesson you have used mathematics to help you understand a flaw in the way we conduct our elections. Write a summary of what you have learned.

11. Mathematics can expose flaws in the way problems are solved. You have used mathematical representations to understand a flaw in the way elections are decided. Mathematics can also find better ways to solve problems. Discuss the topic of elections with several other people. Write down any suggestions they have for conducting elections differently. Perhaps one of these suggestions will lead to a solution that does not suffer from the flaw you have seen.

LESSON TWO

Improving the Election Process

KEY CONCEPTS

Runoff systems

Runoff diagram

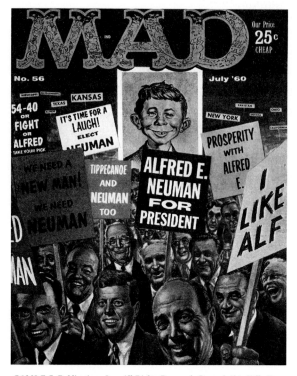

PREPARATION READING

Can the System Be Improved?

*I*f the majority of voters feel that a candidate is the poorest choice, should that candidate win? Most people say no. However, the plurality method, which is used for many elections in the United States, sometimes produces a winner that a majority of voters do not like.

What can be done to improve this situation? Is there a better way to conduct an election in a democratic society?

The plurality method involves very simple mathematics. The votes are counted, and the winner is the candidate with the most votes (or the highest percentage if the totals are converted to percentages). The mathematics you already know can be used to design different voting methods. Would a different method suffer from the same flaw?

In this lesson, you will devise other voting procedures and examine them for flaws.

FINDING A BETTER WAY

An election process has two parts:
(1) the way people vote, and
(2) the way the winner is determined.
The plurality method does those two things in this way:

(1) You vote for only one candidate by making a mark (or punching a hole) on the ballot. (In some cases, such as school board elections, you can vote for two candidates if there are two school board positions open at the same time.)

(2) The votes for each candidate are totaled, and the one with the most votes is the winner. Vote totals are often expressed as percentages.

To change the way an election is conducted, you must consider both parts of the process and describe a new way to do at least one of the two.

In this activity, your task is to describe a different way to conduct an election. To do so, you must describe the way people vote and the way the votes are used to determine the winner.

1. Discuss voting procedures with others in your group. In particular, try to do the following two things.

a) Take a close look at the plurality method's flaw. What feature of the method creates this flaw?

b) Suggest a way to avoid the plurality method's flaw. Select a new voting process and describe it in writing. Remember, you must describe how people vote and how the winner is determined.

ACTIVITY

FINDING A BETTER WAY

2

2. The preferences of voters in the 1970 New York Senate race are shown again in **Figure 1.2**. Apply your new voting method to these preferences. Does your method produce the flaw that the plurality method produces in this case?

Figure 1.2.
Preference diagrams for the 1970 New York Senate race.

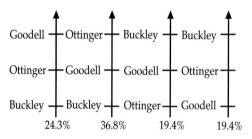

3. Just because a method does not produce a bad result in one case doesn't mean it won't produce it in another. Do you think it's possible that your method might ever produce a winner that is ranked lowest by a majority of voters? Explain.

4. If you feel you have a method that never produces the plurality method's flaw, congratulations! However, it's possible that your method has other problems. Think about difficulties your method might encounter. Write them down. Finally, tell whether you think these problems are more serious or less serious than the plurality method's flaw.

ACTIVITY

TRYING THEM OUT

3

You have examined the plurality method and seen that it has a serious flaw. You have had a chance to propose at least one method of your own.
Whenever you use mathematics to solve real-world problems, you don't get a good feeling for how well your solution works until you give it a try. So, the purpose of this activity is to try your method—to take it for a "test drive," so to speak.

Select a topic of current interest in your class, your school, your community or state, or the nation. It should be one in which there are more than two choices. Have all the members of your class vote in at least two different ways. One way should be plurality and the other should be the method you devised in Activity 2. If your class devised more than one method in Activity 2, try them all.

1. When your class has voted, use the data to find the plurality winner. Also find the winners by the methods you devised in Activity 2.

2. If your class data show the candidate that voters ranked lowest, check the data to see if the plurality method's flaw occurred in your class vote.

3. Based on your test of the method you devised in Activity 2, what problems do you think your method might encounter if it were used in state or national elections?

CONSIDER:

1. How can an election method avoid choosing a winner that is disliked by a majority of voters?

2. Why do you think a flawed method is used at all?

Runoff Systems

Runoff systems are commonly used to decide elections in the United States and other countries. A typical **runoff system** requires a second election if no candidate gets a majority. In the second election, the candidate who received the most votes runs against the candidate who finished second. Runoff systems are sometimes proposed as a way of avoiding the plurality method's flaw.

[REUTERS JULY 9, 1996]

Incumbent Boris Yeltsin won the July 3 runoff round of the 1996 Russian presidential election, beating Communist-nationalist challenger Gennady Zyuganov by approximately 13 percentage points. The runoff was held because no candidate received more than 50% of the vote in the first-round election held June 16. Yeltsin and Zyuganov were the top vote-getters in that round, receiving 35% and 32%, respectively. The third-place finisher, Aleksandr Lebed with 15% of the vote, joined Yeltsin's government as head of Russian security affairs on June 18.

Note:
The percentages in the Reuters report do not total 100 because there were several other candidates receiving votes.

Final Results

Boris Yeltsin	53.82%	40,208,384
Gennady Zyuganov	40.32%	30,113,306
Against Both	4.83%	3,604,550
Estimated Turnout	68.89%	74,815,898

INDIVIDUAL WORK 2

A Popular Election Method

Throughout history, people have proposed alternate ways of conducting elections, just as you did in this lesson. In the items that follow, you will consider an alternate method that may already be familiar to you.

The principle behind the runoff method is simple. If no candidate gets over 50% of the votes, then hold another election, but give the voters only two choices. Only the top two vote-getters are on the ballot in the runoff election.

Voter preferences from the 1970 New York Senate election are shown again in **Figure 1.3**.

1. Which candidate would have been eliminated if the runoff method had been used? Why?

2. In a runoff election, for whom do you think the eliminated candidate's supporters would vote?

3. The preference diagrams you constructed in Lesson 1 for the 1992 presidential election are shown in **Figure 1.4**. Which two candidates would be in a runoff election, and which of the two would win?

4. If a runoff had been held in the 1992 presidential election, could anything have happened between the first election and the second that might have changed the winner?

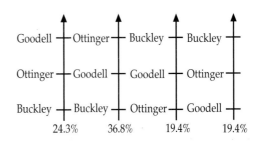

Figure 1.3.
Preference diagrams for the 1970 New York Senate race.

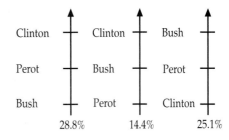

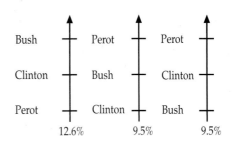

Figure 1.4.
Preference diagrams for the 1992 presidential election.

5. The preference diagrams you constructed in Lesson 1 for the 1968 presidential election are shown in **Figure 1.5**. Which two candidates would be in a runoff election, and which of the two would win?

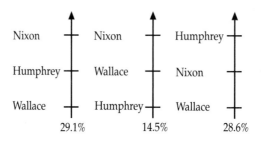

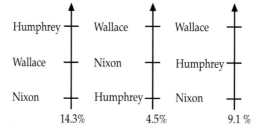

Figure 1.5.
Preference diagrams for the 1968 presidential election.

6. The preference diagrams you constructed in Lesson 1 for the 1892 presidential election are shown in **Figure 1.6**. Which two candidates would be in a runoff election, and which of the two would win?

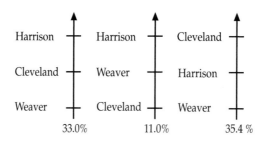

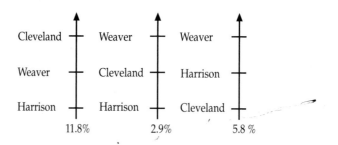

Figure 1.6.
Preference diagrams for the 1892 presidential election.

7. The 1970 New York Senate race is an example of an election in which the majority of voters felt that the plurality winner was the poorest choice. You have seen that a runoff election would not have produced that same flaw in that election. Can the runoff method ever produce that flaw in a three-candidate election?
Explain your answer.

8. Voters do not always vote on candidates for office, sometimes they vote on issues. For example, in 1994, California voters voted on a controversial measure that denied public services like schooling to undocumented immigrants. Although ballot measures like this one are often politically controversial, the election is not mathematically controversial. A runoff is never needed. Why can't a vote on an issue require a runoff?

9. Federal elections, which choose the president and the members of Congress, do not use runoffs. Nor do the voters rank the candidates. Thus most of the preferences you have seen for federal elections were based on polls if polling information was available. Some state and local elections do use runoffs, although the voters do not rank the candidates. Instead, the voters go to the polls a second time if the winner does not get over half the votes the first time. A recent example is the November 7, 1995 election for mayor of San Francisco. Here are the results of that election.

Willie Brown	34.9%
Frank Jordan	33.4%
Roberta Achtenberg	27.0%
Other candidates	4.6%

Since no one received a majority, a runoff was held between Brown and Jordan on December 12, 1995. Here are the results.

Willie Brown	56.6%
Frank Jordan	43.3%

a) Assume that the same people voted each time and that their preferences didn't change between the two elections. Draw preference diagrams for these people. There isn't a single right answer, so you might want to compare your diagrams with those of someone else in your class.

b) Although it didn't happen, suppose that Jordan had been forced to withdraw from the runoff election because of health problems and that the runoff had been between Brown and Achtenberg. According to the preference diagrams you just drew, would Brown still have been elected mayor? If your answer is yes, can you draw another set for which the answer is no?

10. It can be interesting to see who would have won a runoff election if the candidates in the runoff were different. An easy way to do a runoff between two candidates is to cross out all the others. **Figures 1.7**, **1.8**, and **1.9** give an example using the 1970 New York Senate preference diagrams.

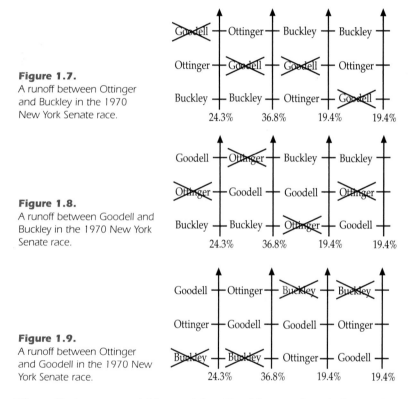

Figure 1.7.
A runoff between Ottinger and Buckley in the 1970 New York Senate race.

Figure 1.8.
A runoff between Goodell and Buckley in the 1970 New York Senate race.

Figure 1.9.
A runoff between Ottinger and Goodell in the 1970 New York Senate race.

Thus, Ottinger would beat either Buckley or Goodell in a runoff; Goodell would beat Buckley, but not Ottinger, in a runoff; and Buckley can't beat either of the other two in a runoff. Sound confusing? Keeping track of what happens in all these runoffs can be tricky, so mathematicians have a way of representing who beats whom in runoffs. They make a **runoff diagram**. In a runoff diagram, a dot is used to represent each candidate and an arrow is drawn from the winner to the loser of each runoff (**Figure 1.10**). Think of the arrow as meaning "beats."

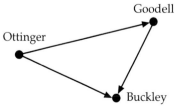

Figure 1.10.
A runoff diagram for the 1970 New York Senate race.

Sometimes the percentage of votes the candidate gets in each runoff is written on the diagram (**Figure 1.11**).

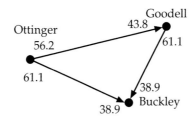

Figure 1.11.
A runoff diagram showing percentages of votes for each runoff.

Go back to the preference diagrams you made for the San Francisco mayoral election. Make a runoff diagram. Include the percentages.

> A runoff diagram is an example of what mathematicians call a graph or a network. Graphs have points called vertices and lines called edges. Sometimes the edges have arrows, and sometimes they do not. Graphs are a common form of representation used to solve many different kinds of problems. For example, graphs are used to route telephone calls every day throughout the world.

11. **Figure 1.12** is a runoff diagram for comic strip preferences in a class of 25 students.

   a) Which comic strip of these four do you think students in this class like best? Why?

   b) Which comic strip is the plurality winner in this class?

   c) Draw a set of preference diagrams that match this runoff diagram. Give *Calvin & Hobbes* as few first-place votes as possible.

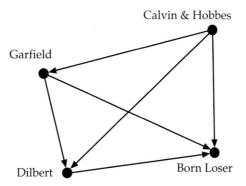

Figure 1.12.
Runoff diagram for comic strip preferences.

   d) Of course, in real elections, only one of the many possible runoffs actually occurs, and that's the one between the two top vote-getters. In your answer to part (c), which of the runoffs in the runoff diagram would be in the real runoff and which comic strip wins it?

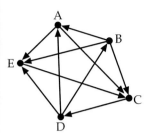

Figure 1.13.
Runoff diagram for breakfast cereal preferences.

12. Companies that develop new products like foods and cosmetics try many different formulations of a new product before deciding on the one to market. Often they test these formulations by having people compare them two at a time. Any time things are compared two at a time, it's like having runoff elections. **Figure 1.13** is a runoff diagram showing how a tester compared various pairs of formulations for a new breakfast cereal. The formulations are called A, B, C, D, and E. When an arrow points from one formulation to another, the arrow is pointing away from the cereal the tester liked.

Based on these tests alone, which formulation would you advise the company to market?

13. a) Write a short summary evaluating the runoff method. Based on the examples you have considered in this lesson, what do you think are the pros and cons of the runoff method?

   b) As you have seen, election methods often have flaws. However, those flaws occur only some of the time. As you continue in your study of election methods, you will occasionally see a new flaw. After a while, it can get hard to keep track of the methods and the flaws. Make a table that shows the methods and flaws you have seen thus far. Keep the table and update it whenever you see a new method or a new flaw. **Figure 1.14** shows how your table might look. Leave plenty of room to add more notes later and to add additional rows as you learn about new methods.

Method	Flaws
Plurality	Winner can be ranked last by a majority of voters.
Runoff	

Figure 1.14.
Sample voting method organization table.

LESSON THREE

Making a Point with Point Systems

KEY CONCEPTS

Point systems

Matrix

Paradox

The Image Bank

PREPARATION READING

Proving a Point

*I*n the United States, national elections are decided by the plurality method. Most local elections are decided by either the plurality or the runoff method. A runoff improves on the plurality method by requiring that the winner of the election receive over half the votes.

The runoff method is an improvement over the plurality method because it cannot produce a winner that is ranked last by a majority of the voters. But how good an improvement is it? What problems does the runoff method have?

When mathematics is used to solve real-world problems or to improve on existing solutions, the new solutions must be tried and tested to be sure they are really better and that they have no serious flaws of their own. In this lesson, you will give the runoff method a close examination and consider an alternative method.

ACTIVITY

ARE RUNOFFS THE ANSWER?

4

The runoff method is commonly used, not only in state and local elections in the United States, but in elections in other countries.

How well does it accomplish the goal of producing a winner that is not disliked by a majority of voters? In this activity, you will give the runoff method a careful examination and draw your own conclusions.

Most election methods work fine at least some of the time. They are more likely to be controversial in close elections with three or more strong candidates. To examine the runoff method, consider a hypothetical election with four strong candidates. The preferences of the voters are shown in **Figure 1.15**.

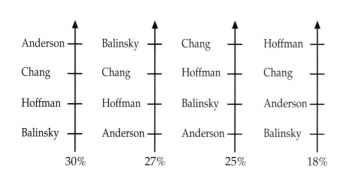

Anderson	Balinsky	Chang	Hoffman
Chang	Chang	Hoffman	Chang
Hoffman	Hoffman	Balinsky	Anderson
Balinsky	Anderson	Anderson	Balinsky
30%	27%	25%	18%

Figure 1.15.
Voter preferences in a hypothetical election.

1. Find the plurality and runoff winners.

2. Discuss how the majority of voters feel about the plurality and runoff winners.

3. How well does the runoff method improve on the plurality method?

4. Sometimes voters can do something to avoid the election of a candidate they do not like very well. One reason they can do this is that opinion polls can give voters a good idea of who will win the election. Suppose that the preference diagrams above are the results of an opinion poll. The supporters of Hoffman know that their first choice is not likely to win the election. How would you advise them to vote on election day?

5. Although voters can try to avoid the problems with the plurality and runoff methods by voting differently, the runoff method can create a different and very unusual problem in some situations.

ACTIVITY

ARE RUNOFFS THE ANSWER?

4

A class of 25 students is trying to decide on a menu for a buffet dinner. They decide to have each member of the class rank the choices, which are burritos (B), chef's salad (C), hamburgers (H), and pizza (P). The results are shown in **Figure 1.16**. Note that the number of students (not the percentage) is at the bottom of each diagram.

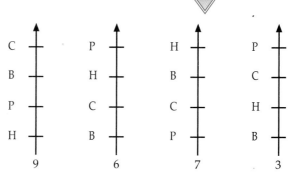

Figure 1.16.
Dinner preferences in a high school class.

a) If the runoff method is used, which foods are in the runoff, and which one wins?

b) Suppose the three students who like pizza best decide to vote for chef's salad because they are concerned about nutrition. To see what happens, reverse the order of pizza and chef's salad on the last preference diagram, and determine the runoff winner.

c) Mathematicians sometimes use the word **paradox** to describe a situation that seems impossible. Why is this runoff example a paradox?

d) Make a runoff diagram for the preferences at the beginning of this question. Does the diagram help you see why this paradox occurred?

e) The preference diagrams at the beginning of this item give the number of students instead of the percentage of students. Do you think questions about elections are easier to answer when numbers of voters are given or when percentages of voters are given?

6. Summarize the problems you have seen with the plurality and runoff methods. What features of these methods cause the problems? How might a new method avoid the problems?

CONSIDER:

1. How does the runoff method improve on the plurality method?

2. Is the runoff method a good alternative to plurality?

INDIVIDUAL WORK 3

Point Systems

Point systems are sometimes called Borda counts. During the 18th century, Jean-Charles Borda (1733–1799) proposed using a point system to decide elections because he was unhappy with the plurality method.

A **point system** gives a candidate points according to how the candidate is ranked. Being ranked first gets the candidate more points than being ranked last. For example, if there are three candidates for an office, a candidate might get three points for being ranked first by a voter, two points for being ranked second, and one point for being ranked last. When using a point system, voters must rank the candidates rather than just vote for the one they like best. Point systems are sometimes proposed as a way of avoiding the plurality method's flaw.

As the news clip about the 1996 college football season shows, one situation in which point systems are used is in the determination of the top college football team. In this poll, college football coaches from around the country are asked to rank the 25 teams they consider best. A point system is applied to the ballots to determine the top 25 teams.

USA Today-CNN Coaches' Poll Jan. 2, 1997

The final USA TODAY-CNN football coaches' poll, with first-place votes in parentheses, record, and total points based on 25 points for a first-place vote through one point for a 25th-place vote:

1. Florida (58)	12-1	1,546
2. Ohio State (4)	11-1	1,466
3. Florida State	11-1	1,408
4. Arizona State	11-1	1,341
5. Brigham Young	14-1	1,261
6. Nebraska	11-2	1,235
7. Penn State	11-2	1,205
8. Colorado	10-2	1,128
9. Tennessee	10-2	1,077
10. North Carolina	10-2	971
11. Alabama	10-3	906
12. Virginia Tech	10-2	791
13. Louisiana State	10-2	746
14. Miami	9-3	636
15. Washington	9-3	622
16. Northwestern	9-3	594
17. Kansas State	9-3	564
18. Iowa	9-3	549
19. Syracuse	9-3	446
20. Michigan	8-4	390
21. Notre Dame	8-3	381
22. Wyoming	10-2	259
23. Texas	8-5	141
24. Army	10-2	106
25. Auburn	8-4	103

OTHERS RECEIVING VOTES: West Virginia (85), Navy (76), Virginia (30), Stanford (24), East Carolina (19), Wisconsin (15), Southern Mississippi (13), Nevada (5), San Diego State (5), Clemson (4), Texas Tech (2).
DROPPED OUT: No. 25 West Virginia

Point systems like the one used to determine the top college football team can also be used in elections. In the items that follow, you will see how to adapt point systems to elections and consider whether point systems have any flaws.

1. The dinner preferences that you saw in this lesson's Activity 4 are shown again in **Figure 1.17**.

   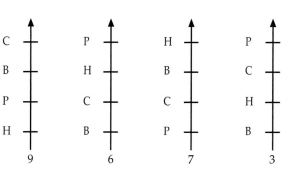

   Figure 1.17.
   Dinner preferences in a high school class.

   a) How many students are in the class?

   b) How many students ranked pizza first?

   c) How many students ranked pizza second?

   d) How many students ranked pizza third?

   e) How many students ranked pizza fourth?

   f) If a first-place ranking is worth 4 points, a second-place ranking is worth 3 points, a second-place ranking is worth 2 points, and a fourth-place ranking is worth 1 point, how many points does pizza receive?

   g) Find the point totals for burritos, chef's salad, and hamburgers.

   h) If the class decides to order only one food, and this point system is used to make the decision, which food is ordered?

   i) Which food is ordered if the plurality method is used? Does the plurality winner get a majority of the votes?

2. As you know, some ways of representing election data are more useful for some purposes than others are. It's not very easy to use a point system with preference diagrams. Tables are better. Mathematicians sometimes call a table a **matrix**. **Figure 1.18** is a matrix that shows the results of the class vote on foods in Item 1. Note that when mathematicians make a matrix, they usually enclose the data in brackets to help separate the data from the row and column labels. The votes for burritos are in the first column of the matrix. Complete the matrix by finding the votes for chef's salad, hamburgers, and pizza.

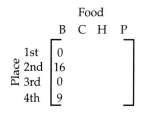

Figure 1.18.
A matrix for the dinner preferences in a class of 25 students.

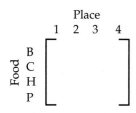

Figure 1.19.
A food-by-place matrix.

A matrix is another representational tool that mathematicians find very useful. Many different kinds of problems can be represented with matrices (the plural of matrix), and representing a problem in a matrix often makes the problem easier to solve. For example, the mathematics of matrices is used in the design of computer animations. You will use matrix representations in several other units in this textbook.

3. In the matrix you just completed (Figure 1.18), the places are in the rows, which go across, and the foods are in the columns, which go up and down. The matrix is called a place-by-food matrix because it is customary to state the row label first and the column label second. The matrix could also be done as a food-by-place matrix. Complete the food-by-place matrix in **Figure 1.19**.

4. Matrices are commonly used in business to keep records. For example, a retail store has the following inventory of a certain kind of shirt: 22 medium, 34 large, and 30 extra-large in blue; 17 medium, 21 large, and 12 extra-large in green; 10 medium, 12 large, and 11 extra-large in red. Write this information in a matrix. State the type of matrix you used and be sure you label the rows and columns.

5. Mathematicians regularly use calculators and computers to help them with their work. Representing election data in a matrix makes point totals easier to calculate. Calculators and computers make point totals even easier. Here are three ways to use calculators and computers to do point totals:

   a) by replaying and editing a calculation on a calculator with a replay feature;

   b) by using matrix multiplication on a calculator with matrix features;

   c) by designing a spreadsheet on a computer that has spreadsheet software.

   Learn at least one of these methods, and apply it to the food-vote data. (The more of the methods you learn, the better you will be at solving problems.) You might be able to learn a method by reading the manual for your calculator or spreadsheet, but you might also need to consult a friend or your teacher. In the future, you should use the method(s) you learn for point system calculations whenever you need to do them.

6. a) How are the results different if you use a 3, 2, 1, 0 point system on the food-election data in Item 1?

   b) Luigi, a member of the class who loves Mexican food, suggests that the class use a 10, 9, 2, 1 point system. He argues that a first- or second-place vote is worth a lot, but a third- or fourth-place vote doesn't mean much. What do you think is Luigi's real motivation?

7. a) Represent the results of the 1970 New York Senate race in a place-by-candidate matrix, and use a 3, 2, 1 point system to determine the winner. Keep in mind that the numbers in the matrix can be percentages rather than vote totals.

   b) Since Buckley received the most first-place votes, a point system that gives a lot of weight to first place might make Buckley the point-system winner. Find a point system that gives the win to Buckley.

8. The preference diagrams you constructed in Lesson 1 for the 1992 presidential election are shown again in **Figure 1.20**.

   a) Represent the election data in a place-by-candidate matrix and use a 3, 2, 1 point system to determine the winner.

   b) Could a different point system have changed the winner? Explain.

9. The preference diagrams you constructed in Lesson 1 for the 1968 presidential election are shown again in **Figure 1.21**. Represent the election data in a place-by-candidate matrix and use a 3, 2, 1 point system to determine the winner.

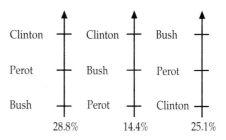

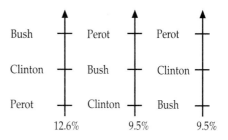

Figure 1.20.
Preference diagrams for the 1992 presidential election.

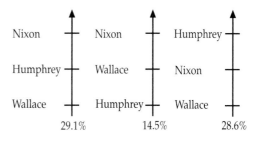

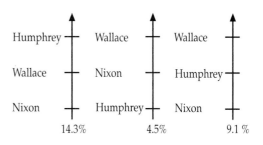

Figure 1.21.
Preference diagrams for the 1968 presidential election.

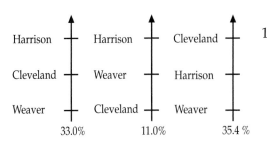

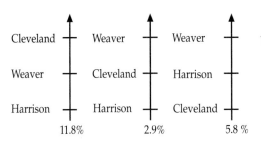

Figure 1.22.
Preference diagrams for the 1892 presidential election.

Candidate

$$\begin{array}{c} \\ \text{Place} \end{array} \begin{array}{c} \text{B} \quad\ \text{S} \quad\ \text{Z} \\ \begin{array}{ccc} 1\text{st} \\ 2\text{nd} \\ 3\text{rd} \end{array} \left[\begin{array}{ccc} 17.2\% & 39.3\% & 43.5\% \\ 45.1\% & 22.5\% & 32.4\% \\ 37.7\% & 38.2\% & 28.5\% \end{array} \right] \end{array}$$

Figure 1.23.
An election matrix.

10. The preference diagrams you constructed in Lesson 1 for the 1892 presidential election are shown again in **Figure 1.22**. Represent the election data in a place-by-candidate matrix and use a 3, 2, 1 point system to determine the winner.

11. Could the matrix in **Figure 1.23** represent the results of an election in which there are three candidates: Brown, Swartz, and Zwang? Explain your answer.

12. The mathematics you use to find point totals is very similar to the mathematics used by other people in a variety of situations. For example, the manager of a clothing store has an inventory of 25 shirts in size medium, 40 shirts in size large, and 50 shirts in size extra-large. A medium shirt costs the store $6.50, a large costs $6.70, and an extra-large costs $6.90. Find the total cost of the store's inventory and explain why the calculation is similar to the calculation of a point total.

13. The preferences that you saw earlier in this lesson's Activity 4, are shown again in **Figure 1.24**. In this election both the plurality winner and the runoff winner are ranked third or fourth by a majority of voters.

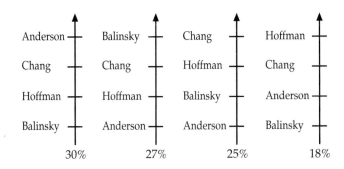

Figure 1.24.
Voter preferences in a hypothetical election.

a) Who is the winner if a 4, 3, 2, 1 point system is used?

b) How do the majority of voters feel about this winner?

c) Change the third preference diagram so that Hoffman is first. Use a 4, 3, 2, 1 point system and determine the winner.

d) What do your answers to the other parts of this question tell you about point systems? Do you think a point system is a good idea?

14. In this lesson's Activity 4, you saw that the runoff method can produce a paradox in which a candidate that can win a runoff election gets more votes and thereby loses the election. If a point system is used, could the winner turn into a loser if some of the voters decide to rank the candidate higher? Explain your answer.

15. In Lesson 2 of this unit you made runoff diagrams to show who would win runoffs between various pairs of candidates.

a) Make a runoff diagram for the preferences in Item 13. Include the option of writing the percentage of votes the candidates get in each runoff.

b) In your runoff diagram you have written three percentages near the dot that represents each candidate. Find the total of the three percentages for each candidate.

c) Do you see any connection between these totals and the point totals the candidates receive when a point system is used? Explain the connection.

16. In the first three lessons of this unit, you have seen several new ways to represent election results. In previous math courses, you have learned many other representations. Select a representation you have learned in a previous course and apply it to the preferences in Item 13. Tell whether you think the representation is a useful addition to those you have learned in this unit.

17. Do you think point systems are an improvement over plurality and runoff? Why?

LESSON FOUR

Other Ways

KEY CONCEPTS

Sequential runoffs

Condorcet method

Pairwise voting

Approval voting

Cumulative voting

The Image Bank

PREPARATION READING

Consider the Alternatives

When mathematics is used to solve real-world problems, the solutions are often not final. Unexpected circumstances sometimes arise in which the solutions do not work well. For example, when there are more than two candidates, the plurality method sometimes produces a winner that is ranked last by most voters.

When a solution to a problem doesn't work well, people look for better solutions. Runoff methods are sometimes used to improve on the plurality method, but some people feel the improvement is not good enough because the winner can be ranked fairly low by a majority of voters.

The search for a good way to conduct elections has led some people to suggest a point system, in which the voters rank the candidates and points are awarded to the candidates according to how highly they are ranked. A point system is less likely to produce a winner that is ranked low by a large percentage of voters because the point system winner is not based on first-place votes only. In fact, a candidate can win with a point system by getting a lot of second-place votes and few last-place votes.

Is a point system the perfect election method or does it have shortcomings of its own? Are there alternatives to point methods?

In this lesson, you will take a closer look at point methods and consider several alternatives.

ACTIVITY

ARE POINT SYSTEMS THE ANSWER?

5

As you have seen, point systems are not likely to produce a winner that is ranked low by a majority of voters, and they do not allow the paradox of turning a winning candidate into a loser when the candidate actually gains votes. Therefore, point systems seem to offer a reasonable alternative to the plurality and runoff methods.

Point systems, however, are not completely free from controversy. In this activity, you will apply a point system to some hypothetical election data to see why.

1. **Figure 1.25** shows preferences of voters for four candidates in an election.

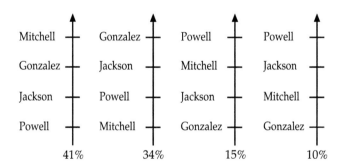

Mitchell	Gonzalez	Powell	Powell
Gonzalez	Jackson	Mitchell	Jackson
Jackson	Powell	Jackson	Mitchell
Powell	Mitchell	Gonzalez	Gonzalez
41%	34%	15%	10%

Figure 1.25.
Voter preferences in a hypothetical election.

a) Use a 4, 3, 2, 1 point system to find the point totals and the winner.

b) Suppose that Powell is considering dropping out of the election because he is running behind in polls and has insufficient funds. If Powell drops out, there are only three candidates, so a 3, 2, 1 point system is used. What are the point totals and who wins?

c) Could Powell use the results you have just seen to any advantage? If so, do you think this is a problem?

ARE POINT SYSTEMS THE ANSWER?

2. Use the preferences you were given at the beginning of Item 1. Suppose that a 4, 3, 2, 1 point system is used to decide the election and polls show Gonzalez the likely winner. Notice that all of the people who rank Powell first like Mitchell better than Gonzalez.

  a) Could these people vote in a way that would prevent Gonzalez from winning? How?

  b) When people vote in a way that is different from how they actually feel about the candidates, they are voting insincerely. Do you think **insincere voting** is a problem?

  c) Can insincere voting occur with other election methods? (See your answer to Item 4 of Lesson 3's Activity 4.)

3. Given the examples you have seen in this activity, do you think that a point system is an improvement over the plurality and runoff systems?

1. Jean-Charles de Borda, the French mathematician who proposed point systems in the 18th century, once said, "My scheme is intended only for honest men." What do you think he meant?

2. Borda made his comment two centuries ago. Do you think it applies today?

Sydney Wins!
2000 Summer Olympics go to Australia

(KANSAS CITY STAR, FRIDAY, SEPTEMBER 24, 1993)

The Image Bank

Sydney, Australia edged out Beijing Thursday for the right to hold the 2000 Summer Olympic Games. Beijing, which was considered the slight favorite, led in each of the first three rounds of voting but could not gain an overall majority. Here's how the International Olympic Committee voted. A simple majority was required to win.

	First round	Second round	Third** round	Fourth* round
Beijing	32	37	40	43*
Sydney	30	30	37	45
Manchester, England	11	13	11*	
Berlin	9	9*		
Istanbul, Turkey	7*			

*Eliminated
**One member did not vote

Sequential Runoffs

ACTIVITY

HOW TO CHOOSE AN OLYMPIC CITY

6

An adaptation of the runoff method is used to choose the city that hosts the Olympics. It is called a **sequential runoff**.

If none of the cities receives a majority of the votes cast, only the city with the fewest votes is eliminated. Then the voters vote again. This process is repeated until one of the cities receives a majority. Sometimes no city gets a majority until only two are left, as happened when the site of the 2000 summer games was chosen.

The difference between a runoff and a sequential runoff is that in a runoff all but the top two candidates are eliminated at once. In a sequential runoff, only one candidate is eliminated at a time.

As is the case with the standard runoff method, it's possible to conduct a sequential runoff without having voters revote if the voters vote by ranking.

An example of a way to conduct a sequential runoff from voter rankings is shown in **Figures 1.26–28**. This example uses hypothetical election data from Lesson 3.

Try the sequential runoff method on some of the preferences you saw earlier in this unit. When you are comfortable with it, answer the following items.

1. Does sequential runoff produce the same winner as runoff? Always, sometimes, or never?

2. Do you think the sequential runoff method is likely to choose a winner that is ranked low by a majority of voters?

3. Some people suggest an adaptation of the sequential runoff method in which the candidate with the most last-place votes is eliminated rather than the candidate with the fewest first-place votes. Do the election at the beginning of this activity by eliminating the candidate with the most last-place votes.

Sequential Runoffs

ACTIVITY

6

HOW TO CHOOSE AN OLYMPIC CITY

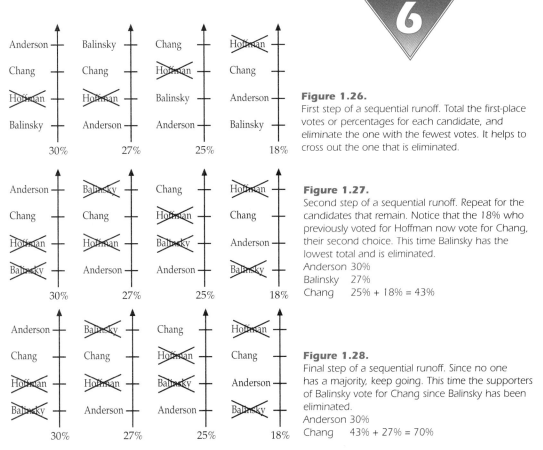

Figure 1.26.
First step of a sequential runoff. Total the first-place votes or percentages for each candidate, and eliminate the one with the fewest votes. It helps to cross out the one that is eliminated.

Figure 1.27.
Second step of a sequential runoff. Repeat for the candidates that remain. Notice that the 18% who previously voted for Hoffman now vote for Chang, their second choice. This time Balinsky has the lowest total and is eliminated.
Anderson 30%
Balinsky 27%
Chang   25% + 18% = 43%

Figure 1.28.
Final step of a sequential runoff. Since no one has a majority, keep going. This time the supporters of Balinsky vote for Chang since Balinsky has been eliminated.
Anderson 30%
Chang   43% + 27% = 70%

4. The runoff method can produce a paradox in which the potential winner loses by gaining more votes. Do you think that the sequential runoff method could produce the same paradox?

5. Summarize what you feel are the pros and cons of the sequential runoff method.

As a project, prepare a presentation to your class on the sequential runoff method. As part of the presentation, have your class vote on something and use the sequential runoff method to determine the winner. If your class ranked the candidates in Lesson 2's Activity 3, you can use those results instead of having the class vote again. You should explain to your class what you feel are the advantages and disadvantages of the method, and you should have the members of your class try the method on several sets of preferences to be sure they understand it.

ACTIVITY

7

The Condorcet Method and Pairwise Voting

ROUND & ROUND WE GO

A French mathematician, the Marquis de Condorcet (pronounced *condorsay*) (1743–1794) once proposed a voting system that requires the winner to be able to beat every other candidate in head-to-head contests. In other words, the winner must be able to beat each of the other candidates in a runoff.

1. The runoff diagrams you made in Lesson 2 are very helpful when analyzing the **Condorcet method**. Here are two sets of preferences you have seen before. Make a runoff diagram for each of them.

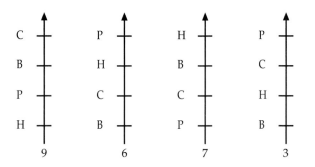

Figure 1.29.
Dinner preferences in a class of 25 students.

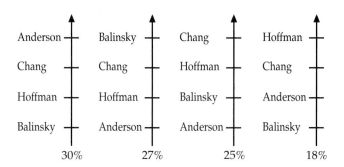

Figure 1.30.
Voter preferences in a hypothetical election.

a) Dinner preferences in a class of 25 students (**Figure 1.29**).

b) A hypothetical set of preferences in an election with four candidates (**Figure 1.30**).

c) Does the runoff diagram you just drew change if Hoffman is moved to first on the third preference diagram?

2. In each part of Item 1, use your runoff diagram to find the candidate that can beat all others in runoffs.

3. Condorcet's method is difficult to implement if voters have to vote on every possible two-way race. As you can see from

The Condorcet Method and Pairwise Voting

ACTIVITY

ROUND & ROUND WE GO

7

these examples, if voters rank the candidates, they need to vote only once. Do these examples indicate any problems from which Condorcet's method might suffer?

4. Do you see any way of resolving the problems you just listed?

5. When Condorcet's method does not produce a winner, interesting things can happen. **Figure 1.31** shows a runoff diagram for the dinner preferences in a different class.

If the choices are voted on two at a time, which is called **pairwise voting**, almost any choice can win. If you're clever, you can manipulate the vote so that your favorite food wins. Here's how to do it.

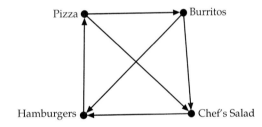

Figure 1.31
Dinner preferences in a class.

Suppose you are a hamburger fanatic and you want to rig the vote so that hamburgers win. Look at the graph. Start at the hamburger vertex and try to move to another vertex. You can't go against the arrows, though—think of them as one-way streets.

There's only one place you can go, that's to pizza.

From pizza you can go either to burritos or chef's salad. Look ahead. If you go to chef's salad next, you won't be able to go to burritos last. The idea is that you have to be able to get to each of the foods once. So, go to burritos next, then to chef's salad.

Here's the order in which you moved around the diagram: hamburgers, pizza, burritos, chef's salad.

a) Now, move through this list in reverse order. Start with an election between chef's salad and burritos. What food wins?

The Condorcet Method and Pairwise Voting

ROUND & ROUND WE GO

b) Continue moving through the list in reverse order. Compare the previous winner to pizza. What food wins?

c) Finally, what food wins between the previous winner and hamburgers?

6. The chairperson of a congressional committee has lots of power. One reason is that the chair decides when bills appear for a vote. Committees often vote on bills two at a time. The winner may later stand a vote against another bill, but the loser usually never gets considered again. If there is no bill that can beat all others, the committee chair can sometimes manipulate the order in which bills appear for a vote to obtain a favorable result.

The runoff diagram in **Figure 1.32** shows how the members of a congressional committee feel about several handgun control bills that have been introduced.

a) Suppose that bill A is the most lenient bill and you favor it. Is it possible to order the vote so that A wins? If so, explain how to do it.

b) Suppose that bill D is the strictest bill and that you favor it. Is it possible to order the vote so that D wins? If so, explain how to do it.

c) Suppose that D and A are voted on first, and that the winner is paired against C. That winner is then paired against B, and, finally, this winner is paired against E. Which bill wins?

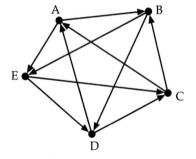

Figure 1.32.
Runoff diagram for handgun control bills.

As a project, prepare a presentation to your class on the Condorcet method. As part of the presentation, have your class vote on something and use the Condorcet method to determine the winner. If your class ranked the candidates in Lesson 2's Activity 3, you can use those results instead of having the class vote again. You should explain to your class what you feel are the advantages and disadvantages of the method, and you should have the members of your class try the method on several sets of preferences to be sure they understand the method.

Approval Voting

ACTIVITY

NO MORE RANKING

8

Approval voting is a simple and recent
system that was first proposed by American
mathematicians in the 1970s. Voters do not rank
the candidates; they simply mark the ballot, as
they do in the plurality method. The difference is that voters can
vote for as many candidates as they like. The winner is the candidate with the most votes. To see how it works, consider the preferences of the voters in the 1970 New York Senate race (**Figure 1.33**).

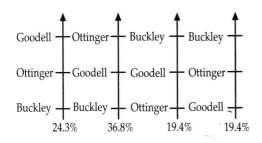

Goodell	Ottinger	Buckley	Buckley
Ottinger	Goodell	Goodell	Ottinger
Buckley	Buckley	Ottinger	Goodell
24.3%	36.8%	19.4%	19.4%

Figure 1.33.
Voter preferences in the 1970
New York Senate race.

Goodell and Ottinger had similar political views, and polls show
that most voters who voted for one of them felt fairly comfortable with the other. However, most Buckley voters did not like
either Goodell or Ottinger very well. Had voters been able to cast
more than one vote, most Goodell and Ottinger supporters
would also have voted for the other. Since approval voting
doesn't require voters to rank, the results would probably look
something like those shown in **Figure 1.34**.

Goodell	Ottinger	Buckley
Ottinger	Goodell	
24.3%	36.8%	38.8%

The Buckley votes can
be combined because
the voters would not
have voted for either
of the other two.

Figure 1.34.
Approval votes in the 1970 New York Senate race.

Approval voting
is quite new, so it
hasn't had time
to achieve
widespread
implementation.
Bills to adopt an
approval voting
system have
been introduced
in several state
legislatures,
including those
of New York,
North Dakota,
and Vermont, but
none has passed
both houses.
Several cities in
the United States
have adopted
approval voting
for local
elections, and
approval voting is
used to elect the
Secretary General
of the United
Nations.

Approval Voting

ACTIVITY

NO MORE RANKING

8

The vote totals are now:

Goodell 24.3% + 36.8% = 61.1%

Ottinger 24.3% + 36.8% = 61.1%

Buckley 38.8%

Notice that in approval voting, the percentages can total more than 100%.

Although the results give Goodell and Ottinger the same total, it isn't likely that the election would actually have been a tie. It probably would have been close.

1. Find approval voting totals for the hypothetical preferences from the 1992 presidential election (**Figure 1.35**). Use the following hypothetical descriptions of the supporters of each candidate:

   Bush supporters were not likely to vote for their second choice.

   Clinton supporters who ranked Perot second were likely to approve of their second choice, but the others were not.

   Perot supporters were likely to vote for their second choice.

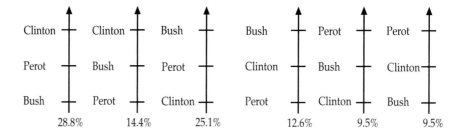

Figure 1.35.
Preference diagrams for the 1991 presidential election.

2. In this lesson's Activity 5, you saw that point systems sometimes encourage voters to vote insincerely. The preferences you saw in that activity are shown again in **Figure 1.36**.

   Using a 4, 3, 2, 1 point system, Gonzalez is the winner. If the voters who support Powell know that Gonzalez is likely to

Approval Voting

NO MORE RANKING

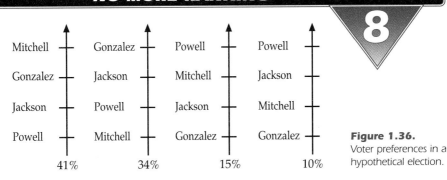

Figure 1.36.
Voter preferences in a hypothetical election.

win, they might decide to rank Mitchell first in an attempt to prevent Gonzalez from winning. This is insincere voting.

You also saw in this same election that the winner changes if Powell drops out. Thus Powell might try to make a deal with one of the other candidates in exchange for either staying in the race or dropping out of it.

Would either insincere voting or a change of winner when a candidate drops out be likely to occur if approval voting is used? Assume that voters always approve of the candidate they rank first and never approve of the candidate they rank last. Make your own decisions about second and third.

3. Do you think approval voting could produce the plurality method's flaw of producing a winner who is ranked low by a majority of voters or the runoff paradox of causing a winning candidate to lose when the candidate gains votes?

4. Do you think approval voting is a good alternative to other methods you have seen in this unit? Explain your reasons. If you think approval voting might have problems, be sure to list them.

As a project, prepare a presentation to your class on approval voting. As part of the presentation, have your class use approval voting to vote on something. You might, for example, have the class vote again on the election you used in Lesson 2's Activity 3. You should explain to your class what you feel are the advantages and disadvantages of the method, and you should have the members of your class try the method on several sets of preferences to be sure they understand the method.

Cumulative Voting

PUT MORE POWER IN YOUR VOTES

The elections you have seen in this unit have one thing in common: several candidates are competing for a single office. In many elections, however, there are several positions to be filled. Most states, for example, have several representatives in Congress. School boards and city councils have several members. In fact, most elections do not choose a president or a governor. They choose members of boards, commissions, and legislatures.

When there are several members of a governing body, should the majority get to elect them all? Many people think that the majority shouldn't get to pick all the members because a minority might go without representation. What do you think?

Suppose that your school board has three members and that a minority group in your school district makes up a third of the population. Should the minority group have a seat on the board?

Consider an example. Suppose there are three candidates from the majority group and one from the minority group. Call the majority candidates A, B, and C, and the minority candidate D. Since there are three positions to be filled, each voter can vote for three candidates. A voter can vote for fewer than three, but not more.

1. Describe the outcome of the election if people tend to vote for candidates from their own group.

There are laws that protect the rights of minorities and require that minorities have representation on governing bodies. These laws are sometimes implemented by dividing a city or state into several districts. For instance, the school district in this example might be divided into three regions so that one of the regions is composed mostly of minority members. Since the minority members are a majority in one region, they will probably succeed in electing one of their own members to the board. The problem with this solution is that it is often difficult to divide a state or city into regions. Sometimes the regions have odd shapes and they are challenged in court.

Cumulative Voting

ACTIVITY

PUT MORE POWER IN YOUR VOTES

9

Another solution to the problem of minority representation is a practice called **cumulative voting**. It is not necessary to divide the state or city into regions. Nor do the voters have to rank the candidates. Cumulative voting is similar to regular voting, but a voter can cast more than one vote for a single candidate. In the school board example, a voter could vote for three different candidates, cast two votes for one candidate and one for another, or cast all three votes for the same candidate. The voter is allowed the same number of votes as there are positions to be filled.

2. In the school board example, suppose the voters prefer the candidates as shown in **Figure 1.37**. The first three preference diagrams represent members of the majority, and the last diagram represents members of the minority.

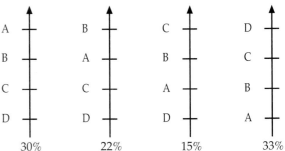

Figure 1.37.
Voter preferences in a hypothetical school board election.

a) Suppose that cumulative voting is not used. Each voter in the majority votes for A, B, and C. Each member of the minority casts a single vote for D, but does not vote for anyone else. Find the vote totals and determine which three candidates are elected to office.

b) Suppose that cumulative voting is used. Each voter in the majority votes for A, B, and C. Each member of the minority casts three votes for D. Find the vote totals and determine which three candidates are elected to office.

c) If cumulative voting is used and the members of the minority cast all their votes for D, is there anything the other voters can do to prevent D's election?

As a project, prepare a presentation to your class on cumulative voting. You should explain the goal of the method and its advantages and disadvantages. Have the members of your class try the method on several sets of preferences to be sure they understand it.

Wrapping Up Unit One

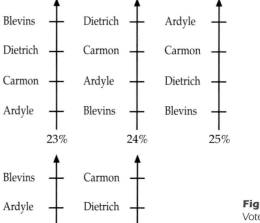

Figure 1.38.
Voter preferences in a hypothetical election.

1. The voter preferences in an election are shown in **Figure 1.38**.

   Here is a list of voting methods. For those that your class studied, determine the winner. Explain how you obtained your answer.

   a) Plurality

   b) Runoff

   c) A 4, 3, 2, 1 point system

   d) Sequential runoff

   e) The Condorcet method

   f) Approval voting if each voter approves of their first two choices

   g) Cumulative voting if there are two seats to be filled, the Blevins voters cast both their votes for Blevins, and all other voters cast a vote for their top two candidates

2. For each method you used in Item 1, write a short description of what you think the method did well or what it did poorly in this election.

3. The state of Louisiana uses the runoff method to select its governor. If no candidate gets a majority of the votes cast, a runoff is held between the top two candidates. In the fall of 1995, there were five strong candidates in the governor's race. Here are the results:

Cleo Fields	279,219
Mike Foster	383,301
Mary Landrieu	271,774
Phil Preis	132,133
Buddy Roemer	262,026
Others	144,530

a) Which candidates were in the runoff, and what percentage of the vote did each of them receive?

b) Is it possible that the two candidates in the runoff were not liked very well by a majority of voters? If so, draw preference diagrams to explain your answer.

c) If many voters had ranked the runoff candidates low, how do you think that would have affected the runoff election?

4. Runoff elections are a little like tournaments because they compare only two candidates at a time. Most tournaments involve only two teams or players competing in any one game or match.

Figure 1.39 is a runoff diagram that shows the results of matches of the final three rounds of the 1996 U. S. Open women's tennis tournament.

a) Who won the tournament?

b) Whom did the winner beat in the final match?

The results of the 1995 Nations Bank Handball USA Cup men's tournament are shown in **Figure 1.40**.

c) Who won the tournament?

d) How is the handball tournament different from the tennis tournament?

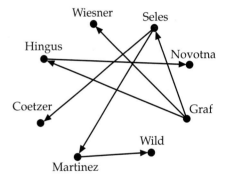

Figure 1.39.
Results of matches in the 1996 U. S. Open women's tennis tournament.

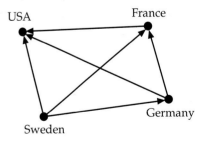

Figure 1.40.
Results of matches in the 1995 Nations Bank Handball USA Cup men's tournament.

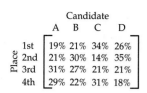

Figure 1.41.
Matrix of candidate rankings in an election.

5. The matrix in **Figure 1.41** shows the first-, second-, third-, and fourth-place rankings for candidates A, B, C, and D in an election.

For each election method, what does this representation tell you? Also tell whether some other representation is more useful.

a) Plurality

b) Runoff

c) A 4, 3, 2, 1 point system

6. Here are the (unofficial) vote totals for the 1996 presidential election.

Bill Clinton	45,628,667
Bob Dole	37,869,435
Ross Perot	7,874,283
Others	1,435,025

(Most of these votes went to Ralph Nader or Harry Browne)

Assume that those who did not vote for either Bill Clinton or Bob Dole were about equally split in their preferences for these two candidates. Would any of the election methods you have studied in this unit have changed the winner of this election? If your answer is yes, explain why.

7. Product ratings of various kinds can be found in books, magazines, and on the Internet. Often a product's rating is based on some type of point system. For example, Midwest Computer Review posts computer software and hardware ratings on its MCR Online Web page. MCR's rating system varies with the kind of product rated. A software game is rated on a scale of 1–100 in each of three categories: sound, graphics, and playability. A 1996 rating of the game ProPinball (Interplay) MCR gave it an overall rating of 250 out of a possible 300. ProPinball received 80 for sound, 95 for graphics, and 75 for playability.

MCR's overall rating gives equal weight to the three categories. Suppose you are a game player who feels that playability is most important, followed by graphics.

a) Suggest a way to modify MCR's overall rating to one more suitable to your tastes.

b) Use your modification to give ProPinball a new rating.

c) Did ProPinball score better or worse in your rating system than it did in MCR's? Explain your answer.

8. Congressional elections in the United States usually do not require runoffs. One reason for this is that primary elections are used to narrow the field of candidates. Thus, the winner of the general election is usually a majority winner. However, runoffs sometimes occur. In August of 1996, federal judges ruled that several Texas congressional districts had been improperly drawn. Since the ruling came after the March primary, the judges threw out the primary results and ordered that all candidates remain on the ballot in the general election. In three districts, the general election winner did not receive a majority, and a special runoff election between the top two candidates was held on December 10. Here are the general election results from the 9th congressional district.

Steve Stockman	88,171
Nick Lampson	83,781
Geraldine Sam	17,886

a) Stockman and Lampson ran in the special election since they were the top two vote-getters. Draw preference diagrams that show how Stockman can win the special election. Use percentages rather than vote totals in your diagrams.

b) Draw preference diagrams that show how Lampson can win the special election.

c) Lampson won the special election. Why do think this happened?

Mathematical Summary

You began this unit by considering the mathematical process for selecting the winner of an election. The plurality method, which is used frequently in the United States, sometimes produces an undesirable result: the winning candidate is rated the worst choice by a majority of voters.

When the mathematics used to solve problems gives inadequate solutions, people search for new ways to use mathematics to achieve better solutions. As you began your search for a better way to determine the winner of an election, you learned new mathematical representations that helped you understand the preferences of voters and the results of an election. Among these are percentages, preference diagrams, and runoff diagrams. You also found that calculators and computers can be helpful tools in applying new election methods.

Your search for a better election method led to an examination of the runoff method. The runoff method never picks a winner that is ranked last by a majority of voters, but it can produce other problems. Since the runoff method can pick a winner that is ranked second-last by a majority of voters, it can be considered a small improvement over the plurality method, but it is certainly not perfect.

The plurality and runoff methods pick winners that are ranked low by a lot of voters because they are based only on first-place rankings. A good way to avoid the flaws of these two methods is to use a method that considers more than just first-place rankings. A point system does just that.

Although point systems don't pick a winner that is ranked low by a lot of voters, they can cause other problems. For example, voters can manipulate the election by voting insincerely, and candidates can demand favors from other candidates in exchange for withdrawing or staying in the election. Voter and candidate manipulation, however, do not occur only with point systems. They can occur with most election methods.

You may have considered alternatives to plurality, runoff, and point methods, such as sequential runoff, the Condorcet method, and approval voting, a recently proposed alternative that some political scientists favor.

The search for ways to improve the election process is an outstanding example of the way mathematics is used to solve real-world problems. Solutions are seldom perfect, and when they are found lacking, the mathematics is changed and new solutions are found. The search for new ways to use mathematics to solve real problems is going on every day in the world around you.

Glossary

APPROVAL VOTING:
A voting method in which the voters are allowed to vote for as many candidates as they like. The winner is the candidate with the most votes.

CONDORCET METHOD:
A method that requires the winner of an election to be able to beat each of the other candidates in a one-on-one race.

CUMULATIVE VOTING:
A voting method sometimes used to elect members of a board or legislature. Each voter is allowed as many votes as there are seats to be filled. The voter can cast more than one vote for a single candidate.

INSINCERE VOTING:
A practice in which voters decide to vote differently from the way in which they actually feel about the candidates.

MAJORITY:
Over half the votes cast in an election.

MATRIX:
A table composed of horizontal rows and vertical columns.

PAIRWISE VOTING:
A practice in which the voters vote on two of the candidates at a time.

PARADOX:
A surprising result—one that runs counter to a person's intuition.

PLURALITY:
An election method that elects the candidate with the most votes.

POINT SYSTEMS:
Point systems assign a number of points for a first-place ranking, fewer points for a second-place ranking, fewer still for a third-place ranking, etc. The winner is the candidate with the highest point total.

PREFERENCE DIAGRAM:
A diagram that shows how voters rank the candidates. It also shows the number or the percentage of voters.

RUNOFF:
An election system that requires the winner to receive a majority of the votes cast. If no candidate receives a majority, another election is held between the top two vote-getters.

RUNOFF DIAGRAM:
A diagram in which a dot is used to represent each candidate and an arrow is drawn from the winner to the loser of each runoff.

SEQUENTIAL RUNOFF METHOD:
A method in which the candidates are eliminated one at a time. The candidate with the fewest votes is eliminated and the voters vote again. The process continues until a candidate receives a majority.

UNIT

Secret Codes and the Power of Algebra

Fiber-optic cables and satellite links carry millions of bits of information every second. Some codes, like the ones used in data compression, help transmit the information quickly and efficiently. Secret codes help preserve the privacy of vital information.

In this unit, you focus on the role of mathematics in designing a coded message that is easy to decode by the receiving party and difficult to crack by an unwelcome interceptor of the message. As you code messages, you learn about linear functions and multiple representations. When you decode messages, you work with inverse operations and solving equations. You will crack codes with the help of statistics.

As you go through this unit, you will compile a mathematical toolbox of skills and concepts that you can use to solve many real-life problems. The most valuable tools are the multiple methods of representation that are introduced in this unit and developed throughout *Mathematics: Modeling Our World*.

SECRET CODES AND THE POWER OF ALGEBRA

People communicate information. Sometimes the information is private or sensitive. If the wrong person intercepts the information, the results can be disastrous.

Perhaps you have written a personal note to a friend and were embarrassed when the note was intercepted by the wrong person. You may have reminded yourself to be more careful next time. Maybe you write in a diary. You might lock your diary in a secure location because you are afraid someone might find it and read it. You do the best you can to protect the information in the diary.

Personal information is not the only material that needs to be protected. This is the Information Age—most information is stored and sent electronically. Wealth is transferred from one bank account to another over phone lines and satellite links. Private messages are sent by e-mail from one computer terminal to another. Important documents, stored on disk, can be retrieved anywhere in the world.

Secret codes are necessary in a civilization that relies on electronic communication. Communication companies and software companies hire cryptographers to design secret codes to protect this vital information.

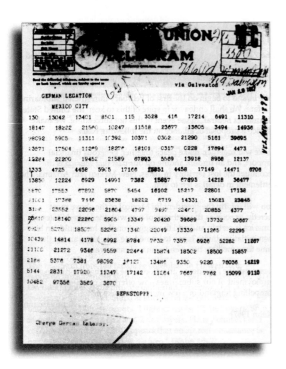

Secret codes also have an important role in history. Military commands were coded during World Wars I and II. The countries with superior cryptographers held an advantage in both wars. The famous Zimmermann telegram, intercepted and deciphered by the British, revealed Germany's attempt to form an alliance with Mexico against the United States in World War I. President Woodrow Wilson asked for a declaration of war not long after this telegram was intercepted.

As you will see in this unit, mathematics is the most valuable tool available to those who make and break codes.

Keeping Secrets

KEY CONCEPTS

Coding

Decoding

Code breaking

Cryptography

Coding methods: substitution, transposition, dictionary

The Image Bank

PREPARATION READING

Codes Are Everywhere

Welcome to the fascinating world of secret codes! MTUJ DTZ JSOTD YMJ HMFQQJSLJ! Codes are everywhere around you. They help ensure that you are charged the proper amount when you buy an item in a store, that your bank account has the correct balance, and that your mail is delivered to the right place relatively quickly.

Many of the codes in use today help speed the processing of information. Secret codes are not used to speed the flow of information, but to keep information secret. Although most people think of secret codes as tools of spies and secret agents, they are used today by many others. Banks, for example, keep their coding systems secret so that computer thieves cannot access accounts. Secret codes are used when a person wants to send a private message to another person and not worry if an unwelcome third person intercepts the coded message.

How would you encode a secret message to a friend? How could you make sure that someone who intercepts the message cannot decipher it? People usually use mathematics to make secret codes. Throughout this unit, you will use mathematics as a tool to design effective secret codes and decode messages.

You can think of the world of secret codes as having three key roles: the coder, the decoder, and the code cracker. At various times in this unit, you will play each role.

The following message contains two pieces of information that can be helpful in making and breaking secret codes. A mathematical process was used to code the message. By the time you finish this unit, you will be able to decode the message.

I Y P K P I I P C F P I Q C B T D F D C P I Y P N T F I J T N N T Q K P I I P C F B Q P Q
V K B F Y W C P W D C P D V C D W Y I T C P O P D K K B Q P D C W D I I P C Q F

Once again, MTUJ DTZ JSPTD YMJ HMFQQJSLJ!

CONSIDER:

1. Name some codes that are familiar to you. How is each used—to speed the processing of information or to keep information secret?

2. Information is often transmitted in coded form so that it cannot be understood by an unwelcome interceptor of the message. What makes one coded message more difficult to crack than another?

3. What makes a code effective?

4. Which role do you think requires the most skill: coder, decoder, or code cracker?

ACTIVITY

THE WORLD OF CODES

1

As you can tell by inspecting the Zimmermann telegram, coded messages are often quite long. Cracking such messages can take weeks or months of effort by a team of trained cryptographers.

In *Mathematics: Modeling Our World*, the problems you solve are as real as is reasonably possible. However, you would not be happy if the messages you are challenged to crack take months of effort. Therefore, the messages in this and other activities are short. They are not actual historical messages. Instead, they convey interesting facts about the use of codes in the world.

This activity has five messages, each coded with a different process. It is unlikely that you will be able to crack all of them now. You will become better at code cracking as you go through this unit. For now, try to determine the meaning of as many of the messages as you can. Later you will have opportunities to revisit these five messages and apply new techniques for cracking codes.

As you attempt to crack coded messages, you might keep in mind the words of Captain Parker Hitt, who wrote the first U.S. Army manual on code cracking: "Success in dealing with unknown ciphers is measured by these four things in the order named: perseverance, careful methods of analysis, intuition, and luck." You will learn "careful methods of analysis." You will have to supply the perseverance.

The process of cracking codes will help you develop ways to build more effective codes. This activity will help you decide what makes one coding process more effective than another.

MESSAGE 1

BPWUIA RMNNMZAWV QVDMVBML I EPMMT TQSM LMDQKM

BW KWLM UMAAIOMA

ACTIVITY

THE WORLD OF CODES

1

MESSAGE 2

6   19 26 18 7 10 23   8 6 17 17 10 9   6   8 13 10 8 16   9 14 12 14 25

14 24   26 24 10 9   25 20   9 10 25 10 8 25   10 23 23 20 23 24   25 13 6 25

6 23 10   20 11 25 10 19   18 6 9 10   14 19   8 20 9 10 24   17 14 16 10

31 14 21   8 20 9 10 24   6 19 9   7 20 20 16   8 20 9 10 24

MESSAGE 3

79 63 55 23   63 27   83 35 23   23 99 35 39 11 39 83 79   7 83   83 35 23

59 7 83 39 63 59 7 51   15 75 103 67 83 63 51 63 31 39 15

55 87 79 23 87 55   59 23 7 75   11 7 51 83 39 55 63 75 23   15 7 59   11 23

79 23 23 59   63 59   83 35 23   95 63 75 51 19   95 39 19 23   95 23 11

MESSAGE 4

HRKND JDNK GR DNOK DBGNIIZGN

UZHGJAND EBHF GR NBAGW BOK GR

UIBV EBHF HRLUBHG KZDFD BAN

BHGJBIIV BEIN GR HRAANHG NAARAD

MESSAGE 5

KBCCAW BVYWAZBWM KMFJENFD RH

LEFRL GTO FFFDVJAJ PVLWJKPGDFQQ

QAIYWZUNX JXERI ULNVOSAZBCCTFP

RH OLSUX OXS NBG

THE WORLD OF CODES

1. Try to crack one or more of the messages. Prepare to present your results to your class.

2. Of the messages you cracked, which were most difficult? Give reasons for your answer.

3. What makes one coding process better than another?

4. Try inventing your own coding process. Code an appropriate sentence using a process you design. Others will be asked to decipher your message if they can.

Before you continue with your coding, decoding, and code cracking, you should learn some vocabulary.

Although **code** and **cipher** have slightly different meanings in some books, they are used interchangeably in this unit. Both refer to the method used to encode a message.

The process of putting a message in coded form is called **encoding**. Taking it from coded form back to its original form is called **decoding** or **deciphering**. The original form of a message is sometimes called the **plaintext**.

Determining the contents of a message without knowing how it was coded is called **code breaking** or **code cracking**.

Cryptography refers to the study of coding, decoding, and code breaking.

INDIVIDUAL WORK 1

Keeping Secrets

Welcome to the world of secret codes. The following items introduce you to more of this interesting topic.

‖‖‖‖‖‖‖‖‖‖‖‖‖‖	17011
‖‖‖‖‖‖‖‖‖‖‖‖‖‖	20078
‖‖‖‖‖‖‖‖‖‖‖‖‖‖	19006
‖‖‖‖‖‖‖‖‖‖‖‖‖‖	05445

Figure 2.1.
The bars for four zip codes.

1. Some codes allow information to be transmitted quickly and efficiently. The U.S. postal service uses zip codes to speed the delivery of mail. ("Zip" stands for "zone improvement program.") The simplest zip code is a five-digit number. (Some codes have nine digits; others have eleven.) The number is converted to a series of short and long bars to be read and sorted by a machine. The bars for several five-digit zip codes are shown in **Figure 2.1**.

a) Determine the code for each digit from 0 to 9.

b) Why do you think the postal service decided to use five characters with three short bars and two long ones? Could they have used two short bars and three long bars instead?

c) If the postal service used six characters with only one long and five shorts in each group, how many numerals could be coded?

d) With six characters, how many longs and shorts would have to be used to cover ten numerals? Could only four characters be used?

e) An effective secret code is easy to encode, easy to decode, and difficult to crack. Do you think that zip-code bars would make an effective secret code?

2. Zip codes and universal product codes (UPCs) are everyday types of codes. Name other situations in which codes are used.

3. a) Write a short message. Encode the message using a simple coding process that you design. (You may use the message your group coded in the activity or create a new coded message.) Explain or diagram the coding process you used.

b) Challenge a member of your family, another student, a friend, or a neighbor to decipher your coded message. Describe the success or failure of the person who tried to crack your code. Note how long it took for the person either to figure out the message or to give up.

4. A **substitution cipher** replaces words or letters with other words or letters. A substitution cipher was used in a telegram sent to a Colonel Pelton in connection with the 1876 presidential race (**Figure 2.2**). The message was coded by using replacements for certain sensitive words.

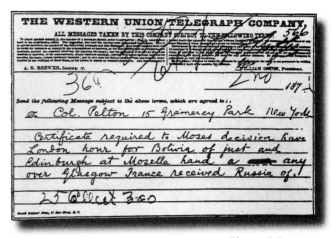

The coded message is "Certificate required to Moses decision have London hour for Bolivia of just and Edinburgh at Moselle hand a any over Glasgow France received Russia of."

Figure 2.2.
The Pelton telegram.

a) Make the following substitutions and write the message that results: replace *Bolivia* with *proposition*, *Russia* with *Tilden*, *London* with *canvassing board*, *France* with *Governor Stearns*, *Moselle* with *two*, *Glasgow* with *hundred*, *Edinburgh* with *thousand*, and *Moses* with *Manton Marble*.

b) Even with the substitutions, the message doesn't make sense. The message was also coded using a **transposition cipher**. In a transposition cipher, the order of the words or letters changes. Below is a list of numbers that tell how to unscramble the Pelton telegram. (Numbers indicate the order in which code words appear in the decoded message. *Governor Stearns, Manton Marble,* and *canvassing board* each count as a single word.)

18, 12, 6, 25, 14, 1, 16, 11, 21, 5, 19, 2, 17, 24, 9, 22, 7, 4, 10, 8, 23, 20, 3, 13, 15

Write the unscrambled message.

The 1876 election between Rutherford Hayes and Samuel Tilden was one of only two times in U.S. history that a candidate won the popular vote but lost in the electoral college. The telegram contained an offer to sell Florida's electoral votes for the amount in the telegram and thereby swing the 1876 election from Hayes to Tilden.

5. One way to set up a coding system is a **dictionary code**. The dictionary matches words or letters to other words, letters, or symbols. Both the sender and the receiver of the secret message must have a copy of the dictionary. In Item 4, the list of replacement words is a type of dictionary.

Use the brief dictionary below to decode the message in the following sentence.

This message is tricky because you would never suspect that, and I would never suspect that, the real words are not written.

and	–	eventually
are	–	amount
because	–	designed
I	–	but
is	–	are
message	–	codes
never	–	their
not	–	of
real	–	a
suspect	–	being
that	–	decoded
the	–	in
this	–	some
tricky	–	not
words	–	reasonable
would	–	avoid
written	–	time
you	–	to

LESSON TWO

UGETGV EQFGU

KEY CONCEPTS

Coding method: shift cipher

Representation of functions

Matrix addition

Function

Coordinate graph

Linear function

Variable

Constant

Domain

Range

PREPARATION READING

Effective Coding

*I*f it's good enough for Caesar, is it good enough for you? Julius Caesar was the emperor of the Roman Empire just before the first century B.C. Caesar used a simple coding method called a **shift cipher** that replaced each letter of the alphabet with a letter three places removed. It's called a shift cipher because you can think of the coding process as moving or shifting the alphabet three places. Did Caesar use an effective code?

In this lesson, you decide whether a shift cipher is easy to encode. In Lessons 3 and 4, you will decide whether the shift cipher is easy to decode and difficult to crack.

Two other considerations are worth keeping in mind when judging a code. One is simplicity. If a simple coding process works well, then you do not need a complicated one. Another consideration is communication. The coder and decoder must be able to tell each other about their method. Communication of the method can be dangerous, particularly in times of war. A method that can be communicated easily is preferable to one that is hard to communicate.

Mathematics plays a primary role in designing effective secret codes. In this lesson, you learn about several key mathematical tools related to codes. Mathematics that is important to secret codes can be used to solve many other types of problems.

As they say in Rome, "GR DV WKH URPDQV GR."

ACTIVITY

2

One of the simplest coding methods is the shift cipher. For example, a shift of +5 codes the letter *D* as the letter *I* because *I* is five letters beyond *D* in the alphabet.

1. How does a shift +5 code the letter *Y*?

2. Suppose you shift the alphabet by first converting a letter to a number that matches its position in the alphabet. (*A*'s **position number** is 1, *B*'s is 2, etc.) Then apply a shift –4 by subtracting 4 from the position number. For example, the letter *J* has a coded value of 6. What is the coded value for the letter *D*? What is the coded value for the letter *B*?

3. a) Use a shift cipher to code a message. Decide the number of spaces to shift each letter. For example, a shift of +3 replaces the letter *A* with *D* and the letter *B* with *E*. Apply your shift cipher to the following message: "Benedict Arnold used a substitution code to send messages to the British in the Revolutionary War."

   b) Describe how to code the letter *Z* with your shift cipher.

   c) Describe how you treated the spaces between words. What is another way to treat the spaces?

   d) You replaced the plaintext letters with different letters. If you replace the letters with numbers, what is your coded message?

   e) Suppose you want someone else to code the message exactly as you did or to understand how you coded the message so that they can decode it. One way to communicate your method is to describe the process in words. What are some other ways? Use several representations to communicate the coding process you used. Create your own diagram if you want.

ACTIVITY

MOVE OVER

2

f) Prepare to present your diagrams or descriptions to your class. Describe the strengths and weaknesses of each diagram or description.

4. Write a brief, appropriate message you would like to send to another student or group. Code the message using another shift cipher. Challenge the student or group to decipher the message. If asked, provide a diagram representing the process you used.

CONSIDER:

1. Suppose you shift the alphabet more than 26 spaces. For example, use a coding process that shifts the alphabet +30. How do you code a letter as another letter?

2. What are the advantages to coding a letter as a number rather than as another letter?

3. Some employers rely on a seniority list when they make decisions about promoting people. If you are number 5 on a seniority list, then only four people have worked at the company longer than you. Suppose the number 1 and 2 people on the seniority list retire from the job. How is the movement of each person on the seniority list similar to a shift cipher?

INDIVIDUAL WORK 2

The Shift

Many codes convert letters to numbers as the first step. Each letter of the alphabet is assigned a position number from 1 to 26. You may want to copy the table of position numbers in **Figure 2.3** into your notebook for future reference.

Plaintext letter	Position number	Plaintext letter	Position number
A	1	N	14
B	2	O	15
C	3	P	16
D	4	Q	17
E	5	R	18
F	6	S	19
G	7	T	20
H	8	U	21
I	9	V	22
J	10	W	23
K	11	X	24
L	12	Y	25
M	13	Z	26

Figure 2.3.
The position numbers of the letters of the alphabet.

A shift cipher gives each letter of a message a position number and adds the amount of the shift to that number. The resulting number is called the **coded value**. For example, coding *K* with a shift of 5 results in a coded value of 16, because *K* is the 11th letter of the alphabet and $11 + 5 = 16$.

1. Find the coded value in each of the following:

   a) Code the letter *J* with a shift of +15.

   b) Code the letter *Q* with a shift of +12.

   c) Code the letter *M* with a shift of –3.

2. a) Can a shift cipher use subtraction? Explain.

   b) Can multiplication or division be used in coding processes? Explain.

3. Is it possible to design a coding process that matches a letter to the number 4.5?

4. Suppose you code by shifting the alphabet +4 followed by a shift of –7.

   a) This is a two-step process. Can you shorten it to one step? If so, how? If not, why not?

   b) If you shift –7 before you shift +4, are the coded values different? Explain.

5. A common practice of coders is to regularly change their method in case a code breaker has cracked the code and is reading the messages. Would changing from shift +20 to shift –6 foil a code breaker who has cracked the shift +20 cipher? Explain.

6. Scientists use two different scales to record temperature: Celsius and Kelvin. You can convert a Celsius temperature to Kelvin by adding 273. For example, 30°C is 303K. (The "degrees" symbol is not used with the Kelvin scale.)

   a) Convert 83°C to Kelvin.

   b) Convert –20°C to Kelvin.

   c) Explain how to convert a Kelvin temperature to Celsius.

   d) How is converting from Celsius to Kelvin (or Kelvin to Celsius) like a shift cipher?

7. Tables and graphs are applied to coding processes later in this lesson. The following questions will help prepare you.

   The table in **Figure 2.4** represents the total cost to send a package based on the weight of the package. (The cost does not account for oversized packages.)

   a) Estimate the cost to send a 14-pound package.

   b) About what does a package weigh if it costs $25.00 to ship?

   The graph in **Figure 2.5** represents the cost to rent a particular power tool, based on the number of hours you rent it.

   c) Use the graph to estimate the cost of renting the tool for 9 hours.

   d) About how many hours did you rent the tool if the cost is $55.00?

Weight (in pounds)	Total shipping cost ($)
1	8.50
5	13.50
10	17.50
15	23.75
20	32.25

Figure 2.4.
Shipping cost.

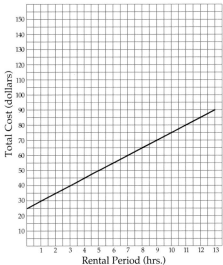

Figure 2.5.
Equipment-rental cost.

8. Tables and graphs appear often in newspapers, magazines, books, and brochures. Find a table or graph in a book, magazine, or newspaper. Draw, photocopy, or cut out the table or graph, and bring it to class to share. Write a sentence or two explaining the information that the table or graph conveys to the reader.

9. In Unit 1, *Pick a Winner: Decision Making in a Democracy*, you used matrices to organize information needed to determine point counts. You can use matrix addition or the list feature of a calculator to speed the process of coding and decoding a message with a shift cipher. The following general directions apply to most graphing calculators. Consult your calculator manual for specifics.

The example codes the word *matrix* with a shift +5 cipher.

- Convert the word *matrix* to position numbers: 13 1 20 18 9 24.

- Enter the numbers into your calculator as a 1 x 6 matrix. The dimensions of the matrix are up to you. You can use any reasonable dimensions. Because *matrix* has six letters, 1 x 6, 2 x 3, and 6 x 1 are good. Name the matrix [A].

- Build a matrix of the same dimensions consisting of all 5s. Name it [B]. (Two matrices may be added only if their dimensions are identical.)

- Perform the matrix addition [A] + [B].

You can also use the list feature of a calculator.

- Convert the word *matrix* to position numbers: 13 1 20 18 9 24.

- Enter these numbers into a list.

- Add 3 to the list.

Use list or matrix features to code *plurality* with a shift +8 cipher.

10. In *The Adventure of the Dancing Men* by Arthur Conan Doyle, secret messages with characters resembling "hieroglyphics" (**Figure 2.6**) are an important part of the plot. What kind of code do you think was used?

Figure 2.6.
A message from *The Adventure of the Dancing Men*.

ACTIVITY

FOLLOW THE ARROW

3

In Unit 1, *Pick a Winner: Decision Making in a Democracy*, you learned that it can be easier to solve a problem in one representation than in another. The same is true of coding, decoding, and code cracking. To be good at all three, you must know a variety of ways to represent a coding process.

In this activity, you use tables, graphs, and a representation called an **arrow diagram** to analyze and communicate coding processes.

Figure 2.7.
Shift +4 arrow diagram.

1. The arrow diagram in **Figure 2.7** represents a coding process.

   a) Describe the coding process.

   b) Draw a new arrow diagram representing the same process, but with letters coded as letters.

   c) Use an arrow diagram to represent a shift of +3 followed by a shift of +7. Copy **Figure 2.8** onto your paper and write the description of each step above its arrow.

Figure 2.8.
A three-step arrow diagram.

   d) What are some advantages of this representation? That is, would the coder, decoder, or code breaker find it useful?

ACTIVITY

FOLLOW THE ARROW

3

Mathematicians use the term **function** to describe a process that transforms items such as letters or numbers into other letters or numbers uniquely. When a message is coded, a plaintext letter or position number is matched with a unique new coded value or letter.

A function produces a single output from an input. In the context of codes, the coded value is the output that results when the coding process is applied to the original position number (the input). In the context of a business, the input may be the number of items sold and the output may be the total profit.

An arrow diagram is a good way to show the steps of a process like a function. A table is another way to represent a function. A table lists some or all of the matched pairs of inputs and outputs.

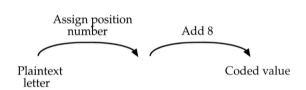

Assign position number   Add 8

Plaintext letter → Coded value

Figure 2.9.
An arrow diagram for a coding process.

2. The arrow diagram in **Figure 2.9** represents a shift coding process.

Plaintext letter	A	B	C	D	E	F	G	H	I	J	K	L	M
Original position													
Coded value													
Plaintext letter	N	O	P	Q	R	S	T	U	V	W	X	Y	Z
Original position													
Coded value													

Figure 2.10.

a) Use the arrow diagram in Figure 2.9 to complete the table in **Figure 2.10.**

b) What are some advantages of this representation? That is, would the coder, decoder, or code breaker find it useful?

FOLLOW THE ARROW   ACTIVITY 3

A **coordinate graph** is another way to represent a coding process. A coordinate graph is based on two lines that intersect at right angles. Each line is called an **axis**. In this unit, the horizontal axis usually represents the plaintext letter or its position number. In general, the horizontal axis is associated with a function's input. The vertical axis represents the coded value or, sometimes, the coded letter. In general, the vertical axis is associated with a function's output. The place where the two axes intersect is called the **origin** (see **Figure 2.11**).

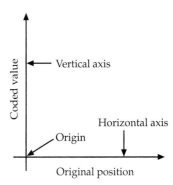

Figure 2.11.
A coordinate graph for coding.

Every location or point on a graph is identified with a pair of numbers called **coordinates**. The numbers in each pair are ordered so the first number represents horizontal position on the graph and the second number represents vertical position on the graph. For example, the point with coordinates (5, 8) identifies the point on the graph with a horizontal position of 5 and a vertical position of 8 (see **Figure 2.12**). In the context of codes, the point with coordinates (5, 8) means the letter with original position of 5 (the input) is matched with the coded value of 8 (the output).

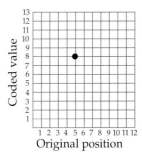

Figure 2.12.
The graph of (5, 8).

Graphs should always have a title or brief description, each axis should be numbered with a consistent scale, and each axis should have a label or description. Note that in Figure 2.12 the brief description is "The graph of (5, 8)," both axes are numbered with scales of one, the vertical axis is labeled "Coded value," and the horizontal axis is labeled "Original position."

ACTIVITY

3

FOLLOW THE ARROW

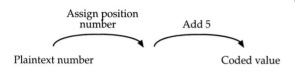

Figure 2.13.

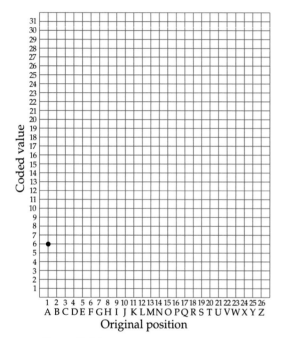

Figure 2.14.
Graph of coded value versus original position.

3. The arrow diagram in **Figure 2.13** represents a coding process.

   a) Use a sheet of graph or dot paper (as in **Figure 2.14**) to graph the coding process.

   b) Describe your graph.

   c) How is the graph different if you shift eight instead of five?

   d) What are some advantages of this representation? That is, would the coder, decoder, or code breaker find it useful?

4. If you want a friend to decode your messages, you must communicate the coding process. You can use an arrow diagram, a table, or a graph to communicate the coding process. Which representation would you choose? Why?

People who exchange coded messages must tell each other how they are coding so the recipient of the message can decode it. Communicating the method is the classic problem of secret codes. Telling someone you are coding by "adding 3" requires brief communication; telling someone an entire table requires a larger communication.

Machines like computers are frequently used as aids by people who code and decode. Computers and calculators work more easily with numbers than with letters.

Symbolic forms, which are the topic of Activity 4, are required by many computers and graphing calculators. Symbolic forms allow the calculator to do much of the tedious and repetitious work. Another advantage of symbolic forms is they are easy to change. Coders often change their process to foil code crackers. Dictionary coding systems are hard to change because it requires writing an entire dictionary that must be transmitted to all parties using the system.

INDIVIDUAL WORK 3

The Arrow Points the Way

1. a) Write a message of ten or fewer words.

   b) Select a number of letters to shift the alphabet and make a table for your shift (as in **Figure 2.15**).

Plaintext letter	A	B	C	D	E	F	G	H	I	J	K	L	M
Code Letter													
Plaintext letter	N	O	P	Q	R	S	T	U	V	W	X	Y	Z
Code Letter													

   Figure 2.15.

   c) Use the table to code your message in letters.

   d) Code your message in numbers.

   e) Do you think coding with letters or with numbers is best? Why?

2. The graph in **Figure 2.16** shows the way a shift cipher codes the first six letters of the alphabet.

   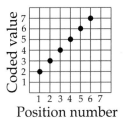

   Figure 2.16.
   Part of the graph of a shift cipher.

   a) Describe the shift cipher in your own words.

   b) How does this shift cipher code the letter *H* as a letter? As a number?

   c) Use this cipher to code the word *codes*.

   d) The word 17 19 16 4 6 20 20 was coded with this shift cipher. Decode it.

3. Arrow diagrams are an excellent way to show a process.

   a) Explain the process in **Figure 2.17** in words.

   Location   Zip code   Bar code

   Figure 2.17.

   b) Explain the process in **Figure 2.18** in words.

   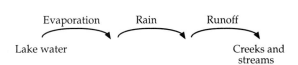

   Evaporation   Rain   Runoff

   Lake water   Creeks and streams

   Figure 2.18.

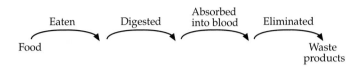

Figure 2.19.

Figure 2.20.

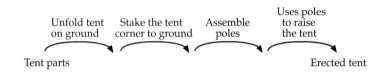

Figure 2.21.
An arrow diagram for setting up a tent.

c) Explain the process in **Figure 2.19** in words.

d) Describe what is wrong with the arrow diagram in **Figure 2.20**.

e) Prepare an arrow diagram to represent the process involved in getting you to school.

f) Think of a familiar process that has several steps. Draw an arrow diagram and write a verbal description for the process.

g) Arrow diagrams can be used to represent mathematical processes. Suppose that each week you receive an allowance of $15.00. From that $15.00 you must pay $5.00 for your gymnastics class. Draw an arrow diagram.

h) Processes represented by arrow diagrams are often reversible. **Figure 2.21** represents the process of setting up a tent with poles and stakes. Refer to the diagram and describe the reverse process of disassembling the tent.

4. The graph in **Figure 2.22** represents part of a coding process.

   a) Describe how to determine where to draw the next two dots on the graph.

   b) What is the meaning of the point (8, 21)?

   c) Use the graph to encode the word *king*.

   d) How would you communicate this coding process to another person without using the graph?

   e) The letter *C* is coded as the number 11. The letter *E* is coded as the number 15. Which letter is coded as the number 31?

   f) Explain why the number 20 cannot represent a letter.

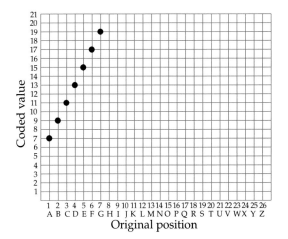

Figure 2.22.

5. Arrow diagrams describe processes. A table shows how symbols and values are paired. Use the arrow diagram to complete the table in **Figure 2.23**.

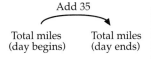

Day	0	1	2	3	4	5	6	7					
Total Miles	0	35											

Figure 2.23
An arrow diagram and the related table.

ACTIVITY

CALCULATOR CODING

4

A graphing calculator is helpful for coding a message. To use it, you must translate a coding process into a symbolic equation and enter the equation into the calculator. This activity shows you how to use the graphing calculator to encode secret messages.

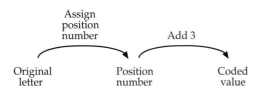

Figure 2.24.
A coding process.

A **symbolic equation** uses symbols to represent mathematical relationships. For example, the arrow diagram in **Figure 2.24** represents a shift of three units. You add 3 to the original position number to obtain the coded value. This statement can be written in the form of a word equation: coded value = position number + 3.

Mathematicians like to keep their equations brief, so they often use a single letter to represent a word or phrase. If the letter c represents "coded value" and the letter p represents "position number," then the equation can be written in symbolic form: $c = p + 3$.

Mathematicians have names for the various parts of symbolic equations. A **variable** represents something that changes or varies. The letters c and p are both variables in the equation $c = p + 3$, because they represent different numbers depending on the letter you are coding.

A **constant** remains unchanged whenever a particular equation is used. The number 3 is the constant in the equation $c = p + 3$. A constant is sometimes called a **control number** because, in a shift code, the constant "controls" the amount and direction of the shift.

ACTIVITY

CALCULATOR CODING

4

1. a) Write a word equation and a symbolic equation to represent a shift of 12.

   b) Write a word equation and a symbolic equation for the arrow diagram in **Figure 2.25**.

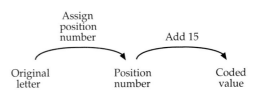

Figure 2.25.

Most of the time you build equations using any letters you choose for the variables. However, most graphing calculators require that you use only the letters X and Y when graphing. The variable X represents the input, which is the position number of the plaintext letter. The variable Y represents the output, which is the coded value. The symbolic equation $c = p + 3$ becomes $Y = X + 3$ when using a graphing calculator. Many calculators allow you to draw several graphs at once. These calculators attach a number to Y so that the graphs can be distinguished. The first equation would be $Y1 = X + 3$. A second equation, $c = p + 10$, might be entered as $Y2 = X + 10$.

Mathematicians use the term **domain** when referring to the collection of all the x-values (inputs) used in a function or coding process. The term **range** refers to the collection of all the y-values (outputs). The function pairs each number from the domain with a number from the range.

For most of the coding processes in this unit, the domain represents the position numbers from 1 through 26. The range represents the coded values and varies with the coding process used.

Some graphing calculators use "range" to describe the highest and lowest values for both X and Y that appear on the graphing screen. Other calculators use the term **window** for the same purpose. "Window" is used in *Mathematics: Modeling Our World*.

ACTIVITY

CALCULATOR CODING

4

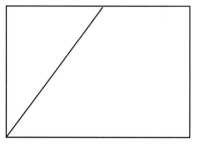

[0, 30] x [0, 30]

Figure 2.26.
Sketch of a calculator graph of Y1 = 2X.

To produce a graph on a calculator, you need to do two things: enter the equation, and set the window.

When you have a graph and you want to sketch the calculator screen, you should indicate the minimum and maximum values for the window. Mathematicians usually enclose the minimum and maximum values in brackets. For example, [0, 30] x [5, 40] means that Xmin is 0, Xmax is 30, Ymin is 5, and Ymax is 40. **Figure 2.26** is a sample sketch of the equation $Y1 = 2X$.

2. Consider the equation $c = p + 3$. You can enter it into a graphing calculator as $Y1 = X + 3$. Since X represents the position number, the lowest position number is 1 (for A) and the highest is 26 (for Z). When you code using a shift +3 cipher, the lowest coded value is 4 (for A) and the highest value is 29 (for Z). It is usually helpful to select a window slightly larger than the actual values you want to graph, to avoid having important points on the edge of the screen.

   a) Enter the equation $c = p + 3$ in the graphing calculator as $Y1 = X + 3$. Set your window to Xmin = 0, Xmax = 30, Ymin = 0, and Ymax = 30.

   b) Graph the equation. Sketch the screen of the calculator on your paper.

The calculator graph looks different from a coding-process graph you draw on paper because the calculator draws a continuous line instead of distinct (or discrete) points for each position number.

Now that you have graphed the equation in Item 2, you will use it to encode letters. For example, to encode the letter P, start with its position number, 16. Use the trace feature of the calculator to trace to the point on the graph where X is as close as possible to 16. The Y-value should be very close to the coded value.

ACTIVITY

CALCULATOR CODING

4

3. What is the numerical value of the encoded
   letter *P* (using the graph from Item 2)?

A graphing calculator can do things that seem a
little strange. It can be hard to get the calculator to
trace to exactly the point you want. For example,
Figure 2.27 is a calculator graph of $Y1 = X + 3$ with
a trace stopped as close as possible to the point
where *X* is 10. If this graph represents a coding
process, then the best you can do is to code *J*, the
tenth letter of the alphabet, as 13. You know that *J*
codes as a whole number, and it seems reasonable
to interpret the coded value as 13.

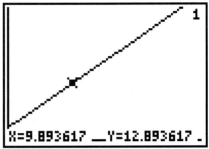

Figure 2.27.
Calculator graph of
$c = p + 3$ or $Y1 = X + 3$.

In a graph like the one in Figure 2.27, the readouts of 9.893617 for
X and 12.893617 for *Y* are the coordinates of the point where the
trace is stopped.

4. Use the graphing calculator and your graph from Item 2 to
   encode the word *window*.

5. Some graphing calculators have a table feature. You need to
   set up the table before you display it. This table should start
   at 1 and reflect increments of 1.

   a) Make a table for the coding process $c = p + 3$. When you
      have the table you want, scroll up and down to see the
      entire table.

   b) Use the calculator table to encode the word *range*.

6. The graphing calculator allows you to graph more than one
   equation at a time. Keep the first equation you entered in
   Item 2 as $Y1 = X + 3$.

   a) Write a calculator equation to represent a shift of 12.
      Call it *Y2*.

ACTIVITY

4

CALCULATOR CODING

b) Adjust the window to include all the values related to a shift +12 cipher. The letter Z has a coded value of 38. Adjust your window to allow 38 to be seen. Enter the equation for $Y2$ in the calculator, and graph it. Draw a sketch of the screen on your paper. You should see two lines in the viewing window of the calculator (one for $Y1$ and one for $Y2$). Label each line in your sketch with its equation.

c) Use the calculator and a shift +12 cipher to encode the word *crypto*.

d) Describe how the graph for $Y1 = X + 3$ is different from the graph for $Y2 = X + 12$.

Both $Y1 = X + 3$ and $Y2 = X + 12$ represent **linear functions**. They are linear because their graphs are lines. They are functions because for each X or original position number (input), there is one Y or coded value (output).

7. a) Will the graphs of shift ciphers always form parallel lines? Explain your answer.

b) What type of coding process might result in a line that is not parallel to $Y1 = X + 3$?

CONSIDER:

1. Tables, graphs, arrow diagrams, matrices on a calculator, and graphs on a calculator are methods that you can use to code messages. When do you prefer each representation?

2. Recall that the first characteristic of a good coding process is that it must be easy to encode. Do you think the shift cipher is easy to encode?

INDIVIDUAL WORK 4

Coding with Graphs

1. The graph representing a shift +5 cipher is not identical to the graph representing a shift +3 cipher. Describe the differences and similarities.

2. To use a graphing calculator for coding, you must convert the process to symbolic form. Convert each of the arrow diagrams in **Figure 2.28** to equations that can be entered into a graphing calculator. The first one is done for you as an example.

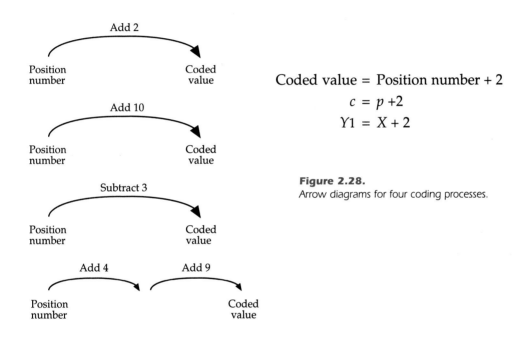

Coded value = Position number + 2

$$c = p + 2$$

$$Y1 = X + 2$$

Figure 2.28.
Arrow diagrams for four coding processes.

3. Suppose you code by shifting the alphabet −8.

    a) Draw an arrow diagram to represent this coding process.

    b) How do you code the letter *B* as a number? As a letter?

    c) Write the equation you would enter into the graphing calculator.

    d) The position number for the letter *A* is 1 and for *Z* is 26. Therefore, when you graph on the graphing calculator, you set the window so Xmin = 1 or lower (usually Xmin = 0) and Xmax = 26 or higher (usually Xmax = 30). For the shift −8 cipher, what are the smallest and largest coded values?

4. Compare a coding process that adds 3 with one that multiplies by 3.

    a) What are the lowest and highest coded values for the addition process?

    b) What are the lowest and highest coded values for the multiplication process?

    c) Describe the differences in the graphs of the two processes.

5. **Figure 2.29** shows four ways to represent the process of computing a sales tax of 6%.

    There are advantages and disadvantages to each representation. Choose the representation you prefer for each question. Explain your choice.

GRAPH

TABLE

Price	Tax
0.50	0.03
1.00	0.06
1.50	0.09
2.00	0.12
2.50	0.15

ARROW DIAGRAM

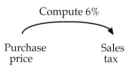

Compute 6%

Purchase price → Sales tax

Figure 2.29.
Four ways to represent a 6% sales tax.

EQUATION

Tax = 0.06 x Price

a) Find the tax on $2.00.

b) Find the tax on $587.16.

c) Show that tax increases as the price increases.

d) Show that tax is computed by multiplying and not by adding.

e) Compare a 6% sales tax with a 8% sales tax.

6. Compare the five representations: word description, arrow diagram, symbolic equation, table, and graph. Which representation do you prefer for each of the following?

a) To code a word or message

b) To transfer the coding process to technology (like a graphing calculator)

c) To communicate your coding process to another person

d) To compare two different coding processes to see if they code one or more letters the same

7. a) What coding process reverses the alphabet? That is, what is the equation that codes A as Z, B as Y, C as X, etc.?

b) Is this a shift cipher?

c) Is any letter left unchanged by this coding process?

d) How is the graph of this coding process different from the graph of a shift cipher like $c = p + 2$?

8. The following message was coded using the process $c = p + 4$. Decipher the message. Draw a diagram or explain the process you used to decode.

C S Y V   T E W W A S V H   L E W   F I I R   G L E R K I H   X S

H S G X S V H I G S H I V.

LESSON THREE

Decoding

KEY CONCEPTS

Inverse of a function

Coding methods: stretch cipher, piecewise cipher

Families of functions

Solving equations

Modular arithmetic

The Image Bank

PREPARATION READING

Reversing the Process

Coded messages must be decoded by the people receiving them. Can you decode a message if someone tells you how it was coded?

One way to decode is to list all the coded letters or words and give a copy to those who receive your messages. If you want the letter *A* to be represented by *Q*, you list *A* and *Q* as the original and coded forms. This can be done with a simple table in many cases.

If the code uses words to replace other words, an entire dictionary must be exchanged. The disadvantage of this type of process is that, as more people are involved, it is more likely that a copy of the dictionary may fall into the wrong hands.

A system that uses a mathematical process gets around this difficulty because the process is simple. Historically, mathematics has often been used to encode messages because mathematical processes are usually easy to reverse.

Decoding is like undoing or reversing a process. Suppose you walk a particular route from your home to a store. Usually you can return to your home by traveling the route in the reverse order. If you have ever had to take apart a machine to fix something, then you probably reversed the steps when you put it back together.

This lesson explores how to decode secret messages by reversing a mathematical process. You have used tables, graphs, equations, and arrow diagrams to *encode* a message. How do you use them to *decode* a message?

CONSIDER:

1. Decoding is the reverse of encoding. What is the reverse of each of the following processes?

   - Wrapping a package

   - Filling a pool with water

   - Putting on roller blades and clamping the buckles

   - Getting into a car, putting the key into the ignition, and turning the ignition switch clockwise

   - Opening a suitcase, packing the suitcase with clothes, and then closing the suitcase

2. What operation reverses each of the following mathematical operations?

   - Addition

   - Subtraction

   - Multiplication

   - Division

3. Sometimes you need to find half of a number or 25% of a number. What is the reverse of finding half of a number? What is the reverse of finding 25% of a number?

ACTIVITY

CODES IN REVERSE

5

An effective coding process makes coding and decoding relatively easy for whoever is sending messages. In this activity, you consider four messages and the method used to encode them. You must be able to reverse the process, or the method doesn't pass the ease-of-decoding test. Judge for yourself. Try to decode all four.

Mathematicians use the term **inverse** to describe a process that converts a coded value to the position number. In order to decode each of these messages, you must determine the inverse of the process used to code them.

Message 1 was coded with the process in **Figure 2.30**.

P F W H W G V   Q F O Q Y W B U   C T   H V S   U S F A O B

S B W U A O   Q C R S   K O G   O   R S Q W R W B U

T O Q H C F   W B   K C F Z R   K O F   H K C

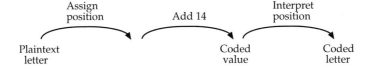

Figure 2.30.
The coding process for Message 1.

Message 2 was coded with the process shown in **Figure 2.31**.

29 17 14   15 18 27 28 29   30 28 14   24 15   12 24 13 14 28

29 24   28 14 23 13   22 14 28 28 10 16 14 28   32 10 28   18 23

28 25 10 27 29 10   10 27 24 30 23 13   15 24 30 27

17 30 23 13 27 14 13   11 12

Plaintext letter	A	B	C	D	E	F	G	H	I	J	K	L	M
Position number	1	2	3	4	5	6	7	8	9	10	11	12	13
Coded value							16						
Plaintext letter	N	O	P	Q	R	S	T	U	V	W	X	Y	Z
Position number	14	15	16	17	18	19	20	21	22	23	24	25	26
Coded value		24											

Figure 2.31.
Part of the coding process for Message 2.

CODES IN REVERSE

ACTIVITY 5

Message 3 was coded with the process shown in **Figure 2.32**.

8 20 9 10 24 6 23 10 24 20 18 10 25 14 18 10 24 26 24 10 9 14 19 8 20
18 21 26 25 10 23 24 25 20 8 20 18 21 23 10 24 24 6 19 9 23 10 8 20
27 10 23 14 19 11 20 23 18 6 25 14 20 19

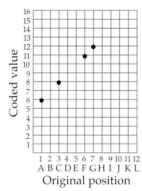

Figure 2.32.
Part of the coding process for Message 3.

Message 4 was coded with the process $c = p + 20$.

25 32 25 42 25 34 24 29 27 29 40 46 29 36 23 35 24 25 39 23 21 34 32
35 23 21 40 25 21 24 25 32 29 42 25 38 45 36 35 29 34 40 32 29 31 25
45 35 41 38 39 23 28 35 35 32 35 38 28 35 33 25 43 29 40 28 35 41 40
21 34 21 24 24 38 25 39 39

CONSIDER:

1. How do you use a table to decode?

2. How do you use an arrow diagram to decode?

3. How do you use a graph to decode?

4. How do you use an equation to decode?

5. Which representation works best for decoding? Why?

INDIVIDUAL WORK 5

Reverse That Process

1. Below are the seven steps for flushing a sand pool filter.

   • Turn off the filter.

   • Plug the drain.

   • Remove the blue filter hose from the water.

   • Loosen the clamp on the green filter hose.

   • Detach the green filter hose.

   • Attach the blue hose to your yard hose.

   • Turn on the water to flush the filter.

   When the water exiting the green hose is clear, the filter is properly backflushed.

   Describe how to reassemble the filter system.

2. Otto is at home and needs to drive to the park to pick up his sister. His mother left the following directions: north on Washington Street, left on Sixth Street, right on Pacific Boulevard, left on Huron Drive, and left on Quincy Place. Describe the route Otto should follow, including the direction he should turn, to return home from the park.

3. a) Suppose you encode by adding 12, as in **Figure 2.33**. Diagram the decoding process.

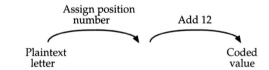

Figure 2.33.
Coding by adding 12.

   b) Suppose you encode by subtracting 8. Describe or diagram the decoding process.

   c) Suppose you encode by multiplying by 3. Describe the decoding process using at least two different representations.

4. When you travel from California to New York, you move your clock ahead three hours. Draw an arrow diagram demonstrating the original and the inverse process.

5. Decode each message.

a) The arrow diagram in **Figure 2.34** represents the coding process for this message:

   19 22 23 31 17 22 15   16 23 31   28 23.

b) The equation $c = p - 2$ represents the coding process for this message:

   0 19 7 10 2   17 3 1 16 3 18   11 3 17 17 −1 5 3 17.

c) The table in **Figure 2.35** represents the coding process for this message:

   46 18 24 24   16 10 24 32   50 30 42.

d) The graph in **Figure 2.36** represents part of the coding process for this message:

   25 6 16 10   25 13 10 18   6 21 6 23 25.

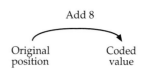

Figure 2.34.
A coding process.

OP	CV
A	2
B	4
C	6
D	8
E	10

Figure 2.35.
Part of a coding process.

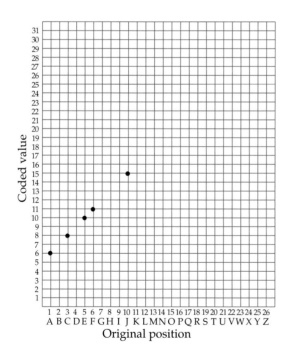

Figure 2.36.
Part of a coding process.

e) Combine the messages from parts (a)–(d).

GRAPH

TABLE

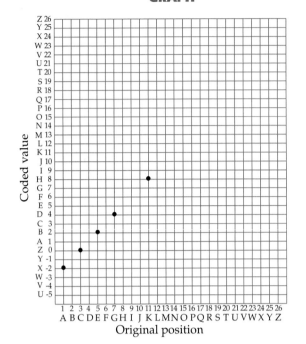

Original position	Coded value
A	−2
B	−1
C	0
D	1
E	2

ARROW DIAGRAM

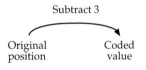

Subtract 3

Original position → Coded value

Figure 2.37.
Four representations of the same coding process.

EQUATION

$$c = p - 3$$

6. a) **Figure 2.37** shows four different representations of the same coding process. Decode the following message:

4 15 24 13 5 16 15 2 19 2 24 9 13 24 17 17 2 0 11 16.

b) Describe this coding process in words.

c) Which representation did you use to decode the message? Explain why you chose one representation rather than the other three.

d) What is the domain for this coding process? What is the range?

e) Does $c = p - 3$ represent a linear function? Explain your answer.

7. Suppose you code using two steps, a shift +8 followed by a shift +3. Describe or diagram how to decode.

8. a) A decoding process is shown in **Figure 2.38**. Diagram the coding process.

   b) Describe the coding process in words.

9. Sometimes plaintext messages are coded as numbers and sometimes as letters. Both words below were coded using a shift +6 cipher. In your opinion, which is easier to decode? Give reasons for your answer.

   22 18 7 15 20 26 11 30 26   V R G O T Z K D Z

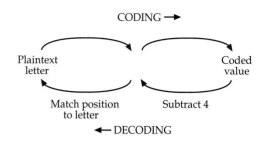

Figure 2.38.
A decoding process.

10. The arrow diagram in **Figure 2.39** describes a new process for encoding messages. The table provides a partial representation of the same process.

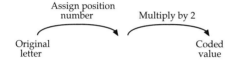

Plaintext letter	A	B	C	D	E	F	G
Original position	1	2	3	4	5	6	7
Coded value	2	4	6	8	10	12	14

Plaintext letter	H	I	J	K	L	M
Original position	8	9	10	11	12	13
Coded value	16	18	20	22	24	26

Figure 2.39.
Arrow diagram and table for a new coding process

a) Decode the word 10 24 2 38 40 18 6.

b) Why is this cipher sometimes called a **stretch cipher**?

c) The equation for this stretch cipher is $c = 2p$. Does this coding process represent a linear function? Explain your answer.

d) How does the graph of the stretch cipher differ from the graph of a shift cipher like $c = p + 2$?

e) Write an equation to represent a coding process that stretches the distance between coded values by more than 2.

11. a) Use a graphing calculator to graph a shift +3 cipher and a stretch 3 cipher on the same graph.

    b) What window did you use for this graph?

    c) Describe how the graphs are similar.

    d) Describe how the graphs are different.

    e) How can you use the graphs to tell if $Y1 = X + 3$ and $Y2 = 3X$ code any letters the same?

12. A coding process uses a shift +2 followed by a stretch of 3.

    a) Draw an arrow diagram to represent the coding and decoding processes.

    b) Code the letter B.

    c) Decode 36. Find the original position and the plaintext letter.

13. For each shift cipher you have studied, you add (or subtract) the same constant number to (or from) each position number. Suppose you invent a coding process in which you add a position number to itself. Describe how this coding process is similar to a shift cipher and how it is different.

14. A family is a group of people who have something in common. For most people, family means being related to one another through birth or marriage or having the same ancestors or house in common. Sometimes members of sports teams or church groups feel like family, having the same goals and struggles in common.

    Mathematicians call a group of functions a family of functions if they have something in common. What do shift ciphers have in common with one another that qualify them as a family?

15. A **piecewise cipher** uses one coding process for part of the alphabet and a different coding process for the rest of the alphabet. Suppose $c = p + 2$ is used to code letters A through K and $c = p + 8$ is used to code letters J through Z.

    a) Code the word *encrypt*.

    b) What are the advantages of a piecewise cipher?

    c) What are the disadvantages of a piecewise cipher?

ACTIVITY

SYMBOLIC DECODING

6

You have used tables, graphs, arrow diagrams, and equations to decode. In this activity you will use symbolic methods to decode messages.
Recall from Lesson 2 that when using a graphing calculator you might not be able to locate the exact point you want because the graphing calculator produces a continuous line. The actual graph of a shift coding process is a series of discrete points that form a linear pattern.

Remember the distinction between continuous and discrete whenever you use a graphing calculator with a coding process. For example, a sidewalk represents a continuous path while a path made of stepping stones is discrete. A steady stream of water represents a continuous flow while a dripping faucet represents a discrete flow.

1. Use a graphing calculator to graph the coding process described by the equation $c = p + 20$. Make a sketch of the graph. Be sure to record the window, too.

2. Use your graph to decode the coded value 30.

In the equation $c = p + 20$, the variable p represents the position number and the variable c represents the coded value. By decoding the coded value 30, you are **solving** the equation $30 = p + 20$. A **solution** to an equation is a number you put in place of p that results in a true statement.

For example, suppose you want to decode 15, and 15 was coded with $c = p + 7$. You decode 15 by solving the equation $15 = p + 7$. Perform the inverse operation by subtracting 7 from 15. The solution is 8. When you replace p with 8, $15 = 8 + 7$ becomes a true statement.

3. When you decode the value 8 that was coded with the process represented by $c = p - 12$, what equation are you solving? What is the solution? How is your solution related to the graph of $c = p - 12$?

ACTIVITY

6

SYMBOLIC DECODING

An equation like $c = p + 7$ really has several solutions. Each solution is a pair of numbers. If p represents the first number of the pair and c represents the second number, then the pair (3, 10) is a solution to the equation $c = p + 7$. Replacing p with 3 and c with 10 produces the true statement $10 = 3 + 7$.

4. Are the pairs (5, 25) and (23, 3) solutions to the equation $c = p + 20$? How can a graph of $c = p + 20$ be used to answer this question?

When you decode, you can write an equation to describe the decoding process. For example, if the coding process is $c = p + 5$, then the decoding process is $p = c - 5$. To decode a coded value like 13, you can replace c with 13 in the decoding equation: $p = 13 - 5$. When you are decoding several coded values, it is quicker to use the decoding equation than to solve a separate equation for each coded value.

5. What is a decoding equation for the stretch cipher $c = 2p$?

CONSIDER:

To decode you can solve an equation.

1. How is solving an equation like decoding with a table?

2. How is solving an equation like decoding with a graph?

3. How is solving an equation like decoding with an arrow diagram?

INDIVIDUAL WORK 6

Step by Step

1. Suppose a coding process replaces a letter with its position number and adds 3. The decoding process must first subtract 3 and then interpret the position number as a plaintext letter. The steps are reversed, and the inverse operation is used (see **Figure 2.40**).

   a) Encode the letter *K*.

   b) Decode the number 15.

   c) What equation are you solving when you decode 15?

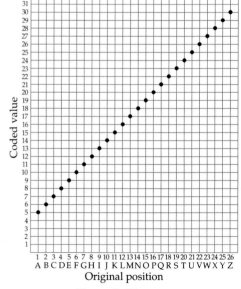

Figure 2.40.
An arrow diagram for both coding and decoding.

2. A coding process is described by the table in **Figure 2.41**.

   a) What are the smallest and largest coded values?

   b) Describe the domain and the range for the coding process.

   c) Encode the letter *P*.

   d) Decode the number 23.

   e) What equation did you solve when you decoded 23?

Plaintext letter	A	B	C	D	E	F	G
Coded value	7	8	9				

Figure 2.41.
Part of a coding process.

3. The graph in **Figure 2.42** represents a shift cipher.

   a) Explain why this coding process is a linear function.

   b) What is the meaning of the point (6, 10) in the coding context?

   c) Decode the number 16.

   d) What equation did you solve when you decoded the number 16?

   e) What equation represents the decoding process?

4. Suppose you graph $c = p + 5$ as $Y = X + 5$ on a graphing calculator.

   a) Is (3, 8) a point on the graph?

   b) Is (3, 8) a solution to the equation $c = p + 5$?

   c) In the context of coding, what is the meaning of the point (3, 8) ?

   d) Is (6, 10) a point on the graph?

Figure 2.42.
A graph of a shift cipher.

e) Is (6, 10) a solution to the equation $c = p + 5$?

f) In the context of coding, what is the meaning of the point (6, 10)?

g) Can a point not on a line be a solution to the equation that was used to create the line?

5. a) What equation do you solve to decode 27 if it was coded with $c = p + 5$?

b) The point (1.7, 6.7) is on the line $y = x + 5$. Check that it is. In the coding context, what is the meaning of the point (1.7, 6.7)?

c) You decide to add five new pages at the beginning of a report you are writing. All the old pages have to be renumbered. For example, the former page 10 is now numbered 15. The equation $y = x + 5$ represents the repagination. Explain the meaning of (16, 21) and (1.7, 6.7).

6. The equation $c = p + 15$ represents a shift cipher.

a) Show the coding and decoding processes in an arrow diagram.

b) Decode 22. What equation do you solve when you decode 22?

c) Is the point (14, 40) a point on the graph of $c = p + 15$? How can you tell without graphing?

7. The word *secret* was coded as M Y W L Y N. Write symbolic equations for the coding process and the decoding process.

8. The following message was encoded using the equation $c = p - 5$:

   –2 13 –4 –2 9 7 12 5 –2 10 –1 0 14 13 0 12 16 4 13 0 14 11 –4 15 4 0 9 –2 0

a) Describe or diagram how to decode the message.

b) Write an equation for the decoding process.

c) Decode the message.

9. Is a shift cipher easy to decode? Explain your answer.

10. Create your own coded message using a shift cipher. Describe or diagram the coding process using an equation, arrow diagram, table, or graph. Give the message and the diagram to another person. Observe while the person attempts to decode the message. Write your observations. Write comments related to how difficult decoding was for the other person, what approaches the other person used, where the person got stuck, questions the person asked, how persistent the person was, and how successful the person was.

CODING FROM ALPHABET TO ALPHABET

7

As you know, coded values sometimes exceed 26, which means you have to subtract 26 from large values if you code as letters. Leaving the coded message as numbers means you do not have to worry about coded values larger than 26. In this activity, you will take a closer look at what happens when you code as letters.

Consider a shift +9 cipher that tries to replace the coded value with a letter (**Figure 2.43**).

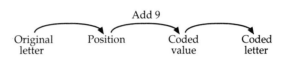

Figure 2.43.
A shift +9 cipher.

A table shows what happens to several letters (**Figure 2.44**).

The letters from *R* to *Z* have coded values of 27 or more. When you subtract 26, the result is the position number for the code letter. The letter *R*, which has coded value 27, is coded as *A* because $27 - 26 = 1$ and 1 is the position number of letter *A*.

Original letter	A	B	C	...	P	Q	R	...	Y	Z
Position number	1	2	3	...	16	17	18	...	25	26
Code number	10	11	12	...	25	26	27	...	34	35
Code letter	J	K	L	...	Y	Z	??	...	??	??

Figure 2.44.
A partial table for a shift +9 cipher.

1. Suppose you are using the coding process $c = p + 32$. Decode the coded values 35, 50, and 54.

When a coded value is very large and you want to find a code letter, you may need to subtract 26 more than once. Suppose the coding process is $c = p + 29$. The letter *A* is coded as the letter *D*, the letter *B* as the letter *E*, etc.

2. Find the code letter that belongs to the original letter *X* using $c = p + 29$.

3. The coding process $c = p + 29$ and the coding process $c = p + 3$ give each letter of the alphabet the same code letter. Can you explain why?

ACTIVITY

CODING FROM ALPHABET TO ALPHABET

7

4. Find three other coding processes that also give the same code letters as $c = p + 29$.

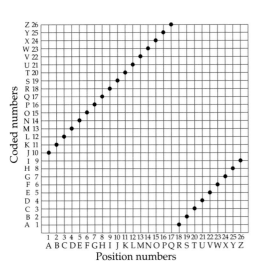

Figure 2.45.
Graph of the coding process $c = p + 9$.

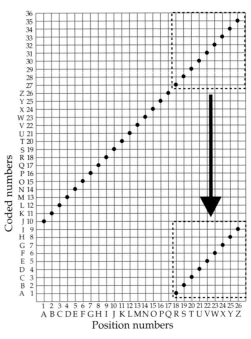

Figure 2.46.
Graph of $c = p + 29$.

Figure 2.45 is a graph of the coding process $c = p + 9$. This graph is different from other graphs you have made in this unit. The first part is a straight line, but the second part, also a straight line, starts at the bottom.

In **Figure 2.46**, you can see how the graph in Figure 2.45 can be found from the ones you have used before. The points that are higher than 26 are pulled down a distance of 26 units.

5. Explain how this action on the graph in Figure 2.46 compares with subtracting 26 from coded values over 26.

ACTIVITY

CODING FROM ALPHABET TO ALPHABET

7

6. The graphs you have just seen belong to the coding process
$c = p + 9$, but also to the processes $c = p + 61$ and $c = p - 17$.
Explain why this is true. Find other coding processes that
have the same graph.

When two coding processes have the same graph, both systems
code each letter of the alphabet with the same code letter. For
example, look at the processes $c = p + 15$ and $c = p - 37$. Here is
how P is coded by these two processes.

$$P \rightarrow 16 \rightarrow 31 \rightarrow 31 - 26 = 5 \rightarrow E$$

$$P \rightarrow 16 \rightarrow -21 \rightarrow -21 + 26 = 5 \rightarrow E$$

Two coding processes that code all letters the same way are
considered identical; whichever process you use, the coded
values are exactly the same.

7. Are the coding systems $c = p - 16$ and $c = p + 11$ identical?

8. How many different coding systems are possible when only
shift ciphers are used?

To find out whether two different shifts create identical coded
messages, calculate the difference between the two shift numbers.
If the difference is a multiple of 26, then the codes are identical.
Mathematicians use a special term to say the difference between
two numbers is a multiple of 26: the two numbers are **congruent
modulo 26**.

The numbers 9 and 61 are congruent modulo 26 because their difference is a multiple of 26: $61 - 9 = 2 \times 26$ or $61 = 9 + 2 \times 26$. For
another way to test this, use the inverse operations. Are 30 and 73
congruent modulo 26? Divide each by 26, and if the remainders
are equal, the numbers are congruent: $30/26 = 1$, remainder 4;
$73/26 = 2$, remainder 21. The remainders are not equal, so 30 and
73 are not congruent modulo 26.

ACTIVITY

7

CODING FROM ALPHABET TO ALPHABET

9. Are these values congruent? Show how you found each answer.

   a) 294 = 8 modulo 26

   b)   26 = 0 modulo 26

   c) −100 = 2 modulo 26

   d)   35 = −42 modulo 26

10. You could use any number as the modulus. The process is the same. For example, 13 modulo 10 = 3. Evaluate each of the following.

    a) 50 modulo 26

    b) 25 modulo 12

    c) 72 modulo 7

    d) 72 modulo 6

    e) 9 modulo 2

You have done a lot of modular arithmetic calculations in your life. You just did them without knowing the mathematical term. The next questions reveal other applications of modular arithmetic.

11. You are making a long trip. It takes you 51 hours to arrive at your destination. You leave home at 9 A.M. on Tuesday. At what time and on which day do you arrive at your destination? What has this to do with modular arithmetic?

CODING FROM ALPHABET TO ALPHABET

As you know, a year has 365 days (366 days during leap year). The days of the year can be numbered in this way:

January 1 is day 1,

January 2 is day 2,

. . . ,

. . . ,

December 30 is day 364, and

December 31 is day 365.

12. What is different in the numbering of the days for a leap year?

In 1993, January 1 was a Friday. **Figure 2.47** is the beginning of a graph that links the day number with the name of that day.

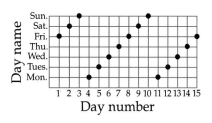

Figure 2.47.
Graph linking day of week with date.

13. Draw the next part of the graph from day 16 to day 25.

14. Which day of the week is day 31? Day 120? Day 275?

15. Draw the part of the graph for days 273 to 279.

16. What can you tell about the day numbers of all days in 1993 that are named Wednesday?

17. Describe the graph for the year 1994. Was it the same as the 1993 graph? If not, what are the differences?

18. Use modular arithmetic to explain why a stretch cipher like $c = 2p$ is not a good coding system if messages are coded as letters.

LESSON FOUR
Cracking Codes

KEY CONCEPTS

Linear patterns

Frequency distributions

David Barber

PREPARATION READING

Cryptography

*S*ecret codes protect the privacy of important information. Coded information is often stored and transmitted electronically. The code cracker gains access, usually illegally, to information protected by codes and passwords, and deciphers the coded information. This is like picking a lock to gain access to a building or file cabinet.

The stakes can be high. During World War II, for example, American and British code crackers worked around the clock to break the Japanese and German coding systems. Their success had a considerable impact on the outcome of the war.

The National Security Agency of the United States is charged with monitoring and protecting information that might compromise the security of the country if leaked to the wrong parties. Cryptographers working for the government design secret codes that deny access to unwelcome people. If an enemy cracked the code and gained access to military secrets, the outcome could be disastrous for the United States.

Almost all instances of code cracking known to history have involved some combination of mathematical knowledge, trial and error, and hard work. Knowing how to build secret codes will help you crack them. Conversely, knowing how to crack codes will help you build better codes and ciphers.

An effective coding process must be difficult to crack. In this lesson, you look at the tools of the code cracker and determine if a shift cipher is an effective coding process.

CONSIDER:

1. Code breakers look for patterns that can expose the method used to code the messages they want to read. What patterns have you seen that might help crack a message coded with a shift cipher?

2. What can the coder do to avoid these patterns?

3. Would adding characters help foil the code breaker?

ACTIVITY

CRACK THE CODE

8

Your task in this activity is to crack the message below, then answer the questions that follow. Use anything you have learned about coding and decoding. Look for patterns that might give away the coding process.

21 9 28 16 13 21 9 28 17 11 17 9 22 27 16 9 30 13 14 23 29 22 12 9 31
9 33 28 23 11 23 12 13 21 13 27 27 9 15 13 27 28 16 9 28 11 9 22 22
23 28 10 13 12 13 11 23 12 13 12 13 30 13 22 17 14 33 23 29 19 22 23
31 28 16 13 11 23 12 17 22 15 24 26 23 11 13 27 27

1. Explain the methods you used to crack the code or, if unsuccessful, the methods you used in attempting to crack the code.

2. What assumptions did you make?

3. If you succeeded in cracking the code, use an arrow diagram to represent the coding and decoding process.

If you didn't succeed in cracking the code, perhaps you could use some more information. Then you could try again. If you did succeed in cracking the code, this part of the activity will provide you with additional information that can help you crack other codes.

ACTIVITY

CRACK THE CODE

8

Code breakers use statistics to help them crack secret codes. They rely on the fact that some letters are used more frequently than others. Which letters do you think appear most often in the English language? Which letters do you think appear least often in the English language?

4. Choose a paragraph from a book or magazine. Tally the number of times each letter occurs in the paragraph, and write the results of your tally in a table like the one in **Figure 2.48.**

A	B	C	D	E	F	G	H	I	J	K	L	M	N	O	P	Q	R	S	T	U	V	W	X	Y	Z

Figure 2.48.
An alphabet tally table.

5. Combine your results with results of other class members. Record the results in another table.

If you did not already crack the code above, use your table of **frequency distribution** to try again. Save the table for future work.

6. Is a shift cipher easy to crack? Is a shift cipher an effective coding process? Explain.

7. Propose a coding process that you think is more difficult to crack than a shift cipher. Remember: it must still be easy to encode and decode.

INDIVIDUAL WORK 7

Code Cracking

1. The following message was encoded with a shift cipher. Crack the code and decipher the message. Explain your strategy and any assumptions you make.

25 30 17 19 25 32 24 21 34 35 41 35 36 21 29 21 17 19 24 35 41 29 18 31 28 31 22 36 24 21 29
21 35 35 17 23 21 39 25 28 28 18 21 34 21 32 34 21 35 21 30 36 21 20 18 41 17 30 31 36 24 21
34 35 41 29 18 31 28 34 17 36 24 21 34 36 24 17 30 18 41 18 28 31 19 27 35 31 22 35 41 29 18
31 28 35 17 35 39 31 37 28 20 18 21 36 24 21 19 17 35 21 22 31 34 17 19 31 20 21

2. Decipher the following message. Explain your strategy and assumptions.

G R Z N U A M N S G Z N K S G Z O I O G T Y N G B K H K K T O T B U R B K J O T I
X E V Z U G T G R E Y O Y L U X G R U T M Z O S K O Z C G Y U T R E O T Z N K Z C
K T Z O K Z N I K T Z A X E Z N G Z S G Z N K S G Z O I Y N G Y H K K T V A Z Z U
C U X Q Y E Y Z K S G Z O I G R R E O T Z N K J K Y O M T U L I U J K Y G T J I O V N
K X Y.

3. a) Crack the code. Decipher the coded portion of the quote below.

"Mathematicians have recently devised a system of coding messages that allows you to tell everyone how to code a message without worrying that it will be deciphered by someone for whom it wasn't intended.

60 24 27 57 57 75 57 60 15 39 33 42 45 69 42 3 57 48 63 6 36 27 9 33 15 75 9 54 75 48 60 45
21 54 3 48 24 75 3 48 48 36 27 15 57 3 6 54 3 42 9 24 45 18 39 3 60 24 15 39 3 60 27 9 57 9 3
36 36 15 12 42 63 39 6 15 54 60 24 15 45 54 75

to allow governments, banks, and others to receive coded messages from anyone."

b) Is this a shift cipher? How do you know?

4. Conduct your own research on frequency. Choose a topic that offers a number of options. You might investigate the types of cars in the student or faculty parking lot at your school (minivans, sport cars, luxury cars, trucks, sedans). You might have the class anonymously write their shoe size on a piece of paper that you collect and then record the distribution. You might survey the types of CDs in a person's collection or the number of titles at a local music store (categories of R&B, country, rock, classical, etc.). Make a prediction before you collect your data. Display your results using a bar graph or circle graph.

5. In word games like Scrabble®, the goal is to build words with available letters and score the highest possible point total. Use the letters below to make a word with the highest score you possibly can. Next to each letter is its point value. (Note that letters that occur most frequently in the English language have the least value.) Compare your score with other students' to see who can find the best word.

A (1) D (2) E (1) I (1) K (5) O (1) P (3) R (1) S (1) T (1) W (4) Y (4)

6. a) Use the arrow diagram in **Figure 2.49** to encode *secret codes*.

   b) Use the equation $c = p + 10$ to encode *are used*.

   c) Use the shift cipher indicated in the table in **Figure 2.50** to encode *to protect information*.

   d) Use the shift cipher indicated in the graph in **Figure 2.51** to encode *that is private*.

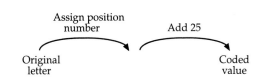

Figure 2.49.
Arrow diagram for a coding process.

Plaintext letter	A	B	C	D	E
Coded value	19	20	21	22	23

Figure 2.50.
Part of a coding process.

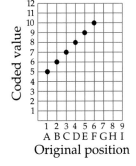

Figure 2.51.
Part of a coding process.

7. a) Use the arrow diagram in **Figure 2.52** to decode V J G U G E T G V X C W N V.

   b) Add the decoding process to the arrow diagram.

   c) The equation $c = p + 5$ was used to encode N X N S Y M J. Reverse the process and decode the message.

   d) Use the shift cipher indicated in the table in **Figure 2.53** to decode J I A M U M V B W N.

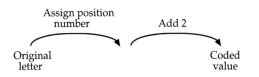

Figure 2.52.
Arrow diagram for a coding process.

Plaintext letter	A	B	C	D	E	F	G	H	I	J
Original position			12					17		
Coded letter			L							

Figure 2.53.
Part of a coding process.

e) Use the graph in **Figure 2.54** to decode
W K H R O G O L E U D U B.

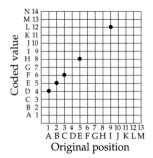

Figure 2.54.
Part of a coding process.

f) Combine the four messages from parts (a), (c), (d), and (e) into one sentence.

g) You have just decoded using four different representations. Which was easiest for you? Why?

8. Everyone in the company is given a weekly raise of $20.00. An arrow diagram can be used to show this simple process (**Figure 2.55**).

Figure 2.55.
Arrow diagram for a salary
raise of $20.00.

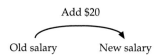

a) Use the arrow diagram to write a simple word equation. Then use n for new salary and s for old salary and write a symbolic equation.

b) Suppose your new salary is $194. What equation can you solve to find the old salary?

c) Suppose the company gives employees the option of taking a 5% raise followed by a $10 increase. Draw an arrow diagram to represent a 5% raise followed by a $10 increase.

d) If you were an employee for this company which option would you choose?

Option A: $20 increase

Option B: 5% raise followed by $10 increase

Explain why your choice is better.

9. In *The Adventure of the Dancing Men* by Arthur Conan Doyle, Sherlock Holmes believes the drawings of dancing men are more than just a child's sketches. Holmes unlocks the secret of the dancing men and astounds everyone involved with the case. Perhaps you can decipher the secret characters as shown in **Figure 2.56**. (In the story, the messages are written by Abe Slaney to Elsie, the wife of Mr. Hilton Cubitt.)

Figure 2.56.
The dancing men
coded message.

10. The following message was coded with a system that uses the letters of the alphabet and a blank space as a 27th character. Try to crack the message. Examine the numbers for patterns similar to those you have seen in previous coding processes. Remember that code breakers succeed by applying their knowledge, and by trial and error. Be persistent!

20 6 54 38 14 46 42 22 52 28 42 16 28 40 44 14 26 28 42 16 28 34 20 36 46 28 22 8 20 48 28 14 36 46 20 48 28 10 22 54 28 36 22 48 18 20 46 28 20 8 20 48 10 46 44 42 52 18 28 10 22 54 28 34 20 14 48 52 20 6 28 42 52 28 16 38 44 22 22 34

LESSON FIVE

Harder Codes

KEY CONCEPTS

Coding method: combination code

Algebraic expression

Order of operations

Equivalent expressions

Distributive property

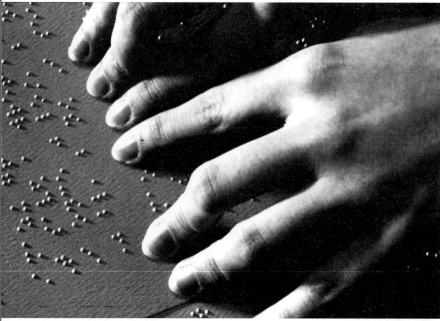

The Image Bank

PREPARATION READING

The Two-Step

You have seen that a shift cipher is easy to encode and decode. Unfortunately it is also easy to crack, because coded messages have patterns that a code breaker can detect.

When a code breaker succeeds in cracking a code, the coder tries to find a better coding process. That's often the way mathematics works in the world. When a process doesn't work quite right, the mathematician fine-tunes it or replaces it.

Friends must still be able to reverse the coding process and decode the message. Unwelcome parties must not find clues that give away the process.

Suppose you add another step. Will a combination of multiplication (stretch) and addition or subtraction (shift) produce an effective coding process? Apply what you learned about multiple representations—arrow diagrams, equations, graphs, and tables—to decide if a two-step coding process is effective.

Don't forget to revisit the five messages in the first lesson. You have tools for cracking codes that you didn't have when you first encountered those messages.

COMBINATION CODES

In this activity, you modify the shift cipher to create a coding process that combines two mathematical operations. You consider whether the new coding process makes life more difficult for the code breaker.

1. A **combination code** combines a stretch and a shift in either order. Code the following message using a process that stretches 2, then shifts +3.

> ## The FBI demands that I remain silent.

2. The following message was coded with the same process. Decode it.

43 19 13 39 13 29 45 41 43 7 13 5 41 13 9 39 13 43 49 5 53 43 33

9 33 29 29 45 31 21 9 5 43 13 43 19 13 43 39 45 43 19

3. Is a message produced by a combination code easy to code and decode?

4. Here is a message coded with a combination code. Crack it.

5   59 62 56 5 44 23 17   14 47 11 65 41 17 44 62   11 5 38 38 17 14

62 26 17   68 47 77 44 29 11 26   41 5 44 65 59 11 56 29 50 62

26 5 59   20 47 29 38 17 14   11 47 14 17   11 56 5 11 35 17 56 59

59 29 44 11 17   29 62 59   14 29 59 11 47 68 17 56 77   47 68 17 56

62 26 56 17 17   26 65 44 14 56 17 14   77 17 5 56 59   5 23 47

5. Do you think combination codes are easy to crack? How do they compare to shift ciphers?

INDIVIDUAL WORK 8

Stretching and Shifting

R ecall that in the coding process $c = p + 3$ the letters c and p are variables representing all pairs of numbers that are solutions to the equation. The number 3 is a constant and acts as a control number. Changing the constant produces a different equation with a parallel graph.

In a coding process with two steps such as $c = 2p + 3$, the numbers 2 and 3 are constants, and both act as control numbers. In Items 1–3, you investigate what happens to the graph when you change one or both of the control numbers.

1. The equation $c = p + 3$ represents a shift of 3 and no stretch. The equation $c = 2p + 3$ represents a stretch of 2 followed by a shift of 3.

   a) Use the graphing calculator and graph both $c = p + 3$ and $c = 2p + 3$.

   b) Describe how the graphs are different.

   c) Predict how the graph of $c = 3p + 3$ differs from the graphs of the two equations in part (a). (Write your prediction, then use the graphing calculator to check it.)

   d) How can you tell from a graph whether a stretch was used in the coding process?

2. Consider two coding processes represented by $c = 3p + 2$ and $c = 3p + 5$.

   a) Without graphing, predict how the graphs are similar.

   b) Without graphing, predict how the graphs are different.

   c) Test your predictions by graphing.

   d) Do these two processes code any letters the same? How do you know?

3. Compare the coding processes $c = 3p + 2$ and $c = 2p + 2$.

   a) Predict how the graphs are similar.

   b) Graph both lines to confirm your answer.

   c) Do these two processes code any letters the same? How do you know?

   d) Do $c = 3p + 2$ and $c = 2p + 2$ represent linear functions?

4. $c = 4p + 1$ represents a coding process.

a) The point (4, 17) is a point on the graph of $c = 4p + 1$. In the context of coding, what is the meaning of the point (4, 17)?

b) Is the point (15, 63) a point on the graph? How can you tell without graphing?

c) The point (1/2, 3) is a point on the graph of $c = 4p + 1$ and is a solution to the equation $c = 4p + 1$. In the context of coding, what is the meaning of the point (1/2, 3)?

d) For $c = 4p + 1$, how much does the value of c change when you change the value of p by 1?

5. a) The message 11 9 16 17 20 3 16 5 7 11 21 23 16 16 7 5 7 21 21 3 20 27 was coded using one of the graphs in **Figure 2.57**. How can you tell which graph was used?

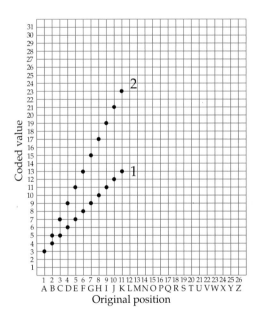

Figure 2.57.
Graphs of two coding processes.

b) The message 39 25 3 29 41 39 3 29 9 39 33 3 7 11 39 was coded using one of the graphs in Figure 2.57. How can you tell which graph was used?

6. A computer on-line service charges a monthly fee of $12.00, with an hourly rate of $2.00.

a) Construct an arrow diagram to represent the process of finding the total monthly bill from the hours used.

b) Write an equation for the billing process. Let c represent total cost. Let h represent the number of hours.

c) Calculate the total monthly bill for service for 12 hours.

d) If you receive a monthly bill for $30, how many hours did you use that month? What equation did you solve?

7. a) Suppose you code using the process $c = -2p + 4$. What is the change in the coded value when you move from one letter to the next?

b) Predict how the graph of $c = -2p + 4$ differs from the graph of $c = 2p + 4$ and write down your prediction. Use a graphing calculator to verify your prediction.

8. Describe how to find all the points that are solutions for both $c = 2p + 3$ and $c = 5p - 4$.

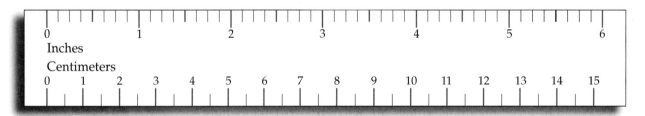

Figure 2.58.
A ruler with centimeter and inch scales.

9. A coding process is a function that matches or maps a plaintext letter to another letter, number, or character. The coding process $c = p + 5$ matches letters from A to Z with numbers from 6 to 31. Every letter is paired with a different number. There are many other processes that map or match a number from one set with one number from another set. Converting from one measurement of length to another is such a function. The ruler in **Figure 2.58** measures length in centimeters along one edge and inches along the other. One inch is matched to 2.54 centimeters.

a) How is 3 cm "coded" as inches?

b) How is 8 in. "coded" as centimeters?

10. a) $C = \frac{5}{9}(F - 32)$ is an equation used to convert a Fahrenheit tempera-
     ture to Celsius. Draw an arrow diagram to represent the process.

   b) Dividing by a fraction is the same as multiplying by its reciprocal.
      So dividing by 5/9 is the same as multiplying by 9/5. Use this
      fact to draw an arrow diagram for converting Celsius tempera-
      tures back to Fahrenheit.

   c) Write an equation to represent the inverse process.

11. Rebecca has her own business. She sells floral arrangements at the
    mall. The mall rental fee is $90 per week. Her materials and supplies
    cost about $6.00 per arrangement.

   a) If c is total cost and a is the number of floral arrangements, write
      an equation to represent Rebecca's total weekly costs based on a,
      the number of arrangements she makes and sells.

   b) How many arrangements did Rebecca sell last week if her total
      cost was $198.00? Write an equation you can solve to answer the
      question. Show the steps needed to solve the equation.

   c) Is this situation represented by a linear function?
      Explain your answer.

12. Distance = rate x time, or $d = rt$, is an equation used to
    find distance, given the rate at which an object travels
    and the time it travels. If an automobile travels at a
    constant rate of 60 mph, the equation is $d = 60t$, where
    d is in miles and t is in hours. The arrow diagram in
    Figure 2.59 illustrates this relationship.

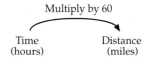

Figure 2.59.
Time and distance arrow diagram.

   a) Graph this equation.

   b) Find d when $t = 3$ hours.

   c) Find t when $d = 720$ miles.

   d) How is the graph different if the speed is 50 mph?

13. a) Use a calculator with a list feature to construct a
       coding table for $c = 4p + 5$. You first need to store
       the original position numbers {1, 2, 3, . . . , 26} as a
       list. Carry out the coding process in two steps by
       first multiplying the list by 4, then adding 5. (See
       Figure 2.60.)

Figure 2.60.
Coding with calculator lists.

b) A computer spreadsheet is a good tool for constructing a coding table. Follow the guidelines below to construct a coding table for $c = 4p + 5$. Here are the basic steps.

- Label the columns (optional, but helpful).
- Enter the plaintext position numbers.
- Define the formula for the coding process.
- Extend the formula to the entire message.

First, of course, open the spreadsheet program.

Label the columns: In the first row of column A, type the letter p and press ENTER or RETURN. (Note the behavior of the cursor. For many spreadsheets, pressing ENTER and pressing RETURN have different effects; try both.) Label column B as $4p$ and column C as $4p+5$.

Enter the plaintext position numbers: In cell A2, type "1" and press ENTER or RETURN. Arrow down to cell A3 if necessary and type "2." Repeat until the position numbers 1–26 have been entered into cells A2–A27.

Enter the formulas for the coding process: In cell B2, type the formula for 4*A2. This formula tells the computer to multiply the value in cell A2 by 4 and put the answer where the formula is, namely, B2. For some spreadsheets, you can type exactly that: 4*A2. For others, you have to let the program know that a formula is coming by starting off with an equal sign or a plus sign. Check with your teacher, your manual, or the help screen of your spreadsheet.

After entering the 4*A2 formula in B2, enter a formula for B2+5 in cell C2; this will complete the coding of the plaintext position number in cell A2.

Extend the formulas from cells B2 and C2 down to code the other position numbers of column A: Spreadsheets differ quite a bit in how they do this process and in what they call it. Commands such as Copy, Copy and Paste, and Fill Down are common. Consult with your teacher or check your manual or help screen for details.

c) In part (b), you typed the plaintext position numbers 1–26 into column A of the spreadsheet. Write a formula that could save you some effort there. (Hint: You still need to enter the number 1 in cell A2. Then find a formula that will get you easily from A2 to A3 and extend it down the column.)

14. a) In Individual Work 2 you coded a message using matrices and a shift cipher. Here is how to use matrices to code the word *matrix* using a stretch of 4 followed by a shift of 5 ($c = 4p + 5$).

- Convert *matrix* to position numbers.

- Enter the position numbers in the calculator as a 1 x 6 or 2 x 3 matrix. Name this matrix [A].

- Create a matrix [B] with the same dimensions as [A] and consisting of all 5s.

- Perform the matrix operation 4[A] + [B], and write the coded values.

b) Use the list feature on a graphing calculator to code the word *matrix* with the process $c = 4p + 5$.

c) Use a spreadsheet to code the word *matrix* with the process $c = 4p + 5$.

Perhaps you have found two-step coding processes a little confusing. That's understandable. Is a stretch followed by a shift necessarily the same as a shift followed by a stretch? Can two processes lead to equivalent results?

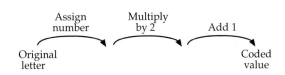

Figure 2.61.
A two-step coding process.

1. Consider two coding methods. Method 1 stretches 2, then shifts +1. (See **Figure 2.61.**)

   Method 2 shifts +1 first, then stretches 2. (See **Figure 2.62**.)

   Does changing the order produce a different coding process? Justify your conclusions.

2. Peter used a graphing calculator to make a table for the process in **Figure 2.63**.

   Sonia used a process with different steps (**Figure 2.64**) and created a table in the same way Peter did.

   Are the two coding processes the same? Explain.

   If you believe they are the same, find another example of two different-looking processes that are really the same.

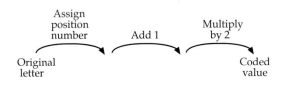

Figure 2.62.
The process in Figure 2.61 in reverse order.

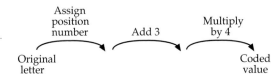

Figure 2.63.
A coding process.

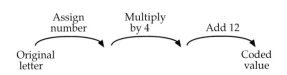

Figure 2.64.
A coding process.

INDIVIDUAL WORK 9

Two-Step Topics

An **algebraic expression** is a collection of variables and sometimes constants, connected by several operations. In the equation $c = 2p + 3$, for example, $2p + 3$ is an algebraic expression. Algebraic expressions are the building blocks of equations. In this individual work, you examine the rules that govern algebraic expressions.

One of the most important things to know about algebraic expressions is whether two expressions are **equivalent**. Two expressions are equivalent if they always produce the same result. For a coding process, equivalence means that both processes code all letters the same way.

Suppose you code a message that changes a plaintext letter to a position number followed by two mathematical operations: add 4 to the position number and then multiply by 3 (**Figure 2.65**).

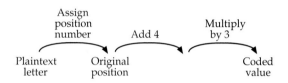

Figure 2.65.
A two-step
coding process.

The symbolic equation reveals the addition and the multiplication operations: $c = 3(p + 4)$.

Notice the parentheses around $p + 4$. Parentheses ensure that addition is performed before multiplication. Without parentheses, the coder would see the equation as $c = 3p + 4$ and would multiply first.

Since order makes a difference, as you saw in Activity 10, it's important to know the rules for the correct order of operations. Mathematicians have agreed that operations in an expression must be performed in the following order:

a) Evaluate expressions within parentheses.

b) Simplify powers (apply exponents).

c) Perform multiplication and division as they occur from left to right.

d) Perform additions and subtractions as they occur from left to right.

Below is an expression involving more than one mathematical operation and two ways to evaluate it. One is correct; the other is not.

$10 - 3 \times 2 = 7 \times 2 = 14$ is incorrect because the subtraction is done before the multiplication, violating the rules.

$10 - 3 \times 2 = 10 - 6 = 4$ is correct.

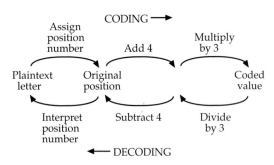

Figure 2.66.
An arrow diagram
for coding and decoding.

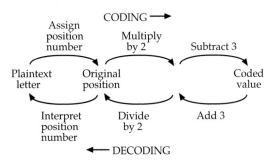

Figure 2.67.
Coding and
decoding processes.

Order of operations is important to the decoding process. An arrow diagram shows how to reverse the operations in the decoding process (**Figure 2.66**).

A symbolic equation to represent decoding is $p = c/3 - 4$. Parentheses are not needed because the division (represented by the fraction bar) is performed first. However, it is not wrong to use parentheses this way: $p = (c/3) - 4$.

The previous example illustrates order of operations with a shift followed by a stretch. The next example illustrates order of operations for a stretch followed by a shift.

Suppose you code a position number by multiplying by 2, then subtracting 3 (**Figure 2.67**).

The equation for coding is $c = 2p - 3$. The equation for decoding is $p = (c + 3)/2$.

In the equation for decoding, parentheses surround $c + 3$. Adding 3 is the first step in the decoding process. The parentheses ensure that addition happens before division.

To illustrate the importance of using parentheses for the decoding process, try decoding the number 15. With parentheses you get $(15 + 3)/2 = 18/2 = 9$. Without parentheses you get $15 + 3/2 = 15 + 1.5 = 16.5$, which is not a position number.

Apply the rules for order of operations as you complete this individual work.

1. The process used to code a message is "Add 1 to a letter's position, then multiply by 2." This process shifts (add) before stretching (multiply).

    a) Draw an arrow diagram for the coding process.

    b) Write a symbolic equation to represent the coding process.

    c) Code the word *variable*.

    d) Add the decoding process to the arrow diagram you drew in part (a).

    e) Write a symbolic equation to represent the decoding process.

    f) Decode 32 38 10 12 38.

2. What is "decor"? The answer to this question is shown below in coded form. It was coded with the process described in Item 1. To make cracking the message more difficult, the coder did not show the blanks between the words.

   a) Decode the message.

      42 18 12 34 4 38 42 32 14 42 18 12 4 34 34 26 12 52 32 44 42 18 38 32
      48 4 48 4 52

   b) Notice that all of the coded values in part (a) are even numbers. Why is that?

   c) Describe other codes with this characteristic.

   d) Give an example of a coding process for which all the coded values must be odd.

   e) Suppose you receive a message that was coded with the process described in Item 1 and one of the characters is a 7. What can you conclude?

3. a) Suppose that you are trying to crack a coded message and you suspect that *A*, *B*, and *C* are coded as shown in **Figure 2.68**. How do you think the letter *D* is coded?

   b) Write an equation for the coding process.

   c) Write an equation for the decoding process.

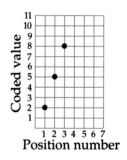

Figure 2.68.
Part of a coding process.

4. a) Draw an arrow diagram for a coding process with a stretch of 2 followed by a shift of +5.

   b) Add the decoding process to the arrow diagram.

   c) Write an equation to represent the coding process and another equation to represent the decoding process.

   d) Use a graphing calculator or spreadsheet to produce a table or graph. Use the table or graph to find the coded value for the letter *R*.

   e) Use the table or graph to decode the number 21. What equation do you solve when you decode 21?

5. a) Do $c = 5p + 1$ and $c = 5(p + 1)$ represent the same coding process or different coding processes? Describe how to confirm or verify your answer.

   b) Represent each coding process with an arrow diagram.

   c) Describe the decoding process for $c = 5p + 1$.

d) For the coding process $c = 5p + 1$, decode the coded value 46. That is, solve the equation $46 = 5p + 1$.

e) Another way to solve the equation $46 = 5p + 1$ is with a graphing calculator. Let $Y1 = 5X + 1$ and $Y2 = 46$. Graph both lines on the same screen. The place where the two lines intersect is the point where $Y1 = Y2$. How do you find the point where the two lines intersect?

f) Describe the decoding process for $c = 5(p + 1)$.

6. One way to explain why two expressions that look different are equivalent involves using a geometric shape to represent the variable and another to represent the constant 1. For example, it is common to use a square for the variable and a circle for 1. Thus the expression $x + 2$ is shown in **Figure 2.69**.

Figure 2.69.
A representation for the expression $x + 2$

a) Use squares and circles to represent a coding process that stretches 2, then shifts three. (Hint: first the variable is doubled, then 3 is added.)

b) Write an equation to represent the coding process.

Suppose you change the order of the math operations in the coding process.

c) Use squares and circles to represent a coding process that first shifts 3, then stretches 2.

d) Write an equation for this coding process.

e) Is a stretch-and-shift cipher the same as a shift-and-stretch cipher with the same control numbers (constants)? Explain your answer.

7. Use squares and circles to show why the coding processes $c = 2(p + 3)$ and $c = 2p + 6$ are equivalent.

Mathematicians use the term **distributive property** to describe a symbolic process for changing $2(p + 3)$ to $2p + 6$. The distributive property may be applied when a constant or variable is multiplied by a sum enclosed in parentheses, as shown in the generalized form

$a(b + c) = ab + ac$.

a, b, and c may represent constants or variables.

In the example $2(p + 3)$, 2 is multiplied by p, and 2 is multiplied by 3. The results are added: $2(p + 3) = 2(p) + 2(3) = 2p + 6$.

Here are two additional examples.

$3(2p + 4) = 3(2p) + 3(4) = 6p + 12$

$5(9 + 3x) = 5(9) + 5(3x) = 45 + 15x$

8. Use the distributive property or circles and squares to create equivalent expressions with parentheses removed.

   a) $5(p + 5)$

   b) $4(2x + 3)$

   c) $2(x + 7) + 2$

9. The squares and circles in **Figure 2.70** represent a coding process.

   Describe or diagram a process for decoding.

Figure 2.70.
A representation of a coding process.

10. Each of the following represents a coding process. Describe the steps for decoding or construct an arrow diagram to represent the decoding process.

    a) $c = 4p + 1$

    b) $c = 2(p + 3)$

    c) $c = 3p - 4$

    d) $c = 5(p + 2) - 3$

11. Use arrow diagrams or symbolic procedures to solve each equation.

    a) $36 = 2x - 10$

    b) $53 = 5x + 8$

    c) $2(x + 7) = 40$

    d) $3(x - 6) = 36$

    e) $3x - 14 = 27$ (Round your answer to the nearest hundredth.)

12. Suppose $m = 5(n + 2) - 3$.

    a) What is the value of m when $n = 8$?

    b) Find the value of n when $m = 67$. What equation do you solve?

13. Every coding process so far has used addition, subtraction, and multiplication. Investigate the following processes involving division to determine if they are equivalent.

Add 6, then divide by 2.

Divide by 2, then add 6.

14. a) Use a graphing calculator to solve the equation $2x + 13 = 27$. Let $Y1 = 2X + 13$ and $Y2 = 27$. Find the coordinates for the point at which the two lines intersect. The x-coordinate is the solution you seek.

b) If two numbers, quantities, or expressions are equal, their difference is zero. That is, if $Y1 = Y2$, then $Y1 - Y2 = 0$. Solve the equation $(2x + 13) - (27) = 0$ by graphing $Y3 = 2X + 13 - 27$ and locating the point on the graph where $Y3$ is 0. Compare your answer to part (a). (Note: On many calculators you can create $Y3$ from $Y1$ and $Y2$ by entering $Y3 = Y1 - Y2$.)

15. Code breakers can crack combination codes by noticing patterns in the graph. Often they can write a symbolic equation from the graph once they notice the pattern.

a) How do you determine an equation of a linear function from the graph?

b) Describe the coding process represented by the graph in **Figure 2.71**.

c) What is the equation for the coding process?

16. The end of the preparation reading for Lesson 1 contains a coded message. You now have the tools you need to crack the code and decipher that message. Do so.

I Y P K P I I P C F P I Q C B T D F D C P I Y P N T F
I J T N N T Q K P I I P C F B Q P Q V K B F Y W C P
W D C P D V C D W Y I T C P O P D K K B Q P D C
W D I I P C Q F

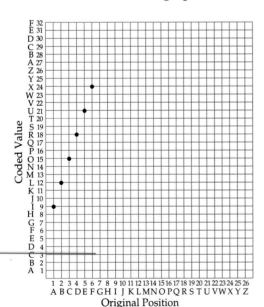

Figure 2.71.

17. The following messages were encoded using two-step coding processes. Crack the codes and decipher the messages. Show all your work to support the steps you take and the decisions you make. Use graphs and tables as needed.

   a) 83 11 89 155 83 95 29 35 113 89 11 101 101 77 59 23 11 125 59 95 89 119
      95 41 23 95 29 35 119 11 113 35 113 35 77 11 125 35 29 125 95
      35 77 35 23 125 113 95 89 59 23 119 83 131 119 59 23 41 95 113
      35 149 11 83 101 77 35 59 119 23 95 29 35 29 95 89 125 95
      23 95 83 101 11 23 125 29 59 119 23 119 11 89 29 83 131 119 125 17 35
      29 35 23 95 29 35 29 17 155 101 77 11 155 17 11 23 71
      35 107 131 59 101 83 35 89 125

   b) L B S Z F W L S M L E W T Z F W R E P S P J A G Z O J N E M K S A
      I A I L S Q P S Z S T P I W T L B S V G M L L B G L I W Q S
      Q G L B S Q G L E M G N Z F W M S I I S I G F S S G I A L W P W J O L
      R S F A B G F P L W O T P W

"Many modern applications of codes are related to electronics. Music, for example, is coded onto compact discs and must be decoded by playback equipment"

(Froelich & Malkevitch 1993, 66)

18. You studied point counts in Unit 1, *Pick a Winner: Decision Making in a Democracy*. Suppose an election among four candidates is decided with a 3, 2, 1 point system. How would the results compare to an election with the same candidates and voters if a 6, 4, 2 point system were used? Use the distributive property to justify your answer.

Through your work with shift and stretch ciphers, you have been introduced to a mathematical concept known as **transformations**. Transformation is another word for function. Since a function can be thought of as moving points on a line (or plane) to other points on the line (or plane), you can think of transformation as moving one set of points, or an object, onto another. Transformations include many different types of operations. Sometimes one such operation is known by more than one term. For example, other words for a shift are a "slide" or a **translation**. Whatever it is called, it is a transformation that moves things without changing their relation to each other. In the case of secret codes, a shift or translation results in a slide of the alphabet along the number line, but the distances between letters and the order among them don't change. Here, translation is accomplished by adding a single constant to all position numbers. A stretch, which uses multiplication, is another type of transformation that is known by more than one term. Another name for stretch is "dilation," which you will explore in more detail in Unit 3, *Landsat*. In fact, you will encounter transformations throughout your work in *Mathematics: Modeling our World*.

LESSON SIX

Illusive Codes

KEY CONCEPTS

Equivalent expressions

Solving equations

PREPARATION READING

Magic

Code breakers are smart people. They look for clues. They know that a two-step process leaves gaps between the coded values, giving away the multiplication in the coding process. A two-step process may slow the code breaking, but it doesn't stop it. It is not effective because it is relatively easy to crack.

Suppose you apply the tricks of the magician to secret codes. Some magicians use number tricks. One goes like this.

> Pick a number.
> Multiply by 3.
> Subtract 1.
> Multiply by 2.
> Add 3.

What did you get?

The magician can quickly tell you the original number. How does he do it? Is it really a trick? Does the magician really reverse all those steps in his head?

The magician's number trick is similar to a multistep coding process. If you increase the number of steps in the coding process, will it be more difficult to crack? Will the increase in the number of steps eliminate the clues that made the two-step process less effective?

You tried the shift cipher, and it passed two of the three criteria for an effective code. You modified the shift cipher, and the resulting two-step process also passed two of the three criteria. Modify the model again. Increase the number of steps, and determine if a multistep coding process is effective. Is it possible that the illusive tricks of the magician may lead to elusive codes for the code cracker?

Remember to check the original five messages from Lesson 1. Perhaps what you learned in Lesson 5 can help you crack one or more of those coded messages.

NUMBER TRICKS

11

The two-step coding process isn't much more difficult to crack than a simple shift cipher. A code cracker can use a frequency distribution or recognize spacing patterns in a graph or table to crack the code.

Perhaps more steps would make codes more difficult to crack. In this activity, you will determine if a multistep coding process is more effective than a two-step coding process.

A magician's number tricks are similar to coding processes. A number trick involves many steps and is designed to confuse the audience. Will a coding process with many steps confuse the code cracker?

One of the questions you seek to answer is "How is the magician able to decode the final number so quickly to reveal the correct original number?" Use your mathematical tools to determine if patterns exist that are useful to the magician. Here is a sample number trick to use in your investigation.

> Choose a number.
> Multiply the number by 3.
> Add 4.
> Multiply the result by 2.
> Add 5.
> Report your result.

1. a) Prepare an arrow diagram for the magic trick, beginning with "Choose a number" and ending with "Final number."

   b) Use math tools to investigate patterns or clues used by the magician.

   c) Using the results of your investigation, describe how a magician can determine the original number quickly.

NUMBER TRICKS

2. a) Pretend the magic number trick in Item 1 is a coding process and code the word *magic* as numbers.

   b) Decode the number 43.

3. The previous items focused on coding and decoding a multi-step process disguised as a magic number trick. Is a multistep coding process difficult to crack? The following message has been coded using a multistep coding process.

   45 19 19 17 47 39 33 39 35 39 37 15 39 17 19 47 11 45 19 51 47 19
   17 13 59 15 39 35 41 11 37 27 19 47 33 27 31 19 47 39 37 59 49 39
   19 37 47 51 45 19 49 25 19 21 27 17 19 33 27 49 59 39 21 35 51 47
   27 15 55 25 19 37 27 49 27 47 41 33 11 59 19 17 13 11 15 31 21 45
   39 35 15 39 35 41 11 15 49 17 27 47 31 47

   a) Try to crack the code and decipher the message. Use tables and graphs when needed.

   b) Describe how you cracked the code or describe the strategies you tried if you were unable to crack the code.

4. Encode a message of your own choosing. Use a multistep process. Challenge another person or group to crack the code and decipher your secret message.

5. Is a multistep coding process effective? Explain your answer.

INDIVIDUAL WORK 10

Number Trickery

1. A magician does a number trick in which the original number is found by subtracting 8. Devise two different tricks that the magician might have used.

2. Here is a number trick.

   > Pick a number between 1 and 10.
   > Add 1.
   > Add your starting number.
   > Subtract 6.

   a) Make a table that shows the results the magician obtained from the starting numbers 3, 4, 7, and 8.

   b) Prepare a graph using the points in your table.

   c) Your four points should lie along a straight line. Use a straight-edge to draw the line. The graph should represent only values from 1 to 10 because the number trick is limited to numbers from 1 to 10.

   d) Suppose that someone tells the magician that the ending number is 7. Use your graph to find the starting number.

   e) Suppose someone tells the magician that the ending number is 2. Locate this on your graph. What is the starting number? Is this possible?

   f) Are your answers to parts (d) and (e) the same if this is a coding process instead of a number trick? Suppose the graph represents a coding process rather than a number trick, and one of the coded values is 7. For which letter could this be the code?

3. a) Model the trick in Item 2 with a symbolic equation suitable for use on a calculator.

   b) Set an appropriate window and graph your equation. Use the trace feature and compare the readouts to the answers you gave in Item 2. Explain any differences between the readouts and your table.

   c) Now use the table feature on your graphing calculator. Make sure the table begins at 1 and increases by 1 so you can compare the table with the answers you gave in Item 2(a).

   d) Suppose someone tells the magician the ending number is 2.4. Find the beginning number. Explain how you determined the answer.

4. a) Describe how to use matrices to code the word *matrix* with this multistep process: add 3, multiply by 2, subtract 5.

   b) Write the word *matrix* in coded form.

5. In Lesson 4, you investigated how often various letters occur in the English language. Code crackers also look for patterns in letter pairs. For example, in the words *crackers*, *pattern*, and *letter* the combination *er* appears. Which letter pairs do you think occur most often in the English language? Make a prediction and write it down. Find a paragraph in a book you are reading. Do a frequency study of letter pairs. Write a summary explaining what you discover.

PICK A NUMBER

12

Multistep coding processes are no harder to crack than two-step coding processes. In fact, a multistep process can be thought of as a two-step process in disguise. The disguise works well for the magician, but not for the coder.

There are many ways to show that a multistep process is equivalent to a two-step process. Here is how to do it symbolically by using the distributive property.

Consider the expression.	$3(2x + 8) - 12$
Distribute 3 by multiplying.	$3(2x) + 3(8) - 12$
Multiply $3(2x)$ to get $6x$ and $3(8)$ to get 24.	$6x + 24 - 12$
Combine $24 - 12$ to get 12.	$6x + 12$

Your job in this activity is to produce a report demonstrating all the methods you have learned to show that two expressions or equations are equivalent. Work with this number trick.

Pick a number.
Multiply by 2.
Add 8.
Multiply by 3.
Subtract 14.

1. Start by using the variable x to represent the starting number. Follow the steps of the magician to create an expression for the result.

2. Use the distributive property to remove the parentheses and find an equivalent expression.

   Use as many additional methods as you can to prepare a report on the equivalence of the expressions you found in Items 1 and 2.

3. Devise a number trick of your own that has several steps. Try it on a member of your family or a friend. Be sure that you know the two-step equivalent of the trick so that you can do it mentally. Try it a few times on your own before doing it for someone else.

INDIVIDUAL WORK 11

The Magic of Algebra

1. Here is a number trick demonstrated with the number 3.

    | Pick a number. | 3 |
    | Multiply by 2. | 6 |
    | Add 8. | 14 |
    | Multiply by 3. | 42 |
    | Subtract 14. | 28 |

    a) Make a table showing several other starting and ending numbers.

    b) What pattern do you notice?

    c) Use the values from the table to create a graph. Label the horizontal axis "Starting number." Label the vertical axis "Ending number."

    d) What two-step coding process produces the same graph?

    e) Represent the steps of the magic trick with an arrow diagram.

    f) Add the decoding process to the arrow diagram.

    g) Suppose the final number is 100. Find the original number.

    h) Does the magician have to do all of the steps?

2. a) Make an arrow diagram to represent this number trick:

    Pick a number
    Multiply by 4
    Subtract 2
    Multiply by 3
    Add 10.

    b) Represent the steps of the trick with a symbolic equation.

    c) Suppose the final number is 64. Find the original number. What equation are you solving?

3. a) Prepare an arrow diagram to represent the following multistep coding process: start with a position number, multiply by 4, add 20, divide by 2, and subtract 3.

    b) Write an equation to represent the coding process.

    c) Add the decoding process to the arrow diagram.

d) Write an equation to represent the decoding process.

e) Suppose the coded value is 31. Find the position number. What was the plaintext letter? How did you find it?

4. Use the distributive property to convert multistep coding processes into two-step coding processes. Use a graphing calculator to compare your answer to the original.

a) $y = 3(5x + 4) - 6$

b) $y = 2(2x + 12) + 15$

c) $y = 5(x + 3) - 9$

d) $y = 8(10x - 3) - 12$

5. Here is a number trick.

Pick a number.
Add 4.
Multiply by 2.
Subtract 7.
Multiply by 3.

a) Model the number trick with squares and circles or with symbols. Write the symbolic equation.

b) What two steps does the magician do to determine the original number from the result?

6. Here is another number trick.

Pick a number.
Multiply by 2.
Add 6.
Divide by 2.
Subtract your original number.

a) Try this trick with three different numbers. Record the results in a table.

b) Model the number trick with squares and circles or with symbols. Write the symbolic equation.

c) How is this trick different from the others you have seen?

7. Here is a number trick.

> Pick a number.
> Double it.
> Add 1.
> Double the result.
> Add 1.
> Double the result.
> Add 2.
> Divide by 8.

a) Model the number trick with squares and circles or with symbols. Write the symbolic equation.

b) What would the magician do to determine the original number from the result?

8. **Figure 2.72** is an arrow diagram of coding and decoding processes.

a) Use x for the position and y for the coded value. Write a symbolic equation that uses the variables x and y.

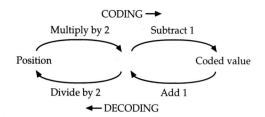

Figure 2.72.
A coding and decoding process.

b) Set a suitable window and graph the equation on a graphing calculator. Sketch your graph.

c) Trace to the point that represents the position number for the letter E. What is its coded value?

d) Trace to a point that lies somewhere between the points that represent E and F. What is the coded value at the point you've chosen? What letter does this represent?

e) Reverse the process. Use x for the coded value and y for the position. Follow the decoding process through the diagram and write a symbolic equation for the decoding process.

f) Set a suitable window and graph the equation on your graphing calculator. Sketch your graph.

g) Use your graph from part (f) to decode this message:
   17 37 9 33 41 1 23 39 29.

h) How could you have used your graph from part (b) to decode the message?

9. Show that the equation $c = 3(p + 1) + 6$ is equivalent to the equation $c = 3p + 9$.

10. A number trick/coding process is represented by $y = 3(5x + 4) - 6$. Write an equation that represents the decoding process.

11. $c = 2(p + 1) + 3p + 4$ represents a coding process.

   a) Draw a two-step arrow diagram to represent the coding process and the decoding process.

   b) Is $c = 2(p + 1) + 3p + 4$ a valid coding process? Give reasons for your answer.

12. a) Find a two-step coding process to match the graph in **Figure 2.73**.

   b) Find a multistep (three or more) coding process to match the same graph.

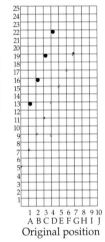

13. $c = p(p + 1)$ represents a coding process.

   a) How do you code 3 using this process?

   b) Decode 72. Explain how you get the answer.

   c) What equation can you solve to decode 72?

   d) Is $c = p(p + 1)$ a valid coding process?

Figure 2.73.
Part of a coding process.

14. Assume that squaring a number is a valid coding process.

   a) Code the word *square* as numbers using a squaring process.

   b) Decode 9 225 16 25.

   c) Is squaring a valid coding process?

15. Write a multistep process that is equivalent to the two-step process $c = 2p + 5$.

16. Can you design a multistep process that makes the resulting number identical to the beginning number? If yes, give an example. If no, explain why not.

17. Describe how to solve the equation $x^3 - 2x^2 = 32$.

18. Suppose you code $c = (16 - p)(1 + p)$. What problems might occur?

19. Write a multistep coding process that codes the word *computer* as 14 50 44 53 68 65 20 59.

20. Suppose you work for a company that invents a state-of-the-art coding machine. Write several sentences explaining how it works. Include diagrams and drawings.

LESSON SEVEN

Matrix Methods

KEY CONCEPTS

Coding method:
keyword codes

Matrix operations

The Image Bank

PREPARATION READING

Frequency Clues

Code breakers know that certain letters and words are used more often than others. No matter how many steps you add to the process, you cannot overcome frequency clues. One-step, two-step, and multistep coding processes are not effective because they do not hide the frequencies of certain letters and words.

Is it possible to design a coding process that does not leave frequency clues?

Throughout this unit, you have been looking for an effective coding process. You started with a simple shift cipher and discovered it is too easy to crack. You modified the process by adding a stretch. However, two-step and multistep coding processes leave frequency clues and other patterns that the code breaker can recognize.

When you complete this lesson, you will be able to design a coding process that is more effective than any you have seen so far. You will also revisit the five messages you encountered in Lesson 1 and try to crack any remaining ones.

MATRIX MAGIC

13

In this activity, you reconsider methods and representations you have already learned to see if they can be adapted to produce a new coding method that foils the code cracker.

Recall that matrices are useful shortcuts for coding and decoding, particularly if you use a calculator with matrix features.

$$\begin{bmatrix} T & H & E & P & A & C & K & A & G & E & I \\ S & I & N & A & L & O & C & K & E & R & A \\ T & T & H & E & A & I & R & P & O & R & T \end{bmatrix}$$

Figure 2.74.
A message stored in a matrix.

For example, suppose you use a shift +5 cipher to code the message "The package is in a locker at the airport." Since the message has 33 characters, you might store it in a 3 x 11 matrix (**Figure 2.74**).

Then you convert the letters to position numbers and add another matrix containing all 5s (**Figure 2.75**).

$$\begin{bmatrix} 20 & 8 & 5 & 16 & 1 & 3 & 11 & 1 & 7 & 5 & 9 \\ 19 & 9 & 14 & 1 & 12 & 15 & 3 & 11 & 5 & 18 & 1 \\ 20 & 20 & 8 & 5 & 1 & 9 & 18 & 16 & 15 & 18 & 20 \end{bmatrix} + \begin{bmatrix} 5 & 5 & 5 & 5 & 5 & 5 & 5 & 5 & 5 & 5 & 5 \\ 5 & 5 & 5 & 5 & 5 & 5 & 5 & 5 & 5 & 5 & 5 \\ 5 & 5 & 5 & 5 & 5 & 5 & 5 & 5 & 5 & 5 & 5 \end{bmatrix}$$

Figure 2.75.
Matrix coding with a shift +5 cipher.

If the first matrix is entered as matrix [A] and the second as matrix [B] on a graphing calculator, the coded values are found quickly. All that's left is to take the message out of the answer matrix and either leave it as numbers or convert it to letters.

Of course, the problem with the coded message is that a code breaker can use knowledge of letter frequencies and linear patterns to crack your code.

1. Can you find a way to alter the process so that the code breaker will find no clues? Look very carefully at the matrices above. The shortcut not only offers a way to make coding easier, but it also offers a way to beat the code breaker if you modify the shortcut slightly. Discuss ways to do so.

2. When you have found a way to modify the shortcut, discuss whether your new method is easy to encode and easy to decode.

3. Also discuss whether your method is easy to communicate between coder and decoder.

4. Discuss the overall merits of your procedure. Is it a good coding method?

Keys to Coding

1. Although a matrix is relatively easy for the coder and decoder to communicate between them, it is possible to make the communication even easier. It can be as simple as sending a single keyword, which is used to code with a **keyword matrix**.

   For example, suppose a coder wants to send the message "Meet me at school" using the keyword *key*. Since *key* has three letters, store the message in a matrix of three columns (**Figure 2.76**).

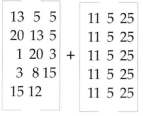

Figure 2.76.
A message stored in a three-column matrix.

   Notice that since the number of letters in the message is not divisible by 3, a blank space is left at the end. You can also add meaningless characters like *X*s at the end.

   Convert the letters of the message to position numbers and add to it a matrix containing the position numbers of the word *key* in each row (**Figure 2.77**).

$$\begin{bmatrix} 13 & 5 & 5 \\ 20 & 13 & 5 \\ 1 & 20 & 3 \\ 3 & 8 & 15 \\ 15 & 12 & \end{bmatrix} + \begin{bmatrix} 11 & 5 & 25 \\ 11 & 5 & 25 \\ 11 & 5 & 25 \\ 11 & 5 & 25 \\ 11 & 5 & 25 \end{bmatrix}$$

Figure 2.77.
Matrix coding with the keyword *key*.

   The message is sent as 24 10 30 31 18 30 12 25 28 14 13 40 26 17.

   a) Explain why the message is hard to crack.

   b) Explain how the person receiving the message would use the word *key* to decode the message.

2. Use a 6 x 5 matrix with keyword *codes* to encode the message

   ## "The password is Captain Codeworthy."

3. Encode the word *package* using a 3 x 3 matrix with *bow* as the keyword. Use a two-step coding process that multiplies the coded values by 3, then subtracts the keyword. (You might get negative numbers in the resulting matrix.)

4. Decode each message. You are given the matrix dimensions, the keyword, and the coding process. [B] is the keyword matrix.

   a) 5 x 6, keyword *Monday*, [A] + [B]

      A T L X D K H Q A I F S V C U M T L B C R E Z

   b) 6 x 3, keyword *SAT*, 2[A] + [B]

      51 37 30 51 3 56 29 13 50 55 41 36 29 41 30 57 41

   c) 6 x 4, keyword *sing*, 5[A] – 2[B]

      27 87 67 31 –23 27 67 51 –33 82 12 11 27 –13 72 31 –23 –13 32

5. The plaintext message "Cryptograms are fun" was coded as S M Y P F T W M A M E F H Z F U Z. Find [B], the keyword matrix, for the coding process [A] + [B].

6. Using a 4 x 6 matrix, encode the message, "No more magic tricks." Choose your own keyword. Keep it a secret. Challenge another student to figure out your keyword.

7. The message is "If you forget the password, you're sunk."

   a) Complete a tally sheet for the frequency distribution of letters in the plaintext message.

   b) Encode the message using a keyword matrix *study* and the process [A] + [B].

   c) Create another frequency distribution, this time with the coded message. Tally the number of times each letter appears in the coded message.

   d) Based on your observations of both frequency distributions, does a keyword matrix defeat the frequency pattern for the most common letters?

8. Conduct an experiment to confirm or refute your conclusions in Item 7.

   a) Select a paragraph from a book or magazine you are reading. Instead of tallying every letter, do a frequency tally based on every fourth letter. Code the entire paragraph using a keyword matrix that you choose. Do a frequency tally for the coded message based on every fourth letter.

   b) Repeat the experiment. Use the same paragraph, same keyword, and the same coded message. This time complete your tally based on every fourth letter starting with the second letter in the paragraph instead of the first letter.

   c) Discuss the likely frequency distribution you would obtain by tallying every fifth letter from a paragraph encoded as in Item 7.

   d) Write your discoveries from the frequency investigations. Conclude whether or not a keyword matrix is an effective way to thwart the frequency distribution of letters in the English language.

9. A keyword dictionary is a type of coding process that defeats the linear pattern of a shift cipher. Choose a word with no repeated letters. *Keyword* is an example of a possible keyword. Match the first seven letters of the alphabet with the letters in *keyword*. Fill in unused letters in alphabetical order (**Figure 2.78**).

Figure 2.78.
Table for a keyword dictionary code.

Plaintext letter	A	B	C	D	E	F	G	H	I	J	K	L	M
Coded letter	K	E	Y	W	O	R	D	A	B	C	F	G	H
Plaintext letter	N	O	P	Q	R	S	T	U	V	W	X	Y	Z
Coded letter	I	J	L	M	N	P	Q	S	T	U	V	X	Z

Will this coding process defeat the statistical clues of a frequency distribution? Explain your answer.

10. Write a one-minute mystery in which the mystery is solved when the ace detective solves the case by cracking a coded message.

ACTIVITY

FINAL MESSAGE

14

In Activity 1, *The World of Codes,* you were given the challenge of cracking five coded messages. All except the last were coded with processes that used a shift, a stretch, or a combination of the two. The last message is different. Here it is again.

KBCCAW BVYWAZBWM KMFJENFD RH

LEFRL GTO FFFDVJAJ PVLWJKPGDFQQ

QAIYWZUNX JXERI ULNVOSAZBCCTFP

RH OLSUX OXS NBG

This message is coded with a keyword. The keyword has something to do with a restaurant.

1. Your task in this activity is to decode the fifth message of Activity 1. Good luck. (Hint: The spaces between words were left in the message when it was stored in the matrix.)

2. When you have finished decoding the message, write a summary of the mathematics you learned in this unit.

Wrapping Up Unit Two

1. The following message was encoded using a shift cipher. Use what you know about patterns created by shift ciphers to crack the code and read the message.

   13 12 5 26 9 5 23 9 7 22 7 24 24 19 24 9 16 16 29 19 25

2. In working with the total cost of electricity based on the number of kilowatt-hours used, the cost reflects a monthly service fee of $10.00 and a charge of $0.15 per kilowatt-hour.

   a) Determine the equation that expresses this situation, graph it, and sketch your graph.

   b) Use the graph to find the total cost of 350 kilowatt-hours.

   c) Use the graph to determine the number of kilowatt-hours used if the bill is $56.00.

3. This message is encoded with the process in **Figure 2.79**.

   28 34 36 52 24 34 20 22 36 52 46 36 10 48 24 30 14 44 16 12 42 16 46 32 16 44 44 8 20 16 44 52 24 30 30 22 16 30 38 56 36 48 46 16 8 42 46 22 16 32 8 38 8 42 46

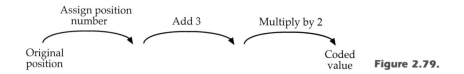

Figure 2.79.

   a) Decode the message.

   b) Describe how you decoded the message.

4. Explain why each of the following is not a good coding process.

   a) Use a telephone keypad to code letters as numbers 1–9.

   b) Use a stretch-2 cipher that matches A to 2, B to 4, and C to 6, and convert each coded value back to a letter.

5. You have two coupons. One is good for $2.00 off the purchase price. The other is good for a 20% discount. Both can be applied to the same purchase. Should you subtract $2.00 and then take off 20%, or take off 20% and then subtract $2.00?

6. Is a stretch of 2 followed by a shift of 3 the same as or different from a shift of 3 followed by a stretch of 2?

7. The following message is encoded with a two-step coding process. Crack the code and decipher the message.

12 48 15 18 60 6 57 18 66 60 18 15 21 48 57 6 69 6 57 30 18 63 78 48 21 51 66 57 51
48 60 18 60 63 48 60 51 18 18 15 6 45 15 63 57 6 12 36

63 27 18 21 39 48 72 48 21 30 45 21 48 57 42 6 63 30 48 45 63 48

27 30 15 18 30 45 21 48 57 42 6 63 30 48 45 63 48 12 48 57 57 18 12 63

18 57 57 48 57 60 63 48 12 48 42 51 57 18 60 60

30 45 21 48 57 42 6 63 30 48 45 6 45 15 63 48

60 78 45 12 27 57 48 45 30 81 18 30 45 21 48 57 42 6 63 30 48 45.

8. Hospital workers are accustomed to identifying the time of day using a 24-hour clock. 8:00 A.M. is written 0800. 11:30 A.M. is written 1130. 3:00 P.M. is written 1500. 11:30 P.M. is written 2330.

   a) Explain how to convert from a 12-hour clock that uses the P.M. notation for hours between noon and midnight to a 24-hour clock. Include additional examples with your explanation.

   b) Describe why converting from a 24-hour clock to a 12-hour clock is a type of modular arithmetic.

9. Here is a number trick.

   Pick a number.
   Multiply by 2.
   Add 8.
   Multiply by 3.
   Subtract 14.
   What is the number?

   a) Prepare a table, graph, and arrow diagram to represent the number trick. Include the inverse in the arrow diagram.

b) Find a two-step symbolic equation for the number trick.

c) Suppose the final number is 100. Find the original number. What equation did you solve to find it?

10. Use the distributive property to convert each multistep process to a two-step process.

a) $y = 3(5x + 4) - 6$

b) $y = 2(2x + 12) + 15$

c) $y = 5(x + 3) - 9$

d) $y = 8(10x - 3) - 12$

11. a) Use squares and circles to model this multistep coding process: add 1, multiply by 3, subtract 2, multiply by 2.

b) Describe how to decode the result using two steps instead of four.

12. Marie forgot the coding process she used for messages that she exchanges with a friend, but she remembers that she decodes with $p = (c + 2)/3$. Describe or diagram the coding process, or write an equation to represent the coding process.

13. Use the keyword *math* to code the word *airport* using shifts only.

14. The coded word below encodes a word using a 4 x 2 matrix and a keyword *it*, using the operation 2[A] + [B]. Decode the word.

41 22 47 58 55 50 45 28

15. Imagine a code in which you take a letter, multiply it by m, and add b.

a) Explain how to decode.

b) Write an equation that represents the coding process.

c) Write an equation that represents the decoding process.

d) Explain why m cannot be 0.

Mathematical Summary

Mathematics is the most commonly used tool for developing an effective coding process. An effective coding process is easy to encode, easy to decode, and difficult to crack.

Mathematicians use the term *function* to describe a process like coding. A function transforms items such as letters or numbers into other letters or numbers. A plaintext letter or position number is matched with a new coded value or letter.

The shift cipher is the simplest coding process. Each letter of the alphabet is assigned a position from 1 to 26. Once you decide the number of spaces to shift the alphabet, there are many ways to represent the shift.

For example, suppose you want to shift the alphabet three spaces.

1. You can use words to describe the process: Shift the alphabet three letters.

2. You can use an arrow diagram to represent the steps in the coding process (**Figure 2.80**).

3. You can write a symbolic equation to identify the relationship between p (the original location) and c (the coded value): $c = p + 3$.

4. You can make a table to match the original or plaintext letter with the coded value or letter (**Figure 2.81**).

Shift 3 letters

Original message   Coded message

Figure 2.80.
An arrow diagram for a coding process.

Plaintext letter	A	B	C	D	E	F	G	H	I	J	K	L	M
Coded value	4	5	6	7	8	9	10	11	12	13	14	15	16
Coded letter	D	E	F	G	H	I	J	K	L	M	N	O	P
Plaintext letter	N	O	P	Q	R	S	T	U	V	W	X	Y	Z
Coded value	17	18	19	20	21	22	23	24	25	26	27	28	29
Coded letter	Q	R	S	T	U	V	W	X	Y	Z	A	B	C

Figure 2.81.
A coding table.

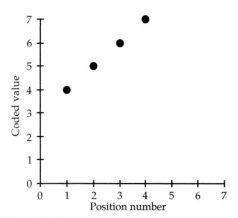

Figure 2.82.
A graph of a coding process.

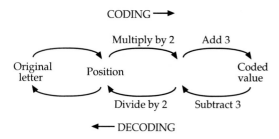

Figure 2.83.
A coding process
and a decoding process.

5. You can make a graph to match the plaintext letter with a coded value or letter (**Figure 2.82**).

Once a message is received, it must be decoded. Decoding reverses the coding process, using inverse operations. The arrow diagram in **Figure 2.83** shows the encoding and decoding process for a two-step cipher.

This arrow diagram shows that a message is coded by first multiplying the position number by 2, then adding 3. You can follow the arrows in reverse order and decode by first subtracting 3 from the coded value, then dividing by 2.

The coding process can be represented with an equation, $c = 2p + 3$. Then each step of the decoding process can be seen in the equation $p = (c - 3) \div 2$: subtract 3 from the coded value, $c - 3$; and divide the result by 2, $(c - 3) \div 2$.

The order of calculations is important when you use a two-step process. $c - 3 \div 2$ and $(c - 3) \div 2$ are not the same. Even though subtraction is written first, it is not done first unless parentheses are used. This is because multiplication and division take priority over subtraction and addition.

When you decode a number like 25 for the coding process $c = 2p + 3$, you solve the equation $25 = 2p + 3$. To solve the equation, you follow the decoding steps of an arrow diagram (Figure 2.83). This process can be written in symbolic form: $2p + 3 = 25$.

Subtract 3: $2p = 25 - 3$, or $2p = 22$.

Divide by 2: $p = 22/2$, or $p = 11$.

Understanding the tools of the code cracker helps build effective codes. The code cracker looks for patterns. Some patterns are revealed through the use of statistics. The value that occurs most frequently in a message probably represents the letter *E* because *E* is the most frequently used letter in the English language. After *E*, the most frequent letters are *T*, *N*, *R*, *I*, *O*, *A*, and *S*. If spaces between words are coded, then spaces occur more often than any letters.

When the coding process is a function, patterns emerge in tables and graphs. The points lie in a line for the graph of a shift cipher and a two-step coding process. Notice the pattern in the coded values for the table representing the two-step process $c = 2p + 3$ (**Figure 2.84**).

The distance between successive coded values is 2.

Plaintext letter	A	B	C	D	E	F
Position number	1	2	3	4	5	6
Code value	5	7	9	11	13	15

Figure 2.84.
Part of a coding process.

Some two-step and multistep coding processes are equivalent.

Process 1: $c = 2(p + 3)$

Process 2: $c = 2p + 6$

The mathematical expressions $2(p + 3)$ and $2p + 6$ are equivalent. No matter what value of p you use, both expressions give the same result.

Another way to see equivalence is to represent the variable p with a shape and the number 1 with a different shape. Use a square for p and a small circle for 1.

To represent $p + 3$, draw a square (p) and three circles (3), as in **Figure 2.85**.

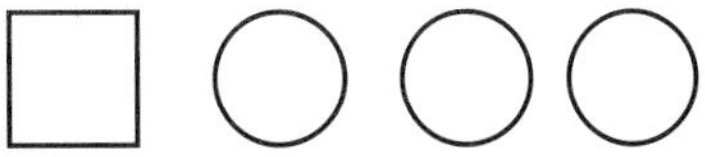

Figure 2.85.
A representation of $p + 3$

To represent $2(p + 3)$, draw the previous figure twice (**Figure 2.86**).

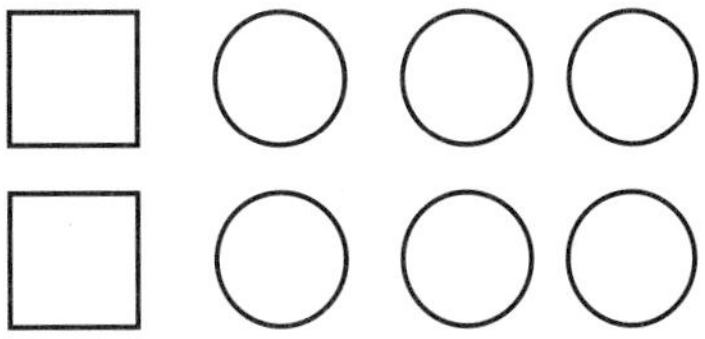

Figure 2.86.
A representation of $2(p + 3)$.

To represent $2p + 6$, start with two squares to represent $2p$. Add six circles to represent $2p + 6$ (**Figure 2.87**).

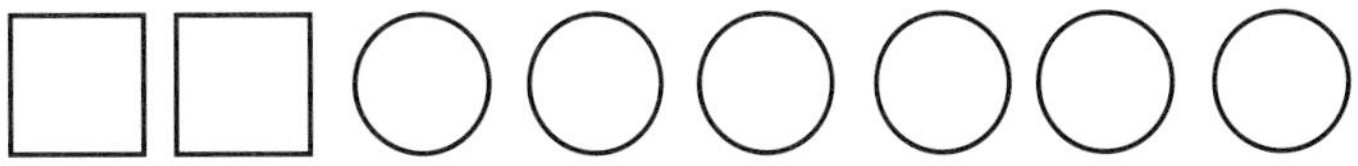

Figure 2.87.
A representation of $2p + 6$.

Mathematicians use the distributive property to describe why $2(p + 3)$ and $2p + 6$ are the same. This property states that multiplication can be distributed to the individual parts of an expression involving addition or subtraction. In this example, multiplying by 2 in $2(p + 3)$ can be distributed to the two parts of the expression $p + 3$, to obtain $2p + 2(3)$.

Increasing the number of steps in a coding process does not make the work of a code breaker more difficult.

Start with the letter's position. p

Multiply by 3.	$3p$
Subtract 2.	$3p - 2$
Add 4.	$3p - 2 + 4$, or $3p + 2$
Multiply by 2 and distribute.	$2(3p + 2)$, or $6p + 4$

The four-step coding process can be done with only two steps. You can also decode with only two steps: subtract 4 and divide by 6.

A matrix is often a labor-saving device. Coders can use it to code an entire message at once rather than a single letter at a time.

$$\begin{bmatrix} M & E & E & T & \_ & M & E & \_ & O & N \\ \_ & F & R & I & D & A & Y & \_ & \_ & \_ \end{bmatrix}$$

Figure 2.88.
A message in a matrix.

To code a message like "Meet me on Friday" with a coding process like $c = 2p + 1$, write each letter in a suitable matrix (**Figure 2.88**).

$$\begin{bmatrix} 13 & 5 & 5 & 20 & 27 & 13 & 5 & 27 & 15 & 13 \\ 27 & 6 & 18 & 9 & 3 & 1 & 25 & 27 & 27 & 27 \end{bmatrix}$$

You choose the size of the matrix. Add blanks at the end if necessary. Replace each letter with its position number (**Figure 2.89**).

Figure 2.89.
A message matrix with position numbers.

Store this as matrix [A] in a calculator, then make a matrix [B] that is the same size as [A] and fill it with 1s.

Use a calculator to compute 2[A] + [B], and the message is coded. Take the numbers out in the same order you put the original letters into the matrix. The coded message is 27 11 11 41 55 27 11 55 31 27 55 13 37 19 7 3 51 55 55 55.

The message can be decoded by reversing the process. You don't have to use the same-sized matrix. All you have to know is the decoding process for $c = 2p + 1$.

The matrix method needs only slight alteration to destroy the patterns recognized by the code cracker. The keyword-matrix method not only disrupts these patterns, but also allows people to send messages to each other using only a single word to communicate the method.

Coding, decoding, and code breaking are good examples of the way mathematics is used in the world. The coder and decoder invent a method, only to have it eventually cracked by a code breaker. Mathematical tools are used by the coder, the decoder, and the code breaker. When the tools no longer do their job well, new ones are invented.

Glossary

ALGEBRAIC EXPRESSION:
A collection of variables and constants connected by several operations. For example, $2p + 3$ is an expression in which the variable p and the constants 2 and 3 are connected by the operations multiplication and addition.

ARROW DIAGRAM:
A diagram that uses labeled arrows to show each step of a process such as coding or decoding.

AXIS:
A line that intersects another line at a right angle in a coordinate plane. The horizontal axis is usually associated with a function's input. The vertical axis is usually associated with a function's output.

CIPHER OR CODE:
A process in which each symbol of a message is represented by another symbol.

CODE BREAKING OR CRACKING:
Determining the contents of a coded message without knowing how it was coded.

CODED VALUE:
A numerical value resulting from a coding process. In the equation $c = 2p + 3$, the letter c represents the coded value.

COMBINATION CODE:
A coding system that combines a stretch and a shift, in either order.

CONGRUENT MODULO K:
Two numbers are congruent modulo k if their difference is an integral multiple of k. Thus, two numbers are also congruent modulo k if both have the same remainder upon division by k. For instance, if values differ by a multiple of 26, as do 56 and 82, they are congruent modulo 26.

CONSTANT:
A quantity that remains the same whenever an equation is used. In the equation $c = 2p + 3$, the numbers 2 and the 3 are constants.

CONTROL NUMBER:
A number that controls some feature of the situation under study. For example, in a combination cipher like $c = 2p + 3$, the constant 2 controls the amount of stretch, and the constant 3 controls the amount and direction of the shift.

COORDINATE GRAPH:
A graph based on two lines (axes) intersecting at right angles.

COORDINATES:
A pair of numbers that identify a location or point on a graph. The order of the numbers in the pair is important: the horizontal location is always first.

CRYPTOGRAPHY:
The study of coding, decoding, and code breaking.

DECODE OR DECIPHER:
To reverse the coding process; to determine the plaintext letter from a coded value or other symbol.

DICTIONARY CODE:
A system in which words or phrases are used for other words or phrases.

DISTRIBUTIVE PROPERTY:
Multiplication can be distributed to the individual parts of an expression involving addition or subtraction. For the expression $2(p + 3)$, the 2 can be multiplied by the p and 3 separately to obtain $2p + 2(3)$.

DOMAIN:
All the possible or meaningful values that originate a process. In the coding context, the position number originates the coding process. Therefore, the domain for a coding process is the counting numbers from 1 to 26. The domain for a group of coordinate pairs is the collection of all the first coordinates. In the number pair (1, 4), the number 1 must be a member of the domain.

ENCODE:
To change a plaintext letter to a coded form.

EQUIVALENT EXPRESSIONS:
Expressions that always have the same value as each other. $2x + 6$ is equivalent to $2(x + 3)$ because, no matter what number you put in place of x, the resulting value is the same for $2x + 6$ and $2(x + 3)$.

FREQUENCY DISTRIBUTION:
A table showing the number of times an event or letter occurs in comparison to related events or letters.

FUNCTION:
A process that transforms items such as letters or numbers into other letters or numbers uniquely.

INVERSE:
The reverse of a process. Decoding is the inverse of encoding. Encoding is the inverse of decoding.

KEYWORD MATRIX:
A matrix in which the numbers of each row represent a single word.

LINEAR FUNCTION:
A function whose graph is a straight line.

ORIGIN:
The point where two axes intersect.

PIECEWISE CIPHER:
A code in which different processes are used for different parts of the alphabet.

PLAINTEXT:
The original language and letters of a message before it is coded.

POSITION NUMBER:
The number from 1 through 26 that is matched to a letter from A through Z. The position number for E is 5.

RANGE:
The numbers that result when a process is applied to the members of the domain. In the coding process, the range is all the coded values.

SHIFT CIPHER:
A coding process that uses addition or subtraction to shift every position number or plaintext letter to another value or letter.

SOLUTION:
The number(s) that, when put in place of the variables in an equation, result in a true statement.

SOLVING:
The process of finding the solution for an equation.

STRETCH CIPHER:
A process that codes a position number to a coded value by multiplication.

SUBSTITUTION CIPHER:
A process that substitutes a number, a symbol, a character of the alphabet, or a word for a plaintext letter or word.

SYMBOLIC EQUATION:
A mathematical statement that two expressions are equal. Variables, constants, and operations may be used in each expression; for example, $c = 2p + 3$.

TRANSFORMATION:
Another word for function. It is typically used in a geometric setting. Since a function can be thought of as moving points on a line (or plane) to other points on the line (or plane), you can think of a transformation as moving one set of points, or an object, into another. Translations (shifts) and dilations (stretches) are examples of types of transformations.

TRANSLATION:
A transformation in which a single constant is added to all position numbers, resulting in a slide of the alphabet along the number line (in the case of a code). A shift cipher is a translation.

TRANSPOSITION CIPHER:
A process that scrambles or rearranges the order of letters, symbols, or words in a message.

VARIABLE:
A quantity that may vary in an equation. In the equation $c = 2p + 3$, c and p are variables.

WINDOW:
The viewing region of a graph, defined by the highest and lowest displayed values for x and y.

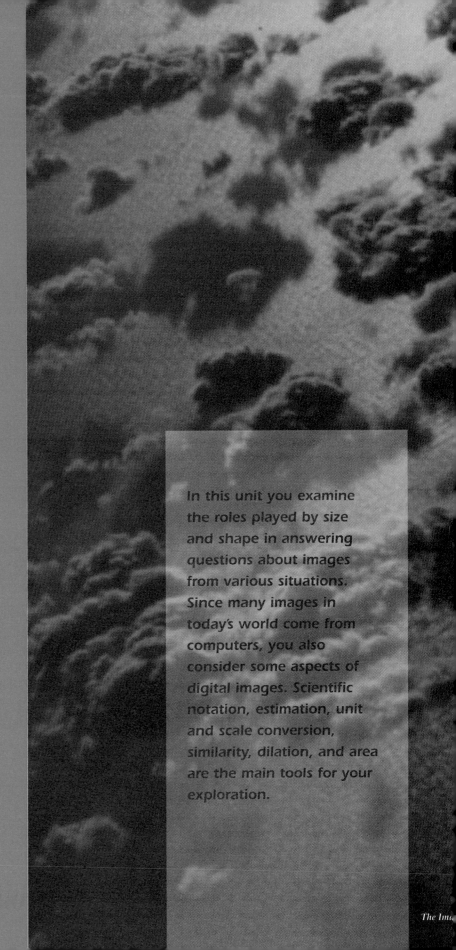

150

In this unit you examine the roles played by size and shape in answering questions about images from various situations. Since many images in today's world come from computers, you also consider some aspects of digital images. Scientific notation, estimation, unit and scale conversion, similarity, dilation, and area are the main tools for your exploration.

How much land has been cleared of forests in a particular region of the world? How can police agencies identify fugitives who have tried to change the way they look? What do these two problems have in common? One common thread is the geometry of size and shape.

Satellite imagery and its use in monitoring Earth's resources are the central context for this unit. Because of their great distance from Earth and their extra-sensitive equipment, satellites such as Landsat 5 can provide information about Earth that would not otherwise be available. In order to determine the extent of deforestation, for example, you will investigate a variety of scales of measure and the nature of computer images. Since images may at first be too small to work with efficiently, you will also examine the geometry of size changes. And, of course, if you are to measure the amount of deforestation in a region, you will need to learn techniques for estimating the area of regions, even when they are not convenient shapes.

But what about police work? Think about visible features of a person that don't change. For example, the size and shape of a person's head depend on the bony structure of the skull; that's not likely to be altered very easily! And, as you know, fingerprints stay the same throughout a lifetime. Whenever a shape does not change, it might play a role in identification.

As you might imagine, geometric reasoning is used in many kinds of work. You are familiar with some computer and satellite images, from videos and movies to weather forecasting. Now you will have a chance to examine some of this work more closely.

LESSON ONE

Seeing from a Distance

KEY CONCEPTS

Distance

Scale

Graphical interpretation

Ratios

Unit conversion

Scale factor

Precision

Significant figures

David Barber

PREPARATION READING

What Resources Can Be Monitored with Landsat Data?

The three articles that follow appeared in the magazines *Science News* and *Discover*. Use the five questions below to guide your reading of each article. Answer the questions with complete sentences.

1. What is the question or problem scientists are investigating?

2. How do the Landsat images help them in this investigation?

3. What are the scientists doing with the images to answer the question they are investigating?

4. How do you think questions of scale affect the scientists' conclusions?

5. Why might scientists also want to visit the regions they are investigating?

Spotting Erosion from Space

RICHARD MONASTERSKY
Science News 136 (4), July 22, 1989, p. 61.

Billions of tons of dirt wash into the oceans each year, making soil one of the world's most endangered resources. Not only does topsoil erosion steal precious nutrients away from fields, it also increases the cost of farming and lowers food production. Now scientists are using satellite images to spot areas facing the greatest erosion danger.

Department of Agriculture researchers exploit the muddy side-effects of erosion in their search. As soil particles wash into lakes and rivers, they change water color by adding red hues, says Jerry C. Richie with the USDA in Beltsville, MD. The earthy tones allow Richie and his colleagues to find soil-filled lakes on images from the U. S. Landsat satellite, which carries several cameras, each recording a specific color of light.

The scientists start by subtracting all land areas from the satellite images, then examine the colors of the remaining lakes and reservoirs. By comparing the colors against a theoretical model they devised, the researchers can estimate with about 90 percent accuracy the total amount of suspended sediment in the water, Richie says. If used several times a year, this technique would quickly tell conservation officers which watersheds suffer the greatest erosion.

The model is based on over a decade's worth of comparisons between satellite images and water samples from a lake in Mississippi and another in nearby Arkansas. Since developing the system, the USDA scientists have tried it out on several more lakes and investigators in Oklahoma are currently testing it on a statewide basis, Richie says.

Frankincense

CHARLENE CRABB
Discover 14 (1), January, 1993, p. 56

According to legend it was a place of immeasurable wealth and, like Sodom and Gomorra, innumerable sins: Ubar, an ancient city that flourished for some 3,000 years as the center of a lucrative trade in the tree resin called frankincense. One of the world's most valued commodities, frankincense was so prized as a perfume, medicine, and religious incense that it once commanded its weight in gold. But sometime between the first and fourth centuries A.D., Ubar vanished—buried under the sands, supposedly by a violent windstorm sent by an angry god.

Ubar remained a city of legend until last winter. Then, in February, a disparate group of adventurers announced that they had located the lost city in the southern Arabian Peninsula, quite near the remote Qara Mountains where the small, scraggly trees that produce frankincense still grow.

The discovery came after a decade of research spearheaded by Emmy Award-winning documentary filmmaker Nicholas Clapp. In his quest to find Ubar, Clapp assembled a team to analyze ancient texts and copies of a map drawn by the Egyptian cartographer Ptolemy around A.D. 200. "Whenever Ptolemy noticed a major trade center, he'd mark it with a little castle with turrets," says the expedition's chief archeologist, Juris Zarins of Southwestern Missouri State University. "One of the places

he marked, in what is today southwestern Oman, was what he called the Omanum Emporium, the marketplace of Oman, in the land of the Ubaritae people."

Clapp had a hunch that the Omanum Emporium was indeed Ubar, but he needed more than the ancient map to pinpoint its location. So he called on Ronald Blom of the Jet Propulsion Laboratory. Ten years ago Blom, a geologist specializing in remote sensing, had used Landsat images to reveal an ancient river system that lay hidden beneath the Sahara. Now Blom and his colleagues began examining satellite images for evidence of Ubar. Using the near-infrared portion of the light spectrum, Landsat, along with the French SPOT satellite, was able to reveal the faint lines of ancient caravan trails. The trails were visible because, over the centuries, their original gravel base was ground down by hundreds of thousands of donkey and camel hooves. The resulting fine grains now have different reflective properties than the surrounding undisturbed sand.

. . . The faint lines converged on the area Ptolemy had labeled the Omanum Emporium on his ancient map. When the team arrived at the site . . . they found evidence of Ubar's end. It was very likely brought about by an earthquake, not a sandstorm, which collapsed the city into a limestone cavern that Ubar's citizens probably never knew existed.

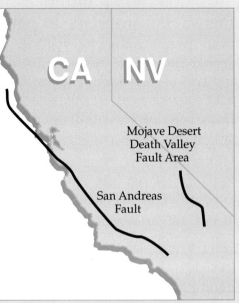

New picture of California Plate Puzzle

RICHARD MONASTERSKY
Science News 137 (11), March 17, 1990, p. 175

As it slowly edges to the northwest, the Pacific plate moves about 48 millimeters a year relative to the North American plate. Researchers think the famous San Andreas fault absorbs a substantial fraction of that slip, on average somewhere between 30 mm to 40 mm each year. But what about the rest of the motion not absorbed by the San Andreas? While scientists five years ago thought that faults west of the San Andreas must absorb the additional slip in southern California, more recent work points to regions to the east of the San Andreas.

and 29 percent over the last several million years, they reported at a meeting of the American Association for the Advancement of Science in New Orleans last month.

This work confirms previous geodetic studies indicating that the Mojave absorbs about 6 mm per year of the slip between the North American and the Pacific, says Wayne R. Thatcher of the U. S. Geological Survey in Menlo Park, Calif.

As part of his study of Mojave faults, Dokka received help from very high sources—the Landsat satellites. John P. Ford and co-workers at the Jet Propulsion Laboratory in Pasadena, Calif., have developed an improved technique to identify faults using Landsat satellite images. This technique allowed the researchers to pinpoint several previously undocumented faults in the Mojave.

Roy K. Dokka and Christopher J. Travis of Louisiana State University in Baton Rouge have spent years of field work studying the faults of the Mojave Desert-Death Valley region. Their work suggests these faults have been taking up a significant amount of the plate motion, between 9 percent

CONSIDER:

1. What is Landsat?

2. What is meant by the terms "macroscopic" and "microscopic" views of Earth?

3. How can scientists use satellite imagery to measure changes on the Earth's surface?

4. What makes satellite images more useful in some cases than images taken closer to the regions being studied?

DEFORESTATION IN THE CZECH REPUBLIC

Many regions of the world have undergone deforestation over a number of years. The reasons vary from place to place and from time to time. At the end of this unit, you will use Landsat images to determine the amount of deforestation that took place between 1985 and 1990 in a particular region of the Czech Republic. **Figure 3.1** shows a portion of the Czech Republic in 1985. The colors have been altered in order to make certain features (in this case, conifer trees) more easily visible. Several forested regions have been marked for you.

Your task is to determine the total area of these forested regions. If you have access to the image on a computer, you may find that helpful, but you can work directly from the images printed here.

As you work, write any questions for which you need answers. If you have ideas about how you might find answers to your questions, record those thoughts, too.

ACTIVITY 1
DEFORESTATION IN THE CZECH REPUBLIC

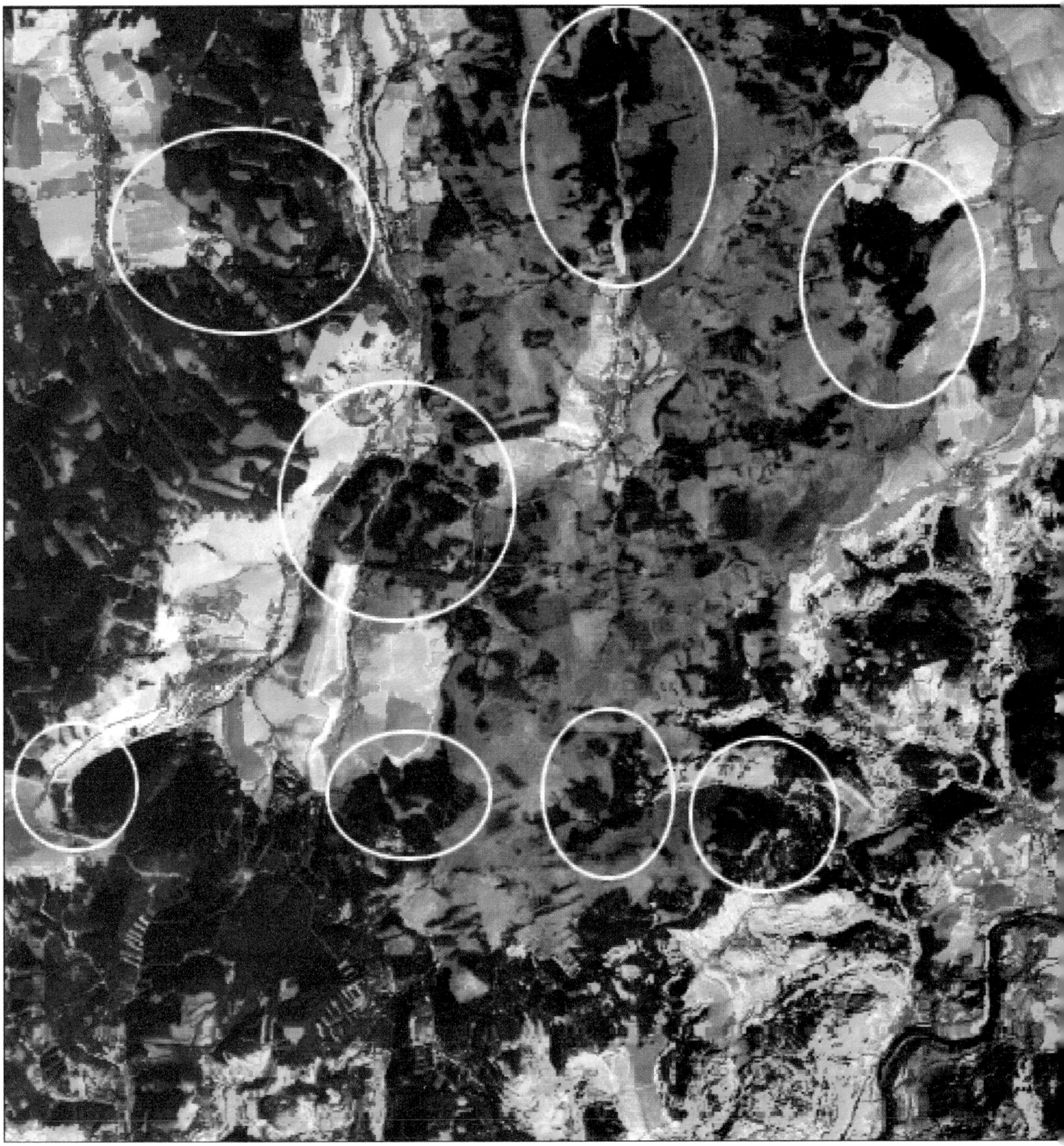

Figure 3.1.
A portion of the Czech Republic in 1985.

INDIVIDUAL WORK 1

Thinking About Distances

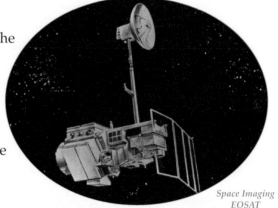

*Space Imaging
EOSAT*

The Landsat 5 satellite orbits the Earth at an altitude of approximately 700 km (440 miles). An airplane typically flies about 9.7 km (6 miles) above the Earth. Use maps to explore the significance of those distances in the following exercises.

1. Locate five features that are about 9.7 km away from your school. These features can be natural (a lake or hill) or constructed (a shopping center or baseball field). Describe how you determined the distance from the school to each feature.

2. Locate five features that are about 700 km from school. Again, describe how you determined the distance. Do you think the distances you determined in this exercise are as precise as those in Item 1? Why?

3. Landsat 5 completes a survey of the earth every 16 days. If it passes overhead this very minute, on what day will it again pass overhead? How close to New Year's Day? How close to midnight on the eve of your birthday?

4. What differences in your community might Landsat 5 detect when it passes overhead 16 days from now?

5. Contact your state geographer or a television meteorologist to find out how often clouds obscure your area on New Year's Day.

VIEWING EARTH FROM DIFFERENT DISTANCES

As you worked to determine the amount of deforestation in the image of the Czech Republic, you probably discovered that you did not have quite enough information to complete your task. From earlier units you know that in order to solve a more complicated problem it is helpful to identify smaller, simpler questions that might move the job along. Look back at your questions from Activity 1 and compare them to the ones below.

What is the scale of the image? If an object measures exactly one cm long on the image, how large is the real thing? Does the direction along which the object is being measured affect the scale?

Can images be enlarged to make measurements easier? If so, how? What does that do to the scale? Does direction matter? Do shapes stay the same? How can you be sure?

How can you measure areas of objects that are not just squares, rectangles, and triangles? Is there more than one method? If so, how do you decide which method is best? What happens to area measurements when you change scales?

All these questions, and others, need to be answered before you will be able to complete the task set forth in Activity 1. Lesson 2 will begin the careful mathematical examination of these issues, but first, let's "set the stage." Satellites travel very high above the earth. How much can be seen from there?

Airplanes typically fly at an altitude of approximately 9.7 km (6 miles). Landsat satellites orbit approximately 700 km (440 miles) above Earth. As you view your town from different distances, think about how your perceptions of your town change. Consider why people might want or need to see objects from different distances.

ACTIVITY

VIEWING EARTH FROM DIFFERENT DISTANCES

2

1. Find your state and the approximate location of your city or town on a globe or map. Take turns looking at your city or town through a cardboard cylinder and create a table, using the headings shown in **Figure 3.2**. As your distances from the map, use 15 cm, 40 cm, 125 cm, and 11 m.

   Work with your partner to check the distances, but answer the items individually. Notice that the questions ask for *perceptions*, not mathematical relationships.

Distance from map	What do you see?	How do you perceive your town?

Figure 3.2.
Sample table headings for Item 1.

2. As you move farther away from your city, how do your perceptions change regarding:

   a) details?

   b) how much of the region you can see?

   c) how your city is related geographically to other cities, other states, other nations?

3. Why might someone want to see an object or place from different distances?

4. How do you think weather satellites have changed the forecasting and reporting of weather as compared to forecasting based only on ground-level observations?

5. What do you think are some of the most important benefits of satellites?

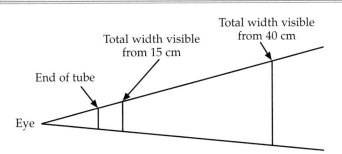

ACTIVITY

VIEWING EARTH FROM DIFFERENT DISTANCES

2

Figure 3.3
Views from a tube.

6. Keep your data from this question for use in Individual Work 2. Construct a table using the headings shown in **Figure 3.5**. Looking through the tube from a distance of 15 cm, help your partner record (in column 2) the total width on the map that you can see at one time. (See **Figure 3.3**.) Then find two clearly visible locations on a map. Have your partner measure the actual distance separating them on the map and record the result in column 4 of the table. Be sure to record units, too.

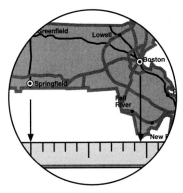

Figure 3.4.
Measuring apparent distance.

Still viewing from a distance of 15 cm, place a small ruler or other scaled device *across the end of the tube* and measure the "apparent distance" between the same two locations. (See **Figure 3.4**.) Record this apparent distance in the third column of the table.

Now move back to a distance of 40 cm, record the total width visible, then locate two new locations that have the *same apparent distance* between them as for the first two objects at 15 cm. Measure their actual map distance and record it in column 4. Repeat for distances of 125 cm and 11 meters.

Both you and your partner should make observations and record information. Note that the numbers in column 3 should not change for any particular person. (But your values may not be the same as your partner's.)

Distance from map	Total width of map visible (units)	Apparent distance (at end of tube)	Distance on map

Figure 3.5.
Table headings for Item 6.

INDIVIDUAL WORK 2

Practice Your Scales

1. Write a coding rule for a simple secret code for which one single equation will serve as a key. Write the same rule in as many different kinds of representations as you can. Then describe the geometric effect of your coding process on the alphabet. For example, does it shift letters, or stretch their spacing, or do something else?

2. **Figure 3.6** shows a portion of a table representation of a coding process. Recall that some codes involve mathematically changing the distances between positions of letters in the alphabet. Similarly, maps present locations drawn with a mathematical change in scale so that they will fit the available space.

   a) Describe the effect of the code on the distances between position numbers for letters, then explain what that means about the equation for the code.

Plain text letters	A	B	C	D	E
Plain text values	1	2	3	4	5
Coded values	15	18	21	24	27

Figure 3.6.
A portion of a coding table.

   b) Use your observation from part (a) together with some specific values from the table to determine the equation for this code.

3. a) Refer back to Item 6 of Activity 2. What was the actual width on the map that was visible when you observed it from a distance of 15 cm? Use the map's scale to determine the distance that width represents in reality.

   b) Repeat part (a) for each of the other viewing distances you recorded in Item 6 of Activity 2.

In your work with maps thus far you have used scales to measure distances, both on the map and in reality. In fact, that's exactly the role of a map's scale: to relate one distance to another distance. Distances, of course, are always associated with *two* locations. That is obvious, but it is sometimes easy to forget. A ruler or other scale may be used to measure a distance in many ways. For example, measuring from the 3-cm mark to the 5-cm mark is just as good as measuring from the end of the ruler to the 2-cm mark; they both indicate a distance of 2 cm. And if a measuring device is broken, or markings are rubbed off in some portions, it may be necessary to do just this kind of adjustment in the location of the ruler.

4. a) Randomly place a meter stick or yardstick on a map of your choice (but be sure the map has a printed scale). Without moving the stick, record the locations (as numbers on the measuring stick) of at least six towns or other features lying along it. Record the names of the features along with their meter-stick locations.

   b) Imagine a giant measuring stick, marked off in kilometers or miles, lying across the region of the world represented by the part of the map you used in part (a) and having the same zero-point as your meter stick. Use the map's scale to convert the locations you recorded into locations along the giant measuring stick.

   c) If you did not do so in part (b), list the locations as values in a two-column table. List the measured locations in the first column (starting with the smallest and going up to the largest) and the corresponding computed locations in the second column.

   d) Use the map's scale to calculate the actual distances between several pairs of these locations. Explain exactly how you get your answers.

5. a) Explain how your table in Item 4(c) is somewhat like the table for coding and decoding messages in Item 2.

   b) List the differences between the process of converting locations to scaled locations and the process of coding, and explain your choices.

6. Use the representations you studied in Unit 2, *Secret Codes and the Power of Algebra*, to write the process for converting direct measurements from your map into actual distances in as many different ways as possible. Comment on which representations seem most helpful, and why.

ACTIVITY

3

CHANGING SCALES

The word **scale** is used in mathematics in several ways. In your work with maps you have already encountered two such meanings. First, and maybe so obvious that you didn't even think about it, is the set of markings on your measuring device (ruler or meter stick). A typical scale on a one-foot ruler is marked in sixteenths of an inch, for example. The same meaning applies to the scale on a graph: you place tic marks at intervals of some convenient separation, such as every five numbers. Thus, this use of the word "scale" indicates a unit of measure.

Another meaning of scale is the one about which you have been thinking. It refers to the use of one size of object (length) to represent another one of an entirely different size. For example, on a city map, 2 cm may be used to represent 1 km. (The reason is clear: a life-size map would be hard to put in your pocket!) It is this meaning of scale that is important in discovering the amount of deforestation in the image of the Czech Republic that you saw in Activity 1. Whenever there is a chance of confusing these two meanings, refer to the first kind of scale as a "ruler."

To analyze the deforestation situation, you need the answers to two questions. First, how do scales work in general? After that, what is the particular scale for the Czech image? In this activity, you will address the first of these two questions.

CONSIDER:

1. Look back at Item 5(a) of Individual Work 2. Then think about a simple code such as you used early in Unit 2, *Secret Codes and the Power of Algebra*; for example, $c = 2p + 3$. Make a table of values of p and c for this coding rule. Pretend your table is two rulers that are matched up. Do a few examples to see how it works, then explain how distances measured with the "p-ruler" can be converted into distances measured with the "c-ruler."

CHANGING SCALES

Because of the importance played by units of measure in most real applications of mathematics and the frequent need to record information about relationships in tables, observe the following convention:

> The first entry in each list is the label for the numbers in that list, together with the units of measure (in parentheses) that those numbers represent. The remaining entries are the numbers themselves, without units displayed.

1. Imagine a map on which 2 cm represent 1 mile. Measurements may be made on the map (in centimeters) or in the real world (in miles). **Figure 3.7** is a table of six corresponding ruler readings for the two measurement systems (map and reality).

map (cm)	1.0	2.0	3.0	5.0	10.0	20.0
reality (mi.)	0.5	1.0	1.5	2.5	5.0	10.0

Figure 3.7.
Corresponding ruler readings.

a) Think of this table as a coding rule, with the map units being the "plain text" and the real locations as being the "coded message." State the coding rule in words.

b) Rewrite your coding rule as an arrow diagram.

c) Use your table or coding rule to represent the process as a graph. Be sure to label the axes of your graph carefully, showing the same labels and units as are recorded in the table.

d) What is the distance from the second to the fifth table reading on the map ruler?

e) What is the distance from the second to the fifth map reading on the reality ruler?

CHANGING SCALES

f) How do you convert from the map-ruler distance to the reality-ruler distance? Write your answer both in words and as an arrow diagram.

g) Repeat Items 1(d–f) for three other pairs of corresponding readings, then summarize your observations. Use both words and arrow diagrams in your summary.

Making a table of corresponding ruler readings is a kind of short cut and space-saving device to take the place of what is perhaps a better visualization of the real situation. Look at **Figure 3.8**, which illustrates the matching of two very different kinds of rulers.

There are several advantages of looking at the map scaling in this way. First, you can use it to make reasonable, quick estimates moving between the two scales, especially if you know how to match up a couple of markings. The second feature illustrated by the figure is the fact that distances are what really matter. That is, the difference between markings matters more than the individual markings themselves.

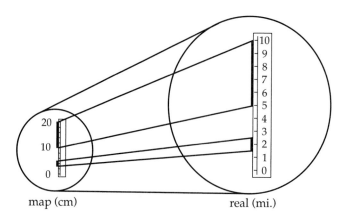

Figure 3.8.
Changing from map units to real units by matching rulers.

map (cm)   real (mi.)

CHANGING SCALES

Figure 3.9 shows another pair of rulers, used by a student in making a "map" of her desk. The markings are related by the equation $D = 2M + 3$, where D is the reading on the desk itself and M is the reading on a ruler on the map. Use this matching to answer Items 2 and 3.

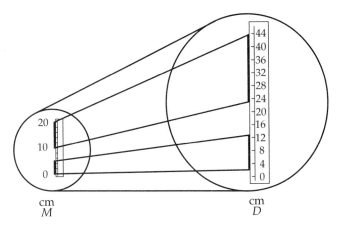

Figure 3.9.
Changing from map units to desk units by matching rulers. $D = 2M + 3$.

2. a) When using a ruler, it is not necessary to begin measuring from the zero marker. Explain how to determine distances using a ruler that has no zero marker.

   b) Based on the equation for the matching illustrated in Figure 3.9, what reading on the "desk" ruler corresponds to the zero marker for the map ruler? Can distances still be measured in this system?

   c) Use the given equation for the matching to construct a table of at least five pairs of corresponding values.

   d) Use your table to repeat the kind of analysis you did in Item 1 parts (d–f).

   e) Write a rule for converting distances from map to desk scales.

ACTIVITY

CHANGING SCALES

3

3. a) Recall from your work with codes in Unit 2, *Secret Codes and the Power of Algebra*, that a graph is another representation of a relationship between two quantities. In that unit you used tables to make graphs, and you learned how to use a graphing calculator, too. Make a graph of the ruler matching illustrated in Figure 3.9. Use a window of –3 to 3 for the map, and –3 and 9 for the desk. Label your axes, and show their units of measure.

  b) If you did not do so in part (a), add dots on your graph to represent those values that you included in your table in Item 2(c). Do they seem to "fit" the pattern?

  c) In Item 2(d) you examined several distances between table values. On your graph, mark as a horizontal segment on the map axis one of the distances between map-ruler locations that you used in Item 2(d). Mark the corresponding desk-ruler distance as a vertical segment on the desk axis. How does the graph relate these two segments?

4. Make up your own scaling situation, and use a table and a graph to explain how to move from map-ruler distances to the reality-ruler distances.

5. What seems to be a general rule for moving from a map-ruler distance to a reality-ruler distance?

6. a) Select one of the scaling situations from Items 1–4. Keep the scaling rule the same. Think of points that are 3 units apart according to the map ruler. How far apart are the corresponding points according to the reality ruler? What happens if the first two points are 6 units apart?

  b) In general, if two towns on a map are twice as far from each other as two other towns, what is true about the corresponding reality-ruler distances? Explain, first by using your answer to Item 5, then by using common sense.

ACTIVITY

CHANGING SCALES

3

SCALING

In Items 1–6, you dealt with two distinct distances that were matched up. For example, in Item 1, a length of 2 cm was matched with a distance of 1 mile. One is small enough to fit on a piece of paper, the other is very large. One represented the other, but they were not really equal distances. The process of converting distances from one system to another, when the corresponding distances are not really equal, is called **scaling**.

You have seen several ways to visualize scaling. First, in looking at a table of corresponding readings, the separations between entries provides a way to "see" the conversion process. For example, if consecutive entries in the map readings differ by 2 cm and the corresponding entries for the reality readings differ by 5 mi., then converting distances from map readings to reality readings requires multiplying by 5/2 and changing the units of measure. Again, the scaling acts on the *distances*, not on the ruler readings themselves.

Use **Figure 3.10** to answer Items 7 and 8.

map (cm)	2.0	4.0	6.0	8.0
reality (mi.)	0.0	5.0	10.0	15.0

Figure 3.10.
A table of locations on two different rulers. Values are not distances but coordinates.

7. Suppose two cities have readings of 4 and 8 on the centimeter ruler placed on the map. How far apart are they in map (cm) units? How far apart are they in reality? Explain your answers.

8. Suppose two cities appear on the map as 15.0 cm apart. Give a pair of possible map scale readings for these cities, then determine their actual distance apart. Can you find the matching scale readings for your hypothetical cm-scale readings?

ACTIVITY

3

CHANGING SCALES

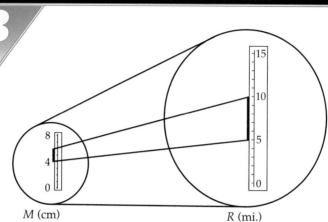

Figure 3.11.
Changing from map
units to real units by
matching rulers.

M (cm)   *R* (mi.)

Another way to represent the same information is by matching the rulers directly.

9. Make a figure similar to **Figure 3.11**, showing the scaling of the distance between the cities mentioned in Item 7 (at cm scale markings 4 and 8).

10. Repeat Item 9 for the scale readings you used in answering Item 8.

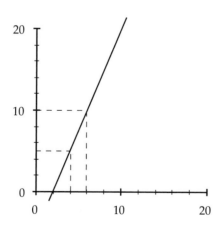

Figure 3.12.
Using a graph to scale.

Yet another way to see the same effect is to look at a graph of the rule relating the ruler readings. The graph allows horizontal segments (map distances) to be turned into vertical segments (reality distances) by "sliding" the horizontal segment up or down to the graph and "reflecting" its endpoints left or right to the vertical axis. (See **Figure 3.12**.)

11. Redraw the graph shown in Figure 3.12 on your paper, then use it to show the scaling of the distance between the cities mentioned in Item 7.

12. Repeat Item 11 for the situation described in Item 8.

CHANGING SCALES

In each of the preceding representations, it is the separation between locations that is important as distance. For that reason, it is customary to use a special symbol to emphasize that you are dealing with distances (differences) rather than just the ruler readings themselves. The symbol is the capital Greek letter **delta**, which looks like a small triangle: Δ. For example, if you use the letter M for a map ruler reading, then ΔM would be a map distance. Likewise, you could use ΔR for real distance.

Distances appear differently in the various representations. In a table, ΔM and ΔR are obtained by subtracting values. In the ruler representation and in the graph, ΔM and ΔR are both seen as line segments. Scaling is the process of moving from ΔM to ΔR or from ΔR to ΔM.

CONVERSION

Remember, "scaling" refers to the change from one measurement system to another when the two represented distances are not really equal, but are associated in a natural way. For example, distances on a map are not equal to real distances between places, but they represent real distances.

There is another familiar situation in which re-expressing measurements needs to be done. Instead of having two honestly different measurements that are matched up, you might have two different-looking measurements that are really identical. This process is called "unit conversion," or just **conversion** for short.

For example, suppose you and a friend each measure the width of a table. Since you are both perfect, both answers are exactly right. But you say the table is three feet wide, and your friend says it's 36 inches wide. A conversion is needed.

This time, you can use real rulers to make a physical model of the table of corresponding readings. Place a yardstick, marked in inches, on the floor or a large table top. Right up against it, place three one-foot rulers end to end, turned face down. Mark the ends of these rulers 0, 1, 2, and 3. (Be sure the zero end of the foot rulers is with the zero end of the yardstick.)

ACTIVITY

3

13. Explain how the physical model depicted in the illustration below is related to a table of readings.

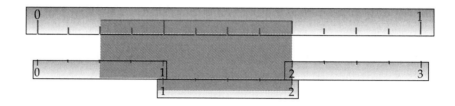

14. a) Devise a method for converting feet to inches. Show several representations, including a graph, of what's going on.

b) If you did not include it in part (a), write a rule for converting feet to inches. Use words, an arrow diagram, and an equation to express your rule.

15. Repeat Item 14 for converting from inches to feet. How do the arrow diagram and equation change?

16. What is the most important feature of your graphs in Items 14 and 15? Show on your graphs the conversion illustrated by the shading in the image shown in Item 13.

INDIVIDUAL WORK 3

Viewing Earth from Space

*I*n Activity 3, you developed several representations for visualizing the process of converting one unit of measure to another. All the representations were based on the idea of matching one set of readings to another. That's the fundamental principle behind all methods, so be sure you understand it well.

Each representation is useful in some instance. One algebraic representation of the conversion from one measurement system to another that you may have discovered in your work with arrow diagrams involves multiplication by something called a **scale factor** or a **conversion factor**. The choice of names depends on whether you have two really different distances that are matched up (scale factor) or two identical distances in different units of measure (conversion factor). But whatever you call it, the process is the same for both situations.

For example, the conversion factor for converting from inches to feet is $\frac{1 \text{ ft.}}{12 \text{ in.}}$. Note that this conversion factor has units. If you multiply an inch measurement (say, 36 in.) by this factor, you get a foot measurement (3 ft.). You might say that the inch units "cancel" in the process since one is in the numerator (36 in.) and the other is in the denominator (12 in.):

$$\frac{36 \text{ in.} \times 1 \text{ ft.}}{12 \text{ in.}} = 3 \text{ ft.}$$

Notice that the choices of what to put in the numerator and the denominator are important, too. The denominator is the "starting" measurement; the numerator is the "final" measurement. So, in the example, since inches are in the denominator, that conversion factor takes inches and turns them into feet.

Using this language, it is now possible to state what is meant by the "scale" of a map.

> The scale of a map is the scale factor for converting from real distances to distances on the map.

Since the scale of a map is a ratio, it may be written in any of several ways: using decimals, fractions, or with a colon. For example, the following are equivalent scales: 0.0125, 1/80, 1:80. Remember, too, that a scale always has units.

1. a) Explain how the conversion factor for converting inches to feet, described above, is related to your work in Items 7–10 of Activity 3.

   b) Find the conversion factor for converting from feet to inches, and explain how it is related to your work in Activity 3.

   c) Re-examine the "conversion rule" graph in Item 10 of Activity 3. If you are converting inches to feet, what are the units on the two axes? How are the axis units related to the units for the conversion factor?

2. a) For a map having a scale in which 2 cm represent 1 mile, what is the scale factor for converting map measurements to real distances?

   b) Use an arrow diagram to write the rule for converting map distances to real distances in this situation.

   c) Repeat parts (a) and (b) for converting real distances to map distances.

   d) Two towns appear to be 3.6 cm apart on the map. How far apart are they, in reality?

   e) How far apart should the map maker place the points representing two cities that are 15 miles apart?

   f) If you have the answer to part (e), explain two ways to decide how far apart to put towns that are 30 miles apart.

   g) Use the delta notation (Δ) to write the relationship between distances on the map (M) and distances in the real world (R) for this situation.

3. a) Suppose you are given an accurate graph relating readings on two scales, but you do not know the equation for the graph. Explain how you could still use the graph, geometrically, to approximate the scale factor for converting distances between the two scales.

   b) What do you infer if a scale factor has *no* reported units?

Figure 3.13.
Fingerprint with key features marked. Scale: 5 cm represent 1 cm.

4. A fingerprint is magnified to a scale in which 5 cm represent 1 cm (**Figure 3.13**). Two of the key features on the magnified print measure 5.4 cm apart. How far apart are they on the actual finger?

5. Pencils sell for 17 cents each. Imagine a barter system in which people are willing to accept pencils instead of cash. Explain how to convert prices from dollars to pencils. Use several representations, including a graph and parallel rulers to illustrate the process.

6. a) Suppose a teacher grades a test using a 0–60 scale. A student's grade using this ruler is called the "raw score" (R) for that student. Grades (G) will be returned to students on a 0–100 scale. Explain how to turn a raw score into a returned score. Use several representations, including a graph and parallel rulers to illustrate the process.

   b) If two students decided to compare scores, what would you know if ΔR is 6? What would you not know?

7. How fast is 50 miles per hour (50 mi./hr.) when expressed in feet per second (ft./sec.)? Hint: Think of two separate conversions, one for distances and one for times.

In Items 8–11, check your estimation skills by writing your estimate of the answer before computing.

8. Astronomers predict that within five billion years, the sun will become a planetary nebula with a radius of 1×10^{10} km. The radius of Earth's orbit is 1.5×10^{11} m. Will Earth's orbit fall inside the radius of the nebula? (Hint: What is the relation between meters and kilometers?)

As you have seen, questions in a number of different settings may be thought of as scaling questions. Distances may be related to each other; time intervals may also be related by scales. In fact, time intervals may be related to distances. For example, if you walk at a constant speed of 5 ft. per second, then a table could relate the distances 5 ft., 10 ft., and 15 ft. to the times 1 sec., 2 sec., and 3 sec. Thus, in the spirit of the work in previous activities, this table might be thought of as a way to convert distances to times (or vice versa). For example, a distance of 97 ft. would correspond to (97 ft.) ÷ (5 ft./sec.) = 19.4 sec.

9. In 1994, fragments of the Shoemaker-Levy comet plunged into Jupiter's atmosphere. The impact of these fragments generated huge flashes of light. If both Earth and Jupiter were aligned on the same side of the sun, and at their mean distance from it, how long would light traveling from Jupiter at 3.00×10^8 meters per second take to reach Earth? (Jupiter's mean distance to the sun is 7.78×10^{11} m.)

10. Space is so vast that distances are frequently measured in light-years—the distance light travels in one year. Light travels at a speed of about 3.00×10^8 m/s. Calculate the number of seconds in a year, and determine the distance of a light-year in meters.

Parsec

The word "parsec" is a contraction of the words "parallax" and "second." It is the distance at which a star would need to be in order for it to have exactly 1 second of parallax as the earth moves from "between" the sun and that star to a point "beside" the sun. See **Figure 3.14**. In this case, a second is not a measure of time. Instead it is a measure of angle; one second is 1/3600 of one degree.

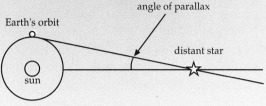

Figure 3.14. Measuring parallax.

11. A parsec (pc) is an astronomical unit used to measure large distances. It is equivalent to 3.26 light-years (lt-yr). Approximately how many km are there in a parsec?

12. Our galaxy, the Milky Way, is about 15,000 parsecs in diameter. Our solar system is located about one-third of the way in from the outer edge of the galaxy. How far is it in light-years from Earth to the center of the galaxy? How far is it from Earth to the edge of the galaxy? How long does light take to cross the galaxy?

Most of the numbers you have used thus far in this unit have been measurements of one kind or another. In working with the Landsat image of the Czech Republic, you have already begun making preliminary measurements for yourself.

No matter what kind of measuring device you use, its accuracy is limited. A typical meter stick has smallest subdivisions in millimeters. Yardsticks have smallest divisions as sixteenths of an inch. Many people are able to see between such marks, and may be able to read such a scale to the nearest half, or even tenth, of the smallest marking. But even then, there is a point beyond which no more precision is possible with that device.

So, every measurement has some built-in level of precision. In general, the number of digits included in a reported measurement will be called the number of **significant figures** for that measurement. Thus, 1.57 centimeters carries three significant figures.

The exceptions to this rule involve zeros and decimals. For numbers with no decimal, all digits are significant except for ending zeros.

For example, 100 has only one significant figure, 120 has two, and 125 has three. If a decimal is displayed and there are non-zero digits to the left of the decimal, then all digits are deemed significant. Thus, 100.0 has four significant figures, and 30. has two, and 1.50 has three. If a decimal is displayed and there are no digits to the left of the decimal, then all digits are significant except for the leading zeros. So, 0.0035 has only two significant figures.

When you get results using a calculator, it is important that you recognize that not all of the decimal places you see on the screen are necessarily meaningful. As a general rule, do not round intermediate calculations. Remember, your final answer cannot become more precise than the least precise measurement at the beginning of the calculation.

EXAMPLE:

> Find the area of a rectangle with a base of 3.6 m and a height of 2.34 m. Remember, these measurements indicate that the base was measured to the nearest tenth of a meter (decimeter), while the height was measured to the nearest hundredth of a meter (centimeter).
>
> $A = b \times h = 3.6$ m $\times 2.34$ m.
>
> $A = 8.424$ m^2 on your calculator display.

Your least precise original measurement (3.6 m) had only two-digit precision. Therefore, your final result should be written as 8.4 m^2.

However, if you were to find the area of the rectangle described above by rounding each dimension to tenths before multiplying (3.6 x 2.3), the answer (8.28 ≈ 8.3 square meters) would not be correct because the rounding was done too soon.

As a general rule, try not to round too soon. Each rounding can decrease the accuracy of your answers. Round your final answer appropriately to indicate the precision of the numbers you were given in the problem.

EXACT NUMBERS AND π

When you do computations, remember that counting numbers are exact. Another example of an exact number is 12 inches in a foot. Do not use the number of digits in exact numbers when deciding how you should round your answer. For example, five boards that each have a length of 3.75 m have a total length of 18.75 m ≈ 18.8 m. Do not round to 20 m just because the 5 has one digit. The 5 is a counting number, so it is exact.

If you have a π key in your calculator, use it instead of a decimal approximation (such as 3.14) when you are calculating.

It should be clear that the level of precision can't get any better as computations are done; the measurements have already been made. Therefore, throughout the remainder of this unit, and in all subsequent work, the following convention will be followed.

No rounding will be done on any intermediate steps of a calculation.

All final answers will be rounded to display the same number of significant figures as the initial measurement having the least number of significant figures.

13. Look back at your work in Item 6 of Activity 2. How precisely were you able to read your measurements? How should those measurements be reported?

14. Apply the rules for significant figures to your answers in Items 8–12.

15. a) Return to the measurements you made in Activity 2. For each set of observations you made (from each viewing distance), determine the scale factor for scaling from ruler-at-the-tube (*T*) distances to map (*M*) distances.

   b) Use the scale factor to write a rule for the scaling, then represent your rule using several methods. Use Δ*T* and Δ*M* in your answers.

16. Repeat Item 15 for scaling from ruler-at-the-tube distances to real-world distances.

17. **Figure 3.15** represents the tube apparatus that you used in Activity 2. Two particular viewing distances are included in the same diagram. Explain the connection between the diagram and your answers to Item 15(a).

Figure 3.15.
Diagram for views through the tube apparatus.

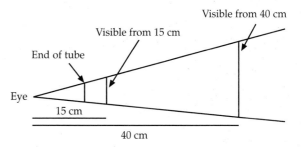

Figure 3.16.
Three students and their shadows.

18. a) Draw a diagram similar to the one in Figure 3.15 that represents the key information in the illustration shown in **Figure 3.16**, which shows three students of different heights casting shadows onto the ground. Label your sketch carefully. Think of "real height" and "shadow length" as two measuring scales. Explain how to use your diagram to convert from real height to shadow length for anyone else in this same situation. (What additional information do you need in order to be able to complete the conversion?)

    b) Interpret the reverse scaling process. Explain how it might be used to determine the height of very tall objects that are not directly measurable.

19. a) Find a map that shows the entire state of Colorado. Measure its length and width on the map, and compute its area on the map. Then use the scale of the map to convert the length and width to real distances, and compute the real area of the state. How are the two areas related to the scale factor for the map?

    b) Check a reference book to find a more precise area of Colorado. How close to this number did you get? Relate your response to what you know about significant figures.

20. If you have access to a computer word-processing program, open any document and check its ruler. Is one inch really one inch? What is the scale of the document window?

LESSON TWO

Satellite Vision

KEY CONCEPTS

Scale

Relative size

raphical interpretation

Pixel

Digitization

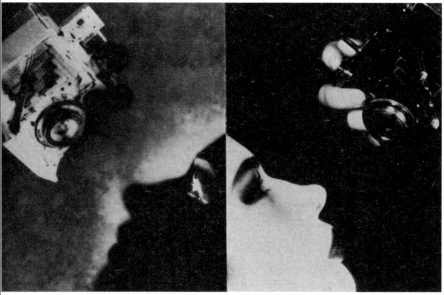

The Image Bank

PREPARATION READING

What Does Landsat 5 See?

You may have learned in Lesson 1 that Landsat satellites are owned and operated by the United States government. The first Landsat satellite was launched in 1972. At that time, it was called ERTS-1 (Earth Resources Technology Satellite-1). In 1975, NASA renamed this the Landsat program. (This is important information if you are trying to research the early Landsat satellites: you have to look under the ERTS project in periodicals dated before 1975.) Scientists are now getting data from the fifth Landsat satellite.

Landsat 5 orbits the earth, passing over each point on the earth (with the exception of the polar caps) every 16 days, taking a complex "picture." This picture captures both what humans can see and certain invisible bands of the infrared spectrum.

We humans have several sensors—eyes, ears, nose, fingers, and tongue—with which we collect data or information about our world. Landsat 5 also has sensors, called **remote sensors** because they do not touch the objects they are sensing. Instead, from about 700 kilometers above Earth, the satellite's sensors collect information transmitted by energy reflected from such things as trees, oceans, and buildings. Landsat 5 then transmits this information back to Earth in the form of complex images, giving us a unique perspective concerning Earth and its resources. Can you name other types of remote sensors?

Users of satellite images may travel to the actual site that appears in a satellite image to compare what they see on the ground with what they see in the image. The information they gather may then be used to interpret other images. In many cases, this process (called **ground truthing**) is the only way to interpret the data transmitted by satellites. For example, an image transmitted by Landsat 5 might show what appears to be a large area of stressed vegetation. Scientists can travel to that area to view the vegetation more closely. The information they gather can then be used to analyze the situation, determine whether there is a problem, and, if so, search for a solution.

Unfortunately, your class probably will not be able to travel to the Czech Republic for ground truthing. However, images typically contain a number of familiar objects, and examining those can tell you a lot.

Recall from Activity 1 in Lesson 1 that for the final project in this unit you will investigate the loss of conifer forest over a five-year period in the Czech Republic. The conifer loss is due to many causes, including acid precipitation and logging. Scientists use satellite images to help them monitor the change in area of forested regions to gain insight into the causes of this situation. To help you understand the image from the Czech Republic that you encountered in Lesson 1, you will also learn about satellite images by studying a Landsat 5 image from the seacoast region north of Boston, Massachusetts. In later lessons, you will compare a Landsat 5 image of the same region of the Czech Republic from 1990 to the one from 1985 that you have already studied. For the final project, you need to determine the change in area of the conifer forests. First, however, you need to learn more about interpreting Landsat 5 images.

On the following pages you see six Landsat images that you will be referring to at different times throughout this unit. The images are in two groups: three Czech images and three Beverly, MA images. **Figures 3.17** and **3.18** show the originals of the Czech and Beverly images. Figures **3.19** and **3.20** are enlargements of the Czech image; **Figures 3.21** and **3.22** are enlargements of the Beverly image.

Figure 3.17.
A region in the Czech Republic
in 1985. Zoom level 0.5.

Figure 3.18.
Beverly, MA. Zoom level 0.5.

Figure 3.19.
A region in the Czech Republic in 1985. Zoom level 1.0.

Figure 3.20.
A region in the Czech Republic in 1985. Zoom level 2.0.

Figure 3.21.
Beverly, MA. Zoom level 1.0.

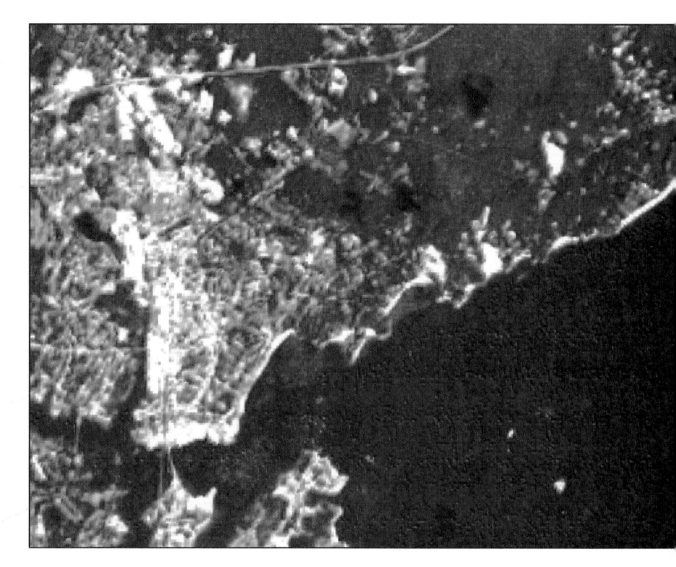

Figure 3.22.
Beverly, MA. Zoom level 2.0.

ACTIVITY

THE ROAD TO PROGRESS

4

1. Examine Figure 3.19, the image of the Czech Republic that you studied in Lesson 1 (Figure 3.1). Identify as many physical features as you can. Look for lakes, streams, buildings, and such.

2. Repeat the identification search using a computer image of Beverly, Massachusetts. You should be able to find roads, bridges, towns, beaches, and more. Be as thorough as you can, and indicate what clues you use to make your decisions. (If you do not have access to a computer, use Figure 3.21.)

3. a) For each image in which you identified objects, determine the relative sizes of those objects or the relative distances between pairs of those objects. That is, select one object or distance to serve as your "ruler" and use it to measure the other items you identified. Record your information in such a way that your classmates can easily interpret your work.

    b) Explain why having good measures of relative sizes might be helpful.

    c) How precise are the measurements in part (a)? For example, what was your unit of measure? Did you measure to the nearest whole unit? Did you estimate fractional units? If so, how closely?

    d) Explain any difficulties you encountered in part (a).

4. Think of ways to make your measurements better, i.e., more precise. When you agree on a method, try it out, and repeat Item 3.

5. Explain any difficulties you encountered in Item 4.

6. Try to find a map of Beverly, Massachusetts that shows some of the same features you were able to find in Item 2. Use its scale to approximate the scale of a particular satellite image on your computer display or in Figures 3.18, 3.21, and 3.22. (Such maps are available on the Internet, for example.)

INDIVIDUAL WORK 4

Information Highway

Your work in Activity 4 focused on measurements within a single image. As you know, working with a single image, especially when it is relatively small, leads to quite a bit of uncertainty in measurements of distance. You probably suggested that one solution to that problem might be to use a larger version of the same image. That might provide greater detail and less uncertainty in individual measurements. That's a great idea, and your knowledge of scaling can help you relate distances in the two images.

Notice that there are two measurement ideas involved here. First, the basis of your work in Activity 4, is the notion of relative size: "How does this distance compare to a *different distance* in the *same image*?" The other idea was examined closely in Lesson 1, the concept of scaling: "How does this distance compare to the *same distance* in a *different image*?"

Items 1–4 refer to **Figures 3.23** and **3.24**.

1. a) Near the bottom of each image is a pair of bridges, running almost vertically. Measure the length of the left bridge in each image. Report the precision of your measurement (for example, "to the nearest half-millimeter"). Then compute the scaling that takes the smaller image to the larger one.

   b) Locate two objects that appear in both images and are fairly far apart. Measure the distance separating them in each image, report your precision, and compute the scaling from the smaller image to the larger one.

   c) Compare your answers to 1(a) and 1(b). Explain any differences in the computed scales.

2. a) Place a small metric ruler on your page so that it lies across both images. If possible, lightly tape the ruler in place so that you will not accidentally move it. Without moving the ruler, find some feature that is visible in both images and record its locations as determined by the ruler readings "in line" with them. For example, the left bridge might appear at the 15-cm mark in the left image and at the 40-cm mark in the right image.

Figure 3.23.
The Beverly-Salem bridges at
zoom level 1.5.

b) Without moving the ruler, repeat part (a) for at least four other features that appear in both images. Record your results in a table such as the one begun in **Figure 3.25**.

c) Sketch a pair of "matched rulers," similar to those used in Lesson 1, that illustrate the matching recorded in your table.

d) Based on your table, write a rule for a scaling that takes distances ΔL in the left image to distances ΔR in the right image.

e) Make a graph from your table. Be sure to label the axes clearly. Show explicitly how it can display the scaling of a distance from one image to the other by marking an appropriate ΔL and ΔR on the axes of your graph.

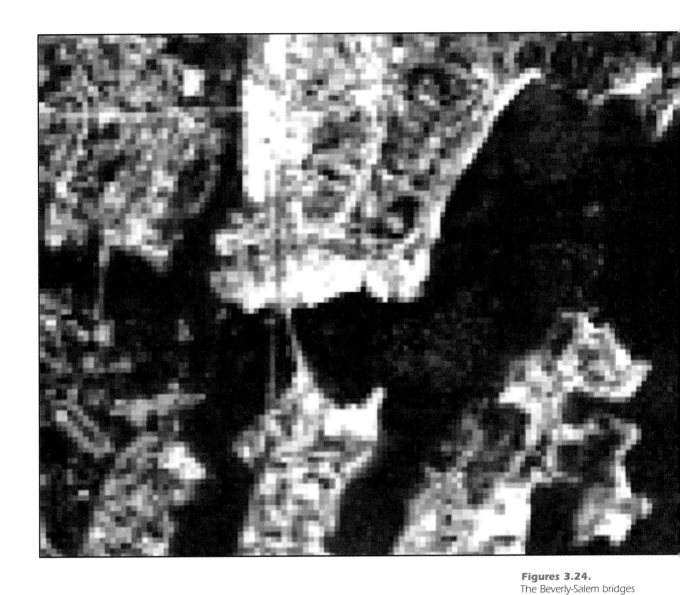

Figures 3.24.
The Beverly-Salem bridges at zoom level 4.0.

3. a) For each image, estimate (separately) the relative widths of the two bridges. Write your answer as a sentence in the form, "In the left image, the right bridge appears to be about ___ times as wide as the left bridge." (You need two such sentences.)

Feature	L, left location (cm)	R, right location (cm)
Left bridge	15	40

Figure 3.25.
Location table for features in Beverly, MA.

b) Repeat part (a) for the lengths of the two bridges.

c) Select another pair of objects that appear in both the images. Compare their relative sizes as you did in parts (a) and (b).

d) Look back at your table of locations from Item 2(b). Use it to determine the relative distances between two pairs of locations in the left image. Compare the relative distances for the same two pairs of locations in the right image.

e) Generalize the results of your comparisons in parts (a–d). (If you don't feel comfortable making a generalization, check another few pairs of objects.) Comment on how the precision of your measurements affects the conclusions you make.

4. a) Find an object or region that appears in both images. For each image, determine (separately) at least three relative measurements within this object or region. ("The left side of the island is twice as long as its bottom; the diagonal is . . . .") Write your answers in a form that your classmates can easily understand and verify for themselves.

b) Discuss the results of part (a) in terms of your generalization in Item 3(e).

5. **Figure 3.26** shows a much-magnified view of the islands near the bottom center of the Beverly image from Figure 3.21. Compare this view to the original appearance of the islands.

6. When you were working on Item 4 of Activity 4, someone in your group may have suggested that you "just keep on zooming" in order to get as much precision in your measurements as you wanted. Based on your work in Item 5 here and on any experimenting you did in Activity 4, discuss the likely effectiveness of this strategy. Be sure to give clear reasons for your assertions.

7. a) Use a magnifying lens to examine a photograph in a newspaper closely. Compare that image to the ones with which you have been working.

b) If you have a home video player that has a "freeze frame" feature, display a single image with it, and use a magnifying lens to examine it. Compare that image to those used in Items 1–7(a).

c) Use a magnifying lens to examine a computer screen and a graphing calculator screen. Add your observations to your comparisons from parts (a) and (b).

In each of the images you have considered up to this point, you have seen that the image is not smooth. Rather, it is composed of lots of tiny dots. These dots are called **pixels**, short for "picture elements." Thus, these images differ from images created by painters, for example. Most devices that display pixels can vary the brightness, and many times even the color, of the pixels. Many calculators, however, simply have pixels either "on" or "off." That makes calculator screens much simpler than other devices. In any case, as you have seen, the existence of pixels in images has some interesting implications for the information you can get from those images.

8. Some calculators use a block of pixels measuring 7 pixels tall and 5 pixels wide as their basic character display when not doing graphs. Without looking at your calculator to see what it does, invent your own alphabetic display system using 7 x 5 pixel blocks to write your own name. That is, decide which pixels within the 7 x 5 block need to be "on" to make each letter in your name.

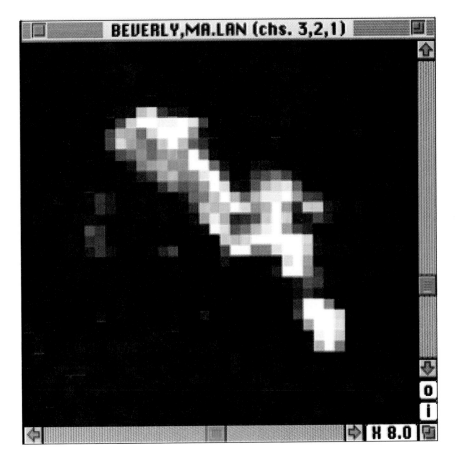

Figure 3.26.
Two islands near Beverly.

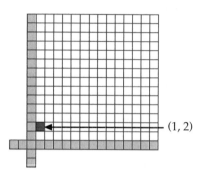

Figure 3.27.
Sample "calculator screen" grid.

(1, 2)

9. Do *not* use a graphing calculator for this item. Instead, use a piece of graph paper. If you have no graph paper, make your own: lay out a large grid of identical squares on a sheet of paper. Mark off three separate regions on your graph paper, each at least 15 squares by 15 squares in size. Now pretend that you are a calculator and the graph paper regions are your screens. Each square of your grid represents one pixel of the screen. If you color it, it represents "on;" otherwise, it is "off." Use a separate screen for each item below. **Figure 3.27** shows an example in which the point (1, 2) is "on."

a) Make a table of values for the equation $Y = X$ with at least ten values. Using equal scales for both axes, record on your graph paper how you would display its graph if you were a calculator. Explain any decisions you make. Be sure that squares represent equal distances in both directions, but they don't have to be one unit per square.

b) Repeat part (a) for the equation $Y = 3X$.

c) Repeat part (a) for the equation $Y = 1.5X$.

d) How does the geometry of the graphs' pixels reflect the numbers in the equation?

WHAT DOES LANDSAT 5 SEE?

This activity is designed to help you understand the imaging process of Landsat 5 and other digital image devices. When you observed the globe or map from different distances in Lesson 1, you noticed a change in perspective. As you moved away from the globe, your ability to identify individual locations on it decreased. Landsat is also limited in this way; the smallest region that Landsat can distinguish is a 30-meter square. This becomes one pixel. This is usually reported by saying that Landsat 5 has "30-meter resolution." In this activity, you will work with pixels that represent much smaller regions.

Your teacher will designate a particular scene—a portion of your school property, a photograph, etc.—for your group and give you blank grids and some gray-scale squares. You will create a gray-scale version of the given scene on your grids.

PROCEDURE:

With the other members of your group, divide your given image into 36 smaller, equal squares. These will be your pixels. Assign these pixels to members of your group; each person should have about the same number of pixels with which to work.

Now examine your pixels. Determine how much light is reflected from each one by matching it as closely as possible to one of the gray levels on your gray-scale squares. Cut a matching square from the squares sheet, and attach it to one of your blank grids in the appropriate place. On the other blank grids, write the number of the gray level in the corresponding place.

When you have finished, each of your 36 pixels should be represented by a digit between 0 and 5, inclusive, representing a single gray-scale level. Give your teacher the grids on which the digits are recorded; keep the grid on which you attached the squares.

ACTIVITY

5

WHAT DOES LANDSAT 5 SEE?

Your teacher will give you another group's grid. Make a rough copy of it for each member of your group to use in Item 2 of Individual Work 5.

To wrap up this exploration, write about the criteria you used to decide upon representative levels of gray at the various stages of this activity. Mention specifically any difficulties you encountered and what your group did to overcome them. What decisions did you have to make? Why? How did you make them?

CONSIDER:

1. How did lighting affect your pixel assignments?

2. Did unusually bright or dark objects contribute heavily to any pixel?

3. Did you and your partners always agree on the level for a given region?

4. How did you deal with situations in which a variety of levels occurred in the same pixel?

5. How did you deal with scenes that were between gray-scale levels?

6. What happened when an object was visible within parts of more than one pixel?

INDIVIDUAL WORK 5

Coding and Decoding Satellite Data

The number assigned to the intensity of reflected light or energy detected for each pixel by each sensor is called the "brightness" or "reflectance" value. Landsat 5 records brightness using a scale of 256 shades of gray. In this activity, you will use only six shades of gray to digitize images, since most people can easily distinguish that many visually but have difficulty with many more levels than that.

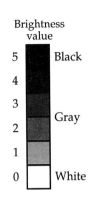

Figure 3.28.
Gray-scale chart.

1. a) A chart of the gray-scale levels you used in Activity 5 has been reprinted as **Figure 3.28**. In **Figure 3.29**, you are given a scene of a house. Using graph paper or a blank grid, assign each square a reflectance value (digit) that corresponds to a gray-scale level on the chart.

   b) How did you decide which reflectance value to assign to pixels that contain more than one shade of gray?

   c) In part (a) above, you coded an image with gray-scale values. This process is known as **digitization**, where numbers are assigned to something that is not inherently numerical or discrete. Now you will decode the image from part (b). Use a pencil to shade each pixel on blank graph paper or on a carefully-drawn grid with the intensity of gray associated with the number in that pixel (its reflectance value).

Figure 3.29.
A simple house.

   d) Your new picture will probably look different from the original. Why? What did you notice about the front door and slanted roof line on the decoded image?

2. Use the grid that you got from the other group at the end of Activity 5 and the gray-scale chart in Figure 3.28 to "color in" a blank grid representing the image that the other group coded. Describe the image you get, and try to determine what the original scene they viewed really was.

3. The sensors of the French SPOT satellite are capable of 10-meter-square resolution, and the resolution offered by some military satellites is even higher. (Higher resolution uses pixels representing smaller regions.) How would higher resolution affect your satellite images of Beverly and the Czech Republic?

4. Each digital satellite image is stored and transmitted by computers, one pixel at a time. For the sake of this question, think of the computer memory as a giant container with a certain number of places to put information. Discuss the relation between storage space, the number of pixels in an image of a region, the size of the full region, and the resolution of the image. For example, what happens if you want very good resolution? What users might be interested in the best resolution, and who might be more interested in other characteristics of the image?

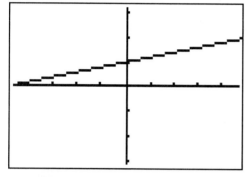

Figure 3.30.
Graph on a
calculator screen.

5.a) When you plot a line on a graphing calculator, the line often looks as it does in **Figure 3.30**. A graphing calculator screen is made up of pixels, and the pixels are only visible when they are lit ("on"). Explain why the line looks as it does.

b) If a graphing calculator had a higher resolution (i.e., the screen's pixels were smaller), how would that affect the line's appearance?

6. You may have seen TV programs or news stories in which the identity of an individual was protected by using "over-sized" pixels over the face. Explain how this method does its job.

7. Recall from the reading that each pixel in a Landsat image represents a 30-meter by 30-meter square. Suppose a bridge measures 40 pixel-widths long in a Landsat image. How long is the real bridge?

8. a) The full-size Landsat image display on your computer measures 512 pixels by 512 pixels. Use this fact, together with the 30-meter resolution of Landsat images, to determine the actual (real-world) width of the region displayed in a full Landsat image.

b) Zoom in or out on your *MultiSpec* image until the "zoom level" box reads x1.0. If you are using the PIC file, it is already at the x1.0 zoom level. If computers are not available, use Figure 3.21.

Use a tape measure to determine (to the nearest tenth of a centimeter), the horizontal distance from one side of the entire image to the other. Be sure to start with the lower scroll bar at the far left. (If the image extends beyond the right side of the screen, you will have to take extra precautions. Measure the distance across the screen. Carefully note an easily identifiable object on the image that rests on the right edge of the screen. Scroll to the right and add in the distance from the object to the right edge of the image. You may also wish to activate the coordinate display and note the coordinates before scrolling, so you can easily find your landmark object.)

Use your direct measurements to determine the actual scale of the screen image. Look back at your answer to Item 6 of Activity 4. How close was your first approximation?

9. a) How long is the left bridge of the pair of bridges at the lower left of the image? How precise is your answer?

   b) How long is the long beach opposite the island near the center of the image? How precise is your answer?

10. Landsat images use pixels measuring 30 meters across. Pixels in SPOT images measure only 10 meters across. Thus, you might say that pixels in Landsat images are three times as large as those in SPOT images. How do the *areas* represented by one pixel in the two kinds of image compare?

11. Cross-stitch is an artistic craft in which small "x"-shaped stitches are used to create a pattern or scene. Patterns are readily available at many crafts stores and consist of what amounts to a piece of graph paper with coded "pixels" indicating the color of the stitch for each pixel. Select a simple image and create a cross-stitch pattern for it.

12. There are numerous computer programs that work with images. You have seen images produced with a program called *MultiSpec*. It was designed specifically to process images using Landsat data and display them on personal computers. If you have access to images from the Internet, you probably have some kind of viewer software. "Paint" and "draw" programs have been used for years, and geometric utility programs are proving very useful. However, among all these programs there may be no common method of displaying an image. Based on your work in Activity 5, discuss the geometric and mathematical decisions that would need to be made, and the possible problems, in converting an image from one kind of software to another.

LESSON THREE

Interpreting Size and Shape

KEY CONCEPTS

Relative size

Corresponding parts

Shape

Similarity

Ratio

Proportionality

Solving proportions

Coordinates

Pythagorean theorem

Dilation

Translation

The Image Bank

PREPARATION READING

Size and Shape

T hink back to the tube-and-map work you did in
Activity 2 (Lesson 1) and to your image measurements
in Lesson 2. By getting closer—moving in—to the map,
you were able to see more. Moving out (farther away) resulted
in a larger region's being visible, but with less detail. When
such movements are done not by actually moving toward or
away from an image, but by changing the size of the image
itself, the process is usually called "zooming."

Zooming in on an image enlarges it. If the image is digitized,
you may able to see its individual building blocks (the pixels).
Knowing the size of a pixel then allows you to measure the
size of an object in the image by counting pixels, but that is an
inefficient method for measuring fairly large objects. It would
be easy to lose count, especially if the magnified object did not
fit on the screen all at once.

Your work in Lesson 2 also led you to determine a scale for the image. With the scale you can measure objects with a ruler of some sort, then convert the measurement into "real-world" distance. How did you determine the scale? Is the scale the same for horizontal and vertical distances? How can you tell?

Scales decrease the need for zooming, but they do not eliminate it. The fact that many computer programs dealing with images have the ability to zoom suggests that many people find that ability useful. (Many calculators have a special zoom key.) With Landsat images, for example, in order to measure a small object, it may be necessary to zoom in to make it large enough for its boundaries to be visible. In order to determine the extent of deforestation in the Landsat image of the Czech Republic, you will need to be able to make the image large enough that you can get accurate measurements.

Before you can analyze the Czech deforestation problem, you need to do the following:

1. Determine exactly what you mean by the size of a forest (or any other object).

2. Find at least one method for measuring that size.

3. Determine how zooming affects the sizes of objects.

4. Determine exactly what you mean when you say that two objects have the same shape.

5. Determine how zooming affects the shape of an object (since that might also affect how you measure its size).

Figures 3.31–3.34 are images of the same region near Beverly, Massachusetts that you first saw in Lesson 2. The first image is the original. Each of the other three is an enlargement of the original.

CONSIDER:

1. Have shapes been preserved in the enlargements in Figures 3.32 –3.34? Explain how you determined your answer.

2. If the shapes of regions change as you zoom in on the image, how might that affect your work on the Czech deforestation problem?

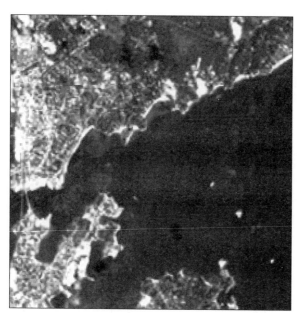

Figure 3.31.
A region near Beverly, MA.

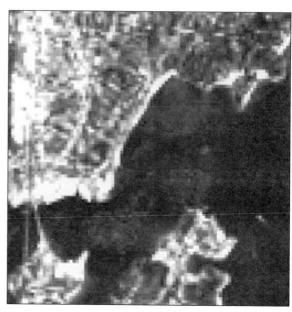

Figure 3.32.
An enlargement of the region near Beverly, MA.

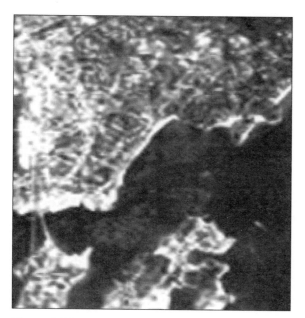

Figure 3.33.
An enlargement of the region near Beverly, MA.

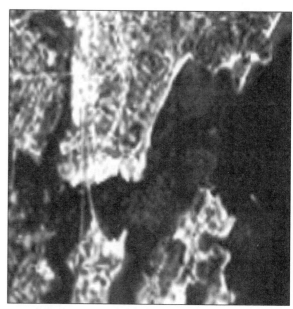

Figure 3.34.
An enlargement of the region near Beverly, MA.

STAYING IN SHAPE

ACTIVITY

6

1. Revisit your work on Items 3 and 4 of Individual Work 4 in Lesson 2. Discuss your results in light of the ideas about shape raised in the Preparation Reading for this lesson.

2. For each pair of figures in **Figures 3.35–3.41**, decide whether the two figures have the same shape.

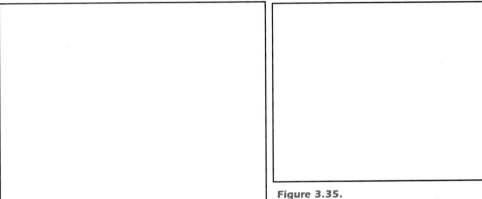

Figure 3.35.
Two rectangles.

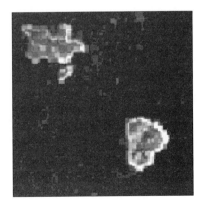

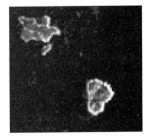

Figure 3.36.
Two islands.

ACTIVITY

6

STAYING IN SHAPE

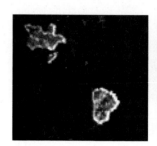
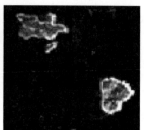

Figure 3.37.
Two islands.

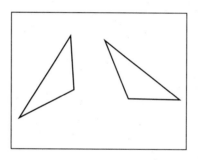

Figure 3.38.
Two triangles.

Figure 3.39.
Two quadrilaterals.

Figure 3.40.
Two faces.

Figure 3.41.
Two stick people.

STAYING IN SHAPE

ACTIVITY

6

3. Use your answers to help explain what you mean by "same shape." Formulate your explanation in a way that someone else could use it to decide whether two objects of their own are the same shape or not.

4. Show how to use your method from Item 3 by justifying your original answers to Item 2.

This activity has focused on determining whether two images have the same shape, even when they are of different sizes. This kind of relationship is needed so often, it has a name. When two objects have exactly the same shape, they are called **similar** objects.

Note that the use of the word "similar" in a mathematical context implies a more strict relationship than when you use it in everyday language. When you say that two shirts are similar, you mean only that they have some features in common. You probably do not mean that their patterns are exactly alike except for size. In most cases, context will let you know which meaning of "similar" is appropriate.

INDIVIDUAL WORK 6

Staying in Shape

1. **Figure 3.42** shows two rectangles.

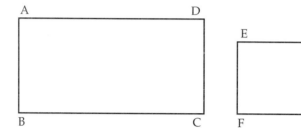

Figure 3.42.
Two rectangles.

a) Collect four corresponding measurements from each rectangle.

b) For rectangle *ABCD*, find the ratio of each measurement to the smallest measurement you collected from that rectangle.

c) Repeat part (b) for rectangle *EFGH*.

d) Do the rectangles seem to have the same shape? Explain.

2. **Figure 3.43** shows two parallelograms.

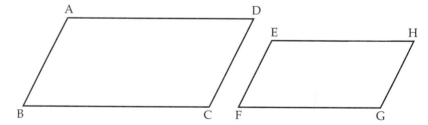

Figure 3.43.
Two parallelograms.

a) Collect four corresponding measurements from each parallelogram.

b) For parallelogram *ABCD*, find the ratio of each measurement to the smallest measurement you collected from that parallelogram.

c) Repeat part (b) for parallelogram *EFGH*.

d) Do the parallelograms seem to have the same shape? Explain.

3. Use your experiences from Activity 6 and from Items 1 and 2 above to state a clear procedure for checking whether two objects have the same shape. How does your method compare to the one you developed in Item 3 of Activity 6?

4. a) **Figure 3.44** is a rectangle. Write a description of the rectangle so that a friend could reconstruct it without seeing the original.

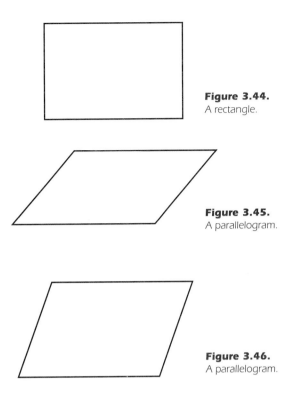

Figure 3.44.
A rectangle.

b) **Figure 3.45** is a parallelogram. Write a description of the parallelogram so that a friend could reconstruct it without seeing the original.

Figure 3.45.
A parallelogram.

c) Repeat part (b) with the parallelogram in **Figure 3.46**.

d) The two figures in parts (b) and (c) have sides of the same relative lengths. Do they have the same shape? Write a convincing mathematical explanation for your answer.

Figure 3.46.
A parallelogram.

5. Suppose a friend has drawn two triangles, *ABC* and *DEF*. Triangle *ABC* is shown in **Figure 3.47**. You are not shown triangle *DEF*.

a) Determine the shape of triangle *ABC* by finding the relative sizes of its sides.

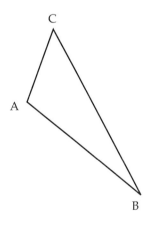

b) You are told that $DE = 2 \times DF$. However, that's all you are told, and you are not permitted to look at triangle *DEF*. From the given information, can you be certain that triangles *ABC* and *DEF* are similar? If yes, explain why. If no, tell what other information you need in order to be sure, then draw two different possible triangles for *DEF* that fit the given conditions.

Figure 3.47.
One of your friend's triangles.

6. On a certain map a measured distance of 2 cm represents a real distance of 10 km.

   a) What distance is represented by a measured distance of 7.3 cm?

   b) What measured distance is needed to represent two towns that are 13 km apart?

   c) Write a rule for converting from measured distance to real distance. Express your rule as an arrow diagram, an equation, and a graph.

   d) What is the scale of this map? (Recall that the scale of a map is the scale factor used to convert actual distance to distance as represented on the map.)

7. Many photocopy machines have a zoom feature that allows you to enlarge or reduce copies of images.

   a) Obtain a map and make a photocopy using a zoom factor of 125%.

   b) Determine whether shape is preserved in this process. Explain your decision.

   c) What does the 125% tell you about the final image?

   d) Look at the printed scale on the original map. Since the copy was made by zooming, that scale is no longer correct. Compute the correct scale for the new map (the copy). Explain your process carefully.

   e) If you did not do so in part (d), make an arrow diagram illustrating the scaling for the copy.

8. **Figures 3.48–3.51** show pairs of images. Are the figures in each pair similar? Justify your answer mathematically.

9. Many communities have underground utility lines (water, phone, cable TV, electricity). Explain why it is extremely important to have accurately scaled maps of utility locations.

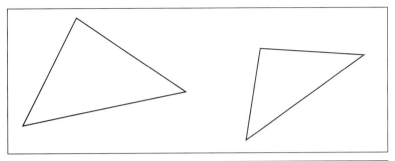

Figure 3.48.
Two triangles.

Figure 3.49.
Two fingerprints.

Figure 3.50.
Two fingerprints.

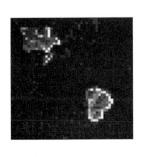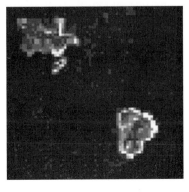

Figure 3.51.
Two islands.

ACTIVITY

7

ZOOM WITH A VIEW

You have spent a fair amount of time thinking about what it means for two objects to have the same shape, to be mathematically similar. One reasonable way to define what you mean by "shape" is to record a number of distance measurements within the object and note the relative sizes of those measurements. If you then make the same measurements in a second object being tested for similarity, the relative sizes should be exactly the same. Then, if every set of relative sizes turns out the same, the objects are similar. This provides an operational definition of similar.

> If the ratio of each pair of distances within one figure is the same as the ratio of the corresponding pair of distances in a second figure, then the two figures are similar.

1. Use the definition of similar to draw two figures that are similar. Carefully label all distances so that a classmate can use the definition to verify that your figures really are similar.

2. Draw two figures having only straight-line sides and meeting the following conditions. First, the ratio of each pair of sides within one figure is the same as the ratio of the "same" pair of sides in the second figure. Second, the figures are *not* similar.

Another word frequently used in connection with the mathematics of similarity is **proportional**. Two variable quantities are said to be proportional if their ratio never changes. For example, whenever the size of a figure changes and the shape stays the same, measurements within the figure are proportional since their ratios don't change. In Figure 3.42 of Individual Work 6, the left side is proportional to the bottom because

$$\frac{AB}{BC} = \frac{EF}{FG} \ .$$

This constant ratio is called the **constant of proportionality**.

ZOOM WITH A VIEW

3. List some proportional pairs in the figures you drew for Item 1, and determine their constant(s) of proportionality.

One drawback in the definition of similarity is that it requires you to check each pair of distances in an object. That can be a lot of checking! However, for now it is a useful method, especially to show that two figures are *not* similar. But, as you may have discovered, there are other methods, too.

Revisit your work on Item 16 of Individual Work 3 in Lesson 1. The viewing-tube diagram is repeated in **Figure 3.52**.

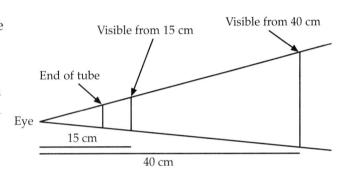

Figure 3.52.
Two views with the tube apparatus.

In Figure 3.52, the regions that are visible are restricted by the tube; only objects lying within a narrow angle may be seen. Since measurements are somewhat difficult to take using this apparatus, the following activity is designed to retain the main geometric restrictions while allowing easier measurement, though only one dimension of the "image" will be represented.

APPARATUS SET-UP:

Have one student in your group hold a meter stick vertically against the floor, with the zero end up. While one person (may be the same person holding the meter stick) holds a string in place at the top of the vertical stick, another should pull the string taut to a point on the floor so that the top angle is between 0° and 45°. This creates angle *A* in **Figure 3.53**.

Have another student hold a second meter stick at a right angle to the vertical stick as shown in Figure 3.53, making sure that one stick is vertical and the other is horizontal. The zero end of the

horizontal stick should be in contact with the vertical stick, and the zero end of the vertical stick should be up. The exact height of the horizontal stick does not matter; it will change during this activity.

Note: Save all your measurements and diagrams for use in Individual Work 7.

Refer to Figure 3.53 to answer the following questions.

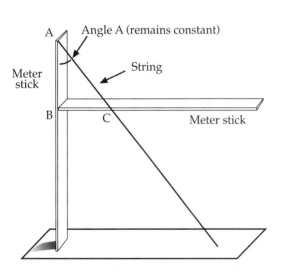

Figure 3.53.

String-and-stick model of viewing from a distance.

4. As stated above, you will be changing the location of the horizontal stick in this set-up. Everything else will remain fixed. Compare this situation to that sketched in Figure 3.52 (the viewing tube). How can the two be thought of as representing the same thing?

5. Assume that the viewing angle never changes, as was the case when using a viewing tube. Two quantities that are of interest in this situation are the distance from which you view a scene and the amount of the scene that you can see. How are these two quantities related? Quantify your results so that you will know how to make numerical predictions. (Hint: Experiment with different locations of the horizontal stick in the apparatus, being sure that the top angle never changes.) Make and label sketches of the different set-ups you use. Keep your figures and measurements for use in Individual Work 7.

INDIVIDUAL WORK 7

Zoom with a View

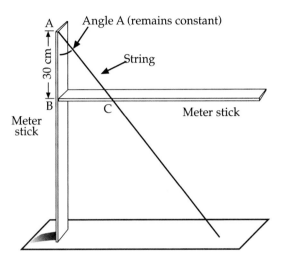

Figure 3.54.   **Figure 3.55.**

1. a) What is the relationship between the corresponding angles of triangle *ABC* (**Figure 3.54**) and those of triangle *ADE* (**Figure 3.55**)?

   b) Repeat the angle comparisons in part (a) for your own figures from Item 5 of Activity 7.

   c) Measure the angles in Figure 3.52, and comment on the relationships within the various triangles.

2. a) What is the ratio of *AD* to *AB* in Figures 3.54 and 3.55? What does this ratio describe in terms of the viewing-tube experiment? (Hint: Recall your answer to Item 1 of Activity 7.)

   b) Predict the ratios of *AE* to *AC* and *DE* to *BC*. Then measure and calculate these ratios. How good were your predictions?

   c) Repeat parts (a) and (b), predicting and comparing your Activity 7 measurements from various heights to those obtained in your initial set-up.

3. What is the viewing-tube interpretation of the lengths corresponding to *BC* in the string-and-stick activity?

4. a) Look back at the table of measurements you collected in Item 6 of Activity 2, Lesson 1. Look at the map distance (corresponding to your single apparent distance) for the 15-cm view and the map distance for the 40-cm view. Compute the ratio of these two distances. Explain the result, using your work in this lesson's Activity 7 as a basis.

   b) Repeat part (a), comparing the remaining tabulated distances to those from the 15-cm view. Explain any discrepancies between actual measurements in Lesson 1 and predictions based on today's work.

5. a) Your work thus far has dealt only with zooming in on a single segment. Based on your measurements thus far, how do you think the appearance of two-dimensional figures would change as you zoom in on them with a fixed angle of view? Use your results from Item 2 or 4 to give a specific numerical example of your conjecture.

   b) Draw a sketch like that in Figure 3.52 (page 211) to illustrate your conjecture.

   c) If you did not address it specifically in part (a), how do you think shapes would change in this situation?

   d) Explain how you could check your answer to part (c).

6. If you accidentally increased the measure of angle A slightly, between Figure 3.54 and Figure 3.55, how would the length of segment DE change? How about segment AE? Which change would be most noticeable?

7. Landsat orbits at a height of approximately 700 km above Earth. What does this mean about the need for precision in setting the angle of Landsat's scanners?

8. **Figure 3.56** shows two regions that are intended to have the same shape. Making that assumption, find the lengths of the sides labeled x and y, and explain how you got your answers.

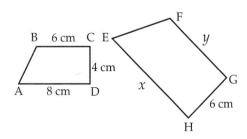

Figure 3.56
Two regions.

9. **Figure 3.57** represents a girl who is five feet tall, casting a shadow that measures six feet in length. A nearby flagpole casts a shadow 24 feet long. How tall is the pole? Explain.

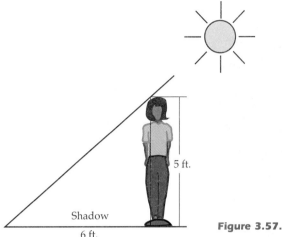

As you saw in Item 1, for the triangles in the string activity and those representing the tube experiment, the angles in the same locations in the triangles are the same whenever the triangles are similar. Remember that similar figures are related by a matching that relates parts of one figure with specific parts of the other figure. These "matched" parts are referred to as **corresponding parts**.

5 ft.

Shadow
6 ft.

Figure 3.57.

In general, for any similar figures, corresponding angles are exactly equal. For triangles, knowing that all corresponding angles are equal is enough to force the triangles to be similar. Thus, at least in the case of triangles, you have another way to test for similarity: check corresponding angles.

> To check for similarity, do one of the following:
> See if all corresponding distance ratios are equal
> (corresponding distances are proportional), or
> see if all corresponding angles are equal.

10. a) In Items 8 and 9 above, identify the corresponding parts of the figures.

   b) Use the above statement about similarity to write and solve as many equations involving the specific information in Items 8 and 9 that you can.

11. During a solar eclipse the moon passes between the earth and the sun. It turns out that the geometry of a solar eclipse is such that during the period of totality (when the entire sun is behind the moon from Earth's perspective) the moon almost exactly fits over the sun; they appear to be the same size.

   a) Sketch a diagram showing the sun, the moon, and a camera on Earth recording the eclipse, that shows this "same apparent size" geometry.

b) The average distance from the earth to the sun is about 1.50×10^{11} m, and the average distance from the earth to the moon is about 3.85×10^8 m. If the moon's diameter is about 3.476×10^6 m, how large is the sun?

12. **Figure 3.58** is from Lesson 1. It illustrates the use of a graph to scale a particular map distance, ΔM, to the real distance that it represents, ΔR. **Figure 3.59** is a variation of Figure 3.58, with the map and real distances drawn "up against the line" instead of on the axes. Note, however, that exactly the same information is displayed. The same map distance, ΔM, is still shown in relation to the real distance, ΔR, it represents.

a) How many triangles do you see in Figure 3.59? Trace the figure and then identify and label each triangle you see.

b) Check the triangles that you found to see which pairs are similar. Explain the methods you use.

c) Redraw Figure 3.59, and add several new triangles showing the scaling of other ΔM distances into their related ΔR distances. Compare all the ΔM-to-ΔR triangles. What do you conclude?

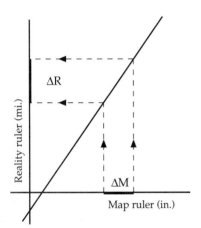

Figure 3.58.
Converting from map distance to real distance.

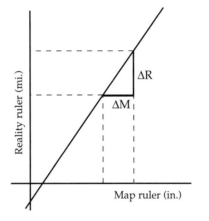

Figure 3.59.
Another view of converting from map distance to real distance.

d) Use the definition of similar stated in Activity 7 and your answer to part (c) to make a general statement about the distances ΔM and ΔR in the triangles you drew.

e) How is the scale factor for map-to-reality scaling related to your work in parts (a–d)?

13. Sketch the graph of a randomly drawn line (not horizontal or vertical). Repeat Item 12 with your new line, and summarize your findings.

ZOOMING INTO MAGNITUDE

In your work with zooming in Activity 7, the zooms were accomplished physically by actually moving closer to the image. This guaranteed that the shape of the image never changed. In fact, the change in the image was just a scaling, and the scale factor was the ratio of the viewing distances. But what about zooming on a computer? Do shapes still stay the same? What is the "geometry" of computer zooming?

For this activity, use **Figures 3.60–3.65** or the *Beverly,MA* image on the computer. If you are using the computer image, select zoom factor x.500 (the number shown in the magnification box near the lower-right corner of the screen).

Place a blank transparency directly on the figure in the textbook or on the computer screen. (Static electricity should hold the transparency in place on the computer monitor.) Use a transparency pen to mark the upper-left corner of the image to help keep the transparency accurately aligned. Move the bottom scroll bar completely to the left and the side scroll bar completely to the top.

1. a) Choose some feature of the image and use a transparency pen to outline it as carefully as you can. Redisplay the image using a zoom factor x1.0, then examine the new image. (If necessary, move the scroll bars on the computer screen back to the top and left, as they were when you started.) Use your edge markings to verify that your transparency is correctly aligned with the image, then trace the new image of the same feature that you outlined. (You should have two images drawn at this point.)

   b) Remove your transparency from the textbook or computer screen, and determine whether the shape of the feature has changed. Justify your conclusion using the definition of similarity.

ACTIVITY

ZOOMING INTO MAGNITUDE

8

Figure 3.60.
Beverly, MA at zoom level 0.5.

Figure 3.61.
Beverly, MA at zoom level 0.6.

ZOOMING INTO MAGNITUDE

Figure 3.62.
Beverly, MA at zoom level 0.7.

Figure 3.63.
Beverly, MA at zoom level 0.8.

ACTIVITY

ZOOMING INTO MAGNITUDE

8

Figure 3.64.
Beverly, MA at zoom level 0.9.

Figure 3.65.
Beverly, MA at zoom level 1.0.

ZOOMING INTO MAGNITUDE

2. Using a straightedge and the "before" and "after" images on your transparency from Item 1, carefully connect corresponding points in the two images with straight lines, and extend those lines across the transparency. For example, connect the tip of the island to the tip of the island, the end of the bridge to the end of the bridge, and so on. Describe the resulting geometric relationship.

3. In Item 1 you measured a number of distances within each separate image of the feature you examined. Now investigate the ratios of corresponding distances (distance after zoom/distance before zoom) between the two images. For example, what is the ratio of the length of the bridge in the final image to the length of the same bridge in the initial image? Relate your answer to the zoom levels displayed at the bottoms of the two images and to your knowledge of scaling.

4. In Item 1 you measured a number of distances within each separate image of the feature you examined, and in Item 3 you investigated the scaling between the two images. Now investigate corresponding distances from the common point of intersection you found in Item 2. For example, measure the distance from the point of intersection to the top-left corner of the feature in each drawing. Repeat for other pairs of corresponding points in your tracing. Relate your answer to your other work.

5. Summarize your findings from Items 1–4 in the form of facts that you believe are true of all zooms on this image. Then explain how someone could test your claims. If you have time, check them yourself.

The magnification of the *Beverly,MA* image (from zoom level x.500 to zoom level x1.0) is an example of a transformation called a **dilation**. The intersection point is called the center of the dilation. The factor by which distances from the center of the dilation are multiplied (from the initial image to the transformed image) is called the **magnitude** of the dilation. The starting image is called the **pre-image** for the dilation, and the final image is called the **image** under the dilation. Thus, the pre-image is the "before" version, and the image is the "after" version. Figures related by dilation are similar. In addition, they have exactly the same orientation in the plane: corresponding segments are parallel.

In this terminology, then, the magnitude of a dilation is the center-to-image distance divided by the center-to-pre-image distance. Note that the order matters.

In the dilation in the items above, objects in the pre-image appear smaller than in the image because the pre-image is closer to the center of the dilation than the image is. Thus the magnitude is greater than 1. The example in **Figure 3.66** illustrates a situation in which the magnitude is between 0 and 1. The image is closer to the center than the pre-image, and it is smaller than the pre-image. Here, *A* is the center of the dilation. Direct measurement shows that the magnitude is approximately 0.56.

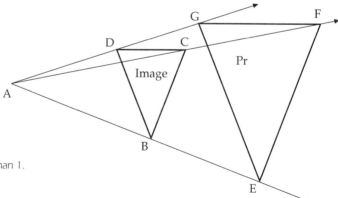

Figure 3.66.
A dilation with magnitude smaller than 1.

INDIVIDUAL WORK 8

Dilations

1. a) Find the center of the dilation in **Figure 3.67** by tracing the figure, then drawing lines from *F* through *C*, from *D* through *A*, and from *E* through *B*, extending the lines so that they intersect.

Your revision of Figure 3.67 should look something like the diagram in **Figure 3.68**. Point *G* is the center of the dilation. If you used the *MultiSpec* software to examine the Landsat images, the zoom level at the bottom of the screen provided enough information to determine the magnitude of the dilation. Zoom levels are not generally available, as is the case in Figures 3.67 and 3.68. However, all is not lost; there are still several ways to find the magnitude.

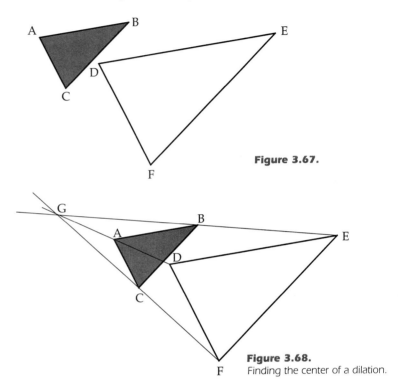

Figure 3.67.

Figure 3.68.
Finding the center of a dilation.

b) Method 1: Use the definition of magnitude. Measure segments *GC* and *GF* in Figure 3.68 to the nearest tenth of a centimeter, and compute the magnitude of the dilation as the after-over-before ratio of distances from the center of the dilation. (Notice that the ratio of the units reduces to 1, so the magnitude of a dilation has no units.) To verify your results, measure the other center-to-point segments and evaluate ratios for corresponding points.

c) Method 2: Use the definition of scale factor. You saw in Activity 8 that the magnitude of a dilation is equal to the scale factor for converting distances in the initial image to distances in the final image. Thus, you can use measurements of lengths within the before and after images to determine the magnitude of the dilation. Measure distances to the nearest tenth of a centimeter within both images, and compute and reduce the after-over-before ratio of corresponding lengths.

d) In similar figures, the measures of the corresponding angles in the two images are the same. Since dilation produces similar figures, the angles here should agree. Verify this by measuring the angles in Figure 3.68.

e) Compare Figure 3.52 (page 211) in Activity 7 to the *GCFEB* portion of Figure 3.68. Relate the geometry of dilations to the geometry of zooming in and to the geometry of scale factors.

f) If triangle *ABC* (Figure 3.68) appears on a computer image that has a zoom level of x.500, what would be the zoom level for an image with corresponding points *DEF*?

In **Figures 3.69–3.71**, the unshaded object is the image of the shaded object in a dilation. Shaded is the pre-image; unshaded is the image. For Items 2–4, do the following four steps.

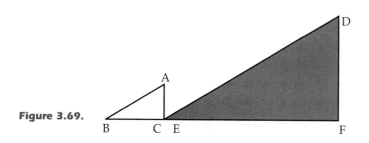

Figure 3.69.

i. Trace the given figure.

ii Find the center of each dilation, and label it *O*.

iii. Find the magnitude of each dilation.

iv. Construct another diagram that is a dilation of magnitude 3 of the original in each figure. Since the shaded object is the pre-image, your drawing should have sides three times larger than the sides of the shaded object. Maintain the same center of transformation.

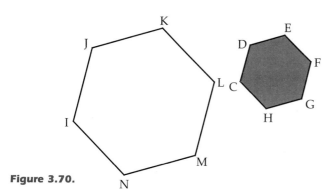

Figure 3.70.

2. Figure 3.69. Triangles.

3. Figure 3.70. Hexagons.

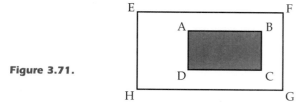

Figure 3.71.

4. Figure 3.71. Rectangles.

5. On graph paper, draw a square with sides 3 cm long. Draw a point (*P*) that will be the center of a size transformation. *P* may be inside the square or outside the square. Draw lines from point *P* to each vertex. Extend them to create an image of the square under a dilation of magnitude 3/2. It may help to think of the dilation as having magnitude 1.5. What are the lengths of the sides of the new square?

6. Look back at Figures 3.35–3.41 (pages 203 and 204) in Activity 6. Determine which of the pairs are related by dilation. For those that are, determine the center and magnitude. For those that are not, explain how you know.

Every dilation produces an image that is similar to its pre-image. Every dilation is a scaling, and the magnitude of the dilation is the scale factor. Corresponding distances within similar figures are related by a common scale factor, so "scaled images" and "similar images" are equivalent expressions. Note that not every pair of similar figures is related by dilation since orientations may not be constant.

> To check for dilation, look for its center by checking for a common point of intersection of lines connecting corresponding points in the two figures.
>
> To check for similarity, look for identical shape ratios in the two figures.
>
> To check for scaling, look for a common scale factor between corresponding distances in the two figures.

EXAMPLE:

Quadrilaterals *ABCD* and *EFGH* in **Figure 3.72** are not related by a dilation.

Known shape ratios in *ABCD* are:
$BC/CD = 1.5$,
$AD/CD = 2.0$;
for *EFGH*,
$FG/GH = 1.5$ and
$EH/GH = 2.0$.

Scale factors (reduced) from *ABCD* to *EFGH* are:
$AD/EH = 2/3$,
$CD/GH = 2/3$,
$BC/FG = 2/3$.

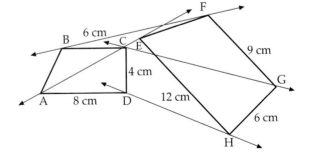

Figure 3.72.
Similar quadrilaterals.

The relationships among some of these distances may be recorded in a special arrow diagram. (See **Figure 3.73**.)

Figure 3.74 shows how such an arrow diagram can be used to describe distances within and between any two similar figures.

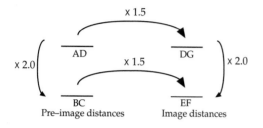

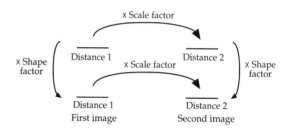

Figure 3.73.
Arrow diagram for lengths in Figure 3.72.

Figure 3.74.
An arrow diagram for distances in similar figures.

7. **Figure 3.75** shows two fingerprints. Do they match? Justify your answer.

Figure 3.75.
Two fingerprints.

THE MATCH GAME

Look back at the questions that were raised at the beginning of this unit. You have made a lot of progress. You know the scale of Landsat images. You can zoom in and out on an image to make more precise measurements of distances and to see features more clearly. You know that zooming preserves shapes. You know how to determine adjustments to scales to allow for size changes due to zooming. In short, you are almost ready to complete the deforestation investigation. What remains is finding efficient methods of measuring area, and you may already have some ideas for that, too.

However, there are other uses for zooming and scale changes than dealing with deforestation, or even working with Landsat images in general. For example, think about the similarities among the following questions.

CONSIDER:

1. Two adjacent communities have local maps printed. Each depicts a region that overlaps a little with the other map. The scales for the two maps are different, but you would like to glue them together so that they can be a single map. How can you accomplish the task?

2. A company maintains extremely confidential records and wishes to restrict access to the portion of the building in which they are stored. They decide to install a new employee-fingerprint-identification system. Thus, the employees who have access to the storage area submit fingerprints that will be stored by a computer and compared to their actual prints when they need access to the records. What does the computer need to do to verify a match?

3. The U. S. Customs check-points in some major airports now have hand scanners. When they return home, citizens who have specially encoded documents slide their

identification into a machine, then insert their entire hand. The scanner checks the hand geometry, and the person's identity is verified. How does the machine determine identity?

4. Builders and architects work from blueprints and physical models of the buildings that they are constructing. How does scale affect their decisions about how much wire to purchase, how much paint will be needed, and how much weight a structure can support?

While not directly involved in solving the Czech deforestation problem, this activity examines some further geometric aspects of dilations that can benefit others in a variety of fields. Think of other situations in which understanding scale and dilation might be valuable.

1. Explain the method that you developed to compare finger-prints in Item 7 of Individual Work 8. Your explanation should be detailed enough that a classmate could carry out your method without your help.

2. Consider question 2 discusses using fingerprints for employee admission to a records room. As you know, law enforcement agencies also use fingerprints in criminal investigations to establish the identities of suspects and victims. Discuss the reliability of the methods you described in Item 1 in terms of your willingness to use it for criminal identification instead of employee access.

3. Suggest as many different methods for matching fingerprints as you can. Identify your choice for the best possible method.

4. Remember that a dilation is a transformation that takes one image and produces a second image, similar to the first. When a dilation is done by hand (pencil and paper), you see both images together. Explain how you could use your knowledge of dilations to help test whether two images match exactly.

THE MATCH GAME

5. Explain any complications that might occur if you were to try to implement your method by using computer software with which you are familiar, or by using a photocopy machine.

Take a look at the ideas you have developed thus far. You have one or more ways to check for matches (mathematical similarity) between prints. No doubt each method is good in theory, but you may have qualms about using them in matters of life and death. Other methods may be perfect, but it's not clear how to get a machine to carry out the details. Since fingerprints are extremely complex images, it is fair to assume that some kind of technology will be used in implementing any method you use. For that reason, the remainder of this activity assumes that some kind of technology will be used by investigators.

6. a) Recall from Lesson 2 that computer images are represented as collections of pixels. That is, each image is really just a bunch of squares on the screen that are lit at different brightnesses. Look back at your work in digitizing a scene in that lesson. Explain how you could record your digitization in a form that would be easy to store in a computer. Think about all the information you would need to monitor.

   b) Suppose the image is 50 pixels wide and 50 pixels tall. How would you modify your recording method to deal with the larger image?

   c) A fingerprint generally does not have much variety in its brightness values. Pixels are mostly just "on" or "off." For the sake of simplification, assume for a moment that that's exactly the situation: each pixel is either "on" or "off," and there is no other possibility. Devise a way of storing fingerprint information as efficiently as possible.

When you digitized an image back in Lesson 2, you used only 36 pixels. A typical computer screen is several hundred pixels wide and tall. There are literally tens of thousands of pixels on such a screen. The computer has to keep track of all those pixels, not only how bright they should be but also where they are on the screen. It uses a coordinate system to record locations.

THE MATCH GAME

If you have been using computer software to work with images in this unit, check your software to see if it includes a coordinate system of some sort. Many generic image programs, for example, have rulers that can be turned on or off to supply information about the location of a particular point. Different programs use different systems. Some measure "up" while others measure "down." Some use "real" units such as inches or centimeters; others use screen units or pixels.

7. Describe the coordinate system used by your computer software. Be as specific as you can. How do you determine coordinates for a particular location? For example, where is (0, 0)? How do you know? What would you mean by (5, 0)?

Your goal is to develop a method of matching images so that a computer can carry out the task of deciding (or at least helping decide) whether two images are alike.

In mathematical investigations, often the best line of attack is to do what you know how to do and then "look back." That is, you compare the results you end up with to those you wanted. Maybe they are not exactly the same, but perhaps you can see a pattern in how they differ. Seeing how they differ may help you fix any problems you introduced by not knowing exactly how to solve the original problem.

This idea is important enough to repeat. Even when you do not know exactly how to solve a problem, start by doing what you *do* know how to do that may be related to what you want. Many times, a problem will "solve itself" once you get started with some part of the problem. Here, for example, you know a lot about size and shape.

That said, it's time to try your methods out. Again, the goal is to produce a method that will allow you to carry out a physical match to verify image similarity—to line two images up exactly, or to overlay the images. Remember, even though you may be working with paper images, you need to be sure that computers could carry out your method, so focus on coordinates as well as on images.

THE MATCH GAME

8. a) Return to the two fingerprints in Figure 3.75 in Individual Work 8 (page 228). Determine by direct measurement the magnitude of a dilation that can be used to make the smaller image the same size as the larger. Explain your reasoning.

   b) Carry out a dilation of the proper magnitude. (If you have access to computers with image software, use the files *Print1* and *Print2* to work with the prints separately, or use the file *Prints* to have them together.) Compare the resulting image to the one you are trying to match. Discuss the match in terms of both size and location. Quantify your observations as fully as you can. If the images do not match, exactly how do they miss? What instructions would you give to friends to allow them to complete the match? Note: Do not actually match the images if the dilation missed in some way. Focus instead on *describing* the results.

   c) Using your instructions for a friend from part (b), make an arrow diagram for the operations that take you from the original configuration of the two images to the final matched version.

9. Create two figures related by dilation. Remove all your construction lines and points so that only the two figures are visible. Then exchange with a classmate. Repeat Item 8 for this pair of figures.

10. Summarize the methods that you found useful in Items 8 and 9. Write instructions, as detailed and precise as you can, that would help someone complete an "image matching" for images related by dilation without additional help from you. Ideally, your method should provide coordinate information so that it can be applied to images containing thousands of points.

INDIVIDUAL WORK 9

Coordinated Transformations

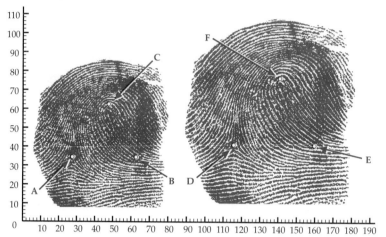

Figure 3.76.
Two fingerprints on a coordinate system.

1. **Figure 3.76** shows the images of the two fingerprints you examined in Figure 3.75. You have already established the relationship between them. However, notice now that the fingerprints have been placed onto a coordinate system. Several distinguishable features have been marked and labeled.

   a) Use a large sheet of paper to construct an image of parts of the left print using a dilation having the same magnitude as the one you used in Item 8(b) of Activity 9. However, this time use the point (0, 0) as the center of your dilation. Record the coordinates of the labeled points following the dilation, together with their coordinates in the pre-image.

   b) Describe any pattern relating the coordinates you recorded for the image to those of the pre-image. That is, what did "dilation centered at (0, 0)" do to the original coordinates? Where would this dilation take the point (40, 20)?

   c) Complete the matching by converting the coordinates of the dilated image to those of the final target print. What mathematical operation on the coordinates does this job?

   d) Summarize what your procedure, starting with dilation, does to the coordinates of the original image, first in words, then using equations, arrow diagrams, or graphs. (You may need to describe the horizontal and vertical actions separately.)

Geometrically, dilation "stretches" a figure. If the magnitude of the dilation is larger than 1, then the image is larger than the pre-image. If the magnitude is between 0 and 1, then the image is smaller than the pre-image. In fact, each distance in the pre-image is multiplied by the magnitude of the dilation.

Looking back at Item 1, an even stronger statement may be made. When a coordinate system is used, a dilation centered at (0, 0) actually multiplies *every coordinate* in the pre-image by the magnitude. Thus, dilation, at least when it's centered at (0, 0), is easy: it's just the multiplication of coordinates.

2. **Figure 3.77** shows two images related by dilation.

   a) Determine the scale factor taking *EFGH* into *ABCD*. Then apply a dilation centered at (0, 0) and having that magnitude to the pre-image *EFGH* to produce an image the same size as *ABCD*. Record the coordinates of each vertex of the image you obtain.

   b) Complete the matching. What do you need to do to the coordinates of the dilated image to make them match *ABCD*?

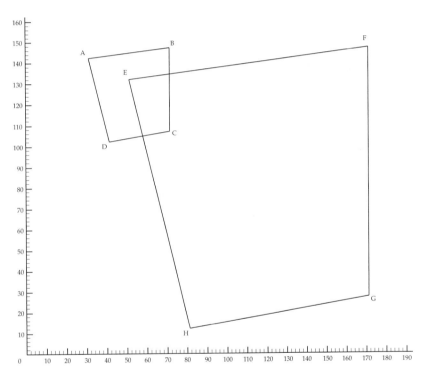

Figure 3.77.
Quadrilaterals related by a dilation.

   c) Summarize the overall process that takes *EFGH* into *ABCD*, first in words, then using at least one other representation.

Notice that your work in Items 1 and 2 are very similar to the work you did with codes in Unit 2, *Secret Codes and the Power of Algebra*. The *x*- and *y*-coordinates have to be treated separately, but coordinates of the pre-image are like a plain text message, and the coordinates of the image are like a coded message. The transformation that turns the pre-image into the image plays the role of the code itself.

That means that the same methods that worked to break codes can be used to discover equations for transformations matching two figures. Just make a table of pre-image x-values and image x-values, then do the same with y-values. Each table leads directly to an equation.

3. **Figure 3.78** shows coordinates for key points in a particular figure. The corresponding coordinates for a figure to match are shown in **Figure 3.79**.

x original	y original
–2	10
0	8
1	9
5	0
2	–4
4	6

Figure 3.78. Initial points.

x target	y target
–1	15
3	11
5	13
13	–5
7	–13
11	7

Figure 3.79. Matching points.

a) Write a formula that will take the points from Figure 3.78 into the points whose coordinates are shown in Figure 3.79.

b) If you have access to a graphing calculator or spreadsheet program on a computer, enter the original coordinates into a pair of lists on your calculator, or into two columns of a spreadsheet. Apply your formula(s) to the coordinates to verify your work; produce two new lists (or columns) that match the target coordinates.

c) Graph the original set of coordinates on a coordinate system. Then, on the same system, graph the target coordinates. Geometrically verify that the figures are related by a dilation.

d) Express your formula(s) in arrow diagram form.

e) Use your arrow diagrams to write, then solve, equations representing the center of the dilation. (Hint: The pre-image of the center is also the image of the center.)

One additional advantage of using a coordinate system, beyond making it easier for computers and calculators to help with the work, is that it makes it possible to measure diagonal distances indirectly.

You have probably studied the Pythagorean theorem in an earlier course. It says that the sum of the squares of the lengths of the two legs of a right triangle is the same as the square of the length of the hypotenuse. The legs are the two sides that form the right angle; the hypotenuse is the remaining side, the long one. See **Figure 3.80**.

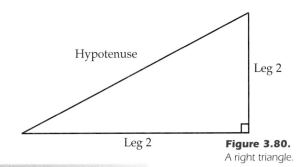

Figure 3.80.
A right triangle.

> The Pythagorean Theorem:
> $(\text{leg } 1)^2 + (\text{leg } 2)^2 = (\text{hypotenuse})^2$

4. Verify the assertion of the Pythagorean theorem by measuring the distances in Figure 3.80, squaring them, and adding the two smaller values.

5. Return to Figure 3.77. Look at the segment labeled *AD*. Imagine a right triangle having *AD* as its hypotenuse and legs parallel to the axes, one vertical and the other horizontal. You may wish to sketch such a triangle lightly into the diagram (on a separate piece of paper).

   a) How many such triangles are possible?

   b) Use the coordinates of *A* and *D* to determine the lengths of the legs of your new triangle. Explain your reasoning clearly.

   c) Combine your results from part (b) with the Pythagorean theorem to determine the length of *AD* without actually measuring it. Then check your answer using direct measurement.

6. The images in **Figure 3.81** show the long beach near Beverly, MA, whose length you computed in Lesson 2. Notice that in each image there is a small cursor at one end of the beach. The location of the cursor is reported in the small box to the right of the image.

   The coordinate system measures in pixels, with (0, 0) in the upper-left corner, (512, 0) in the upper-right corner, (0, 512) at the lower left, and (512, 512) at the lower right. In a *MultiSpec* display, lines of pixels run horizontally and can be thought of as rows in a matrix. Columns of pixels run vertically and can be thought of as columns in a matrix.

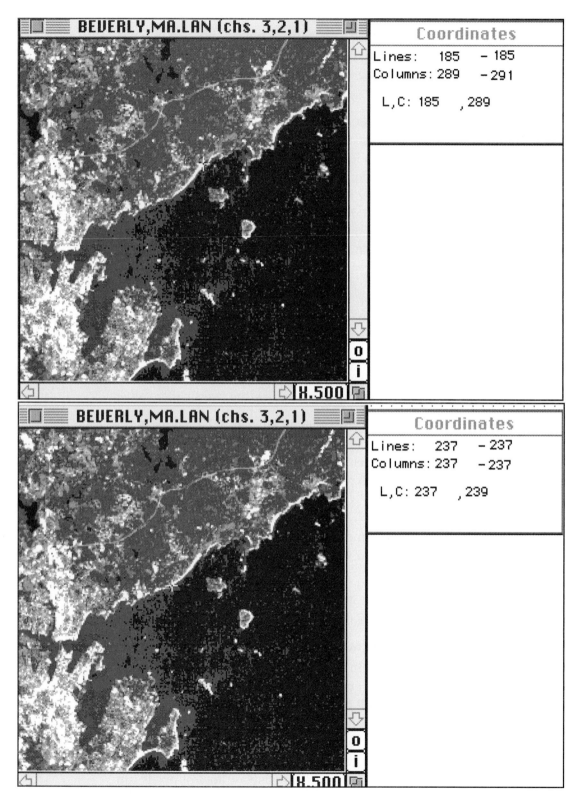

Figure 3.81.
A beach near Beverly, MA, with pixel coordinates displayed.

This coordinate system indirectly defines a unit for measuring distances. For lack of a better name, call the new unit a "pixel-width." One pixel-width is the distance from the center of one pixel to the center of the next one, measured exactly horizontally. Since pixels are square, one pixel-width is also the distance from the center of one pixel to the next, exactly vertically, too. However, you know from experience that the distance from the center of one pixel to the next one, corner-to-corner, is larger than one pixel-width. In fact, the Pythagorean theorem implies that it is about 1.4 pixel-widths in that case.

Use this new measurement system to determine the approximate length of the beach. Compare this answer to your earlier results. Remember that one pixel represents a region measuring 30 meters square. Thus, one pixel-width scales to 30 meters.

7. a) The two quadrilaterals in Figure 3.77 are related by dilation. By directly measuring the corresponding distances within the figures, verify that they really are similar. Explain your work.

    b) Add appropriate lines to the diagram to demonstrate that the figures are indeed related by a dilation. Explain your reasoning.

    c) Determine the magnitude of a dilation that takes *ABCD* onto *EFGH*.

    d) Using pencil and paper, carry out a dilation of *ABCD*, using a randomly chosen center and the same magnitude you obtained in part (c). Record the coordinates of the dilated image as accurately as possible, and relate the resulting figure to *EFGH*.

    e) Use your results from part (d) to explain how to make the two images match exactly if someone else chooses the center of any dilation. Be clear and precise in your description. Include information about coordinates if that is appropriate.

In your investigations in Item 6 of Activity 9, you were asked first to carry out a dilation on an image and then to adjust the location of the result in order to complete the match. How well does the reverse process work? That is, can you align a point first, then dilate for size, and be done? To find out, carry out as many of the Item 8 investigations as you can. Think back to your work with secret codes to get a preview!

8. This item includes directions for working with a computer, a calculator, and paper and pencil. Your teacher may direct you to use one or more of these approaches.

If you have access to a computer with image software, use the fingerprint images from Item 6 of Activity 9. Before doing anything else, make a backup of the file you wish to use, then work only with the backup. Use either the pair of files *Print1* and *Print2* or the file *Prints*, which contains both prints. Move the first print so that one pair of corresponding points is actually exactly overlaid. Then carry out a dilation to produce an image of the correct size. Comment on your results.

If you are using a computer or calculator:

a) Using the coordinate lists from Item 3 above, first carry out an addition to all the original coordinates so that one of the target points will be matched. Sketch or plot the resulting set of coordinates to verify that the image is still its original size, and has the desired point overlaid.

b) Now carry out a dilation centered at (0, 0). Use the algebraic rule you developed in Items 1–3 to assist you. Sketch your results to verify that the dilation was completed correctly. Compare the dilated figure to the original larger figure, both in size and in location. Does "slide then dilate" do the same thing as "dilate then slide?"

If you are using paper, use the quadrilaterals in Figure 3.77 for this item.

a) First, redraw the smaller quadrilateral so that one of its points is exactly on its corresponding point in the larger one. It should still be its original (small) size. Explain how to use the coordinates to assure that your drawing is accurate.

b) Now select a random point in the diagram to serve as a center of dilation. Use it to carry out a dilation of the new small image, using the magnitude you determined in 7(c). Compare the dilated figure to the original larger figure, both in size and in location. Does "slide then dilate" do the same thing as "dilate then slide?"

9. Two students decided to make a secret code for passing notes. They agreed to use multiplication by 3 and addition of 2. However, they forgot to decide which to do first. Each made a key, Jensen using "multiply then add" and Lequita using "add then multiply." Will they be able to use their own keys to decode the other's messages? Explain carefully.

10. a) Return to your work in Items 1 and 2. Record the coordinates of the key points in the pre-image as a list or as columns in a spreadsheet. Then use your multiply-then-add operations from your answers to Items 1 and 2 on the table to verify the match.

    b) Now reverse the order of operations: add, then multiply. Describe the effects.

11. Describe some possible drawbacks to the fact that fingerprints stored on a computer are displayed as pixels.

12. Two adjacent communities have local maps printed. Each depicts a region that overlaps a little with the other map. The scales for the two maps are different, but you would like to glue them together so that they can be a single map. Explain how the geometry you have been studying could be used to help you accomplish the task. Include methods that could be used with computer images and with photocopy machines.

13. Explain how the mathematics you have been studying could be used to help identify missing persons from photographs, even though their appearances may have changed.

14. Look back at the images of Beverly, MA that you have been using in this unit (Figure 3.22 on page 187 is a good example.) Three clouds are in a central portion of the image. You can tell because they have similarly shaped shadows. Using methods you have studied, verify that they really are similar. Then explain the mathematics of how you could overlay the clouds and shadows.

LESSON FOUR

Areas

KEY CONCEPTS

Area

Length-Area
relationship

Approximation

Conversion factors

Solving proportions

Monte Carlo
methods

The Image Bank

PREPARATION READING

Areas

*I*n Lesson 1 you began investigating deforestation in a
region of the Czech Republic. Since then you have
developed ways to enlarge the image in order to see the
regions of interest better, and you have investigated the
geometric implications of zooming.

The final piece of the puzzle is the computation of area. Several
issues need to be addressed. You know what a zoom does to
lengths of objects in an image, but you still need to know what
it does to area. You probably already know how to calculate
areas of familiar shapes, but you need to modify them or
invent new ways for the areas of irregularly shaped regions.

These ideas are fundamental. Each will have to be addressed
before you can complete your work on the Czech question.
Ideally, methods that you develop will work both for computer
images (with pixels) and for smoothly drawn images.

ACTIVITY

FROM A TO Z

10

Figures 3.82–3.84 are different enlargements of the same region near Beverly, MA.

If you do not have access to a computer, use the images in the text to answer the following questions. Figure 3.82 shows zoom level x64.0, Figure 3.83 shows zoom level x32.0, and Figure 3.84 shows zoom level x16.0. Note that a two-pixel-by-three-pixel rectangular region is outlined with dotted lines on each image.

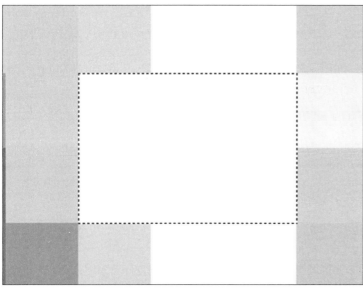

Figure 3.82.
A region near Beverly, MA, at zoom level x64.0.

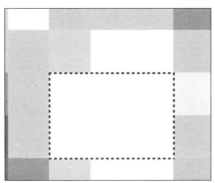

Figure 3.83.
A region near Beverly, MA, at zoom level x32.0.

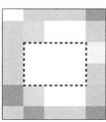

Figure 3.84.
A region near Beverly, MA, at zoom level x16.0.

Ideally, you should get together at a computer with up to three of your classmates. Bring the *Beverly,MA* image up on your computer screen. If you are using *Multispec* software, under the Options menu, pull down to Show Selection Coordinates. A coordinate box will appear in the upper-right corner of your screen; it displays the cursor location by line and column. Position the cursor on the image so that the coordinates (337, 72) show in the coordinate box, then click once at that point.

1. Zoom to a level of 64 (or refer to Figure 3.82), and measure (in millimeters) the actual (screen) width and length of the entire two-pixel-tall-by-three-pixel-wide white region. Record the length and width measurements and calculate the screen area of the arrangement.

2. Zoom to a level of 32 (Figure 3.83), again measuring the width and length of the entire two-pixel-by-three pixel white region and calculating the screen area.

3. What would you predict for a zoom factor of 16 (Figure 3.84)? Zoom to the factor of 16 and check your predictions.

4. At a zoom level of 4, the white block is too small to be measured accurately. Nevertheless, based on your understanding of similar figures and zooms, you can predict the width, length, and area of the white block. What is your prediction?

5. As you changed zoom levels, did the actual size of the region in Massachusetts change?

6. Did the image size change?

7. Did the shape of the image change? Justify using your measurements.

8. If you did not do so earlier, make a table of values from your work above, using columns labeled "Zoom level" and "Area." Then graph Area versus Zoom level using your table. Comment on any pattern you see.

INDIVIDUAL WORK 10

From A to Z

1. Find the areas of each of **Figures 3.85–3.93**.
   Explain your methods (formulas) and measurements.

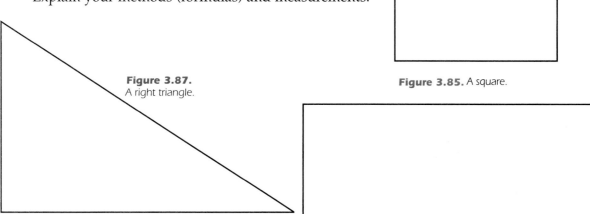

Figure 3.87.
A right triangle.

Figure 3.85. A square.

Figure 3.86. A rectangle.

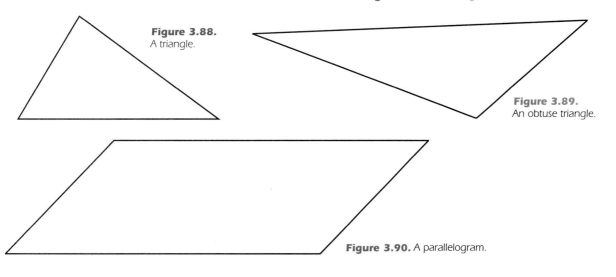

Figure 3.88.
A triangle.

Figure 3.89.
An obtuse triangle.

Figure 3.90. A parallelogram.

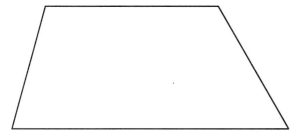

Figure 3.91. A trapezoid.

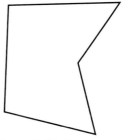

Figure 3.92. A pentagon.

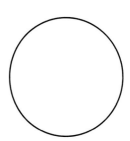

Figure 3.93. A circle.

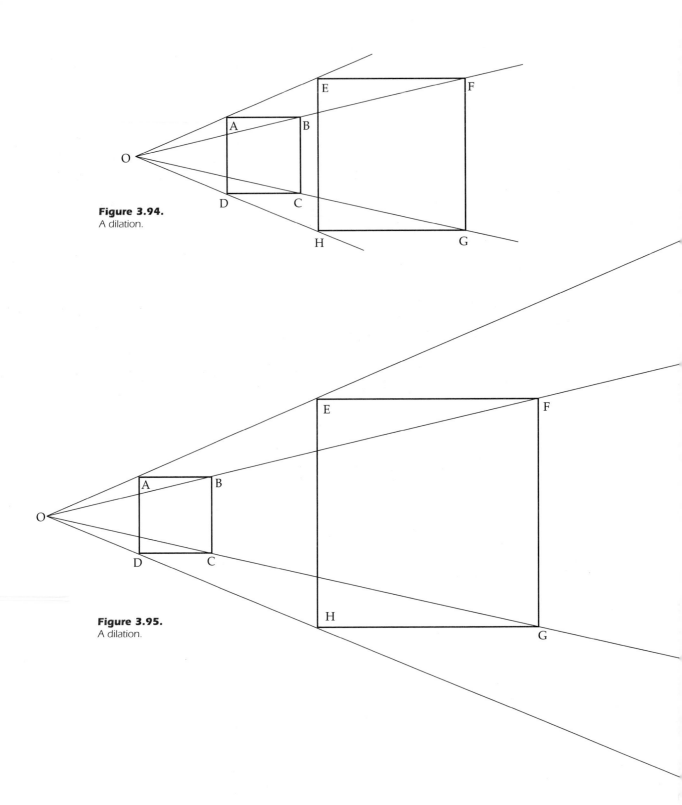

Figure 3.94.
A dilation.

Figure 3.95.
A dilation.

2. In the dilations shown in **Figures 3.94** and **3.95**, square *EFGH* is the image of square *ABCD*. For each figure, answer the following questions. You will need to measure the lengths of the sides and determine the areas of the pre-image and image squares. Measure lengths in centimeters to the nearest tenth of a centimeter.

a) What is the magnitude of the size transformation?

b) What is the ratio of the length of one side of the image square to one side of the pre-image square?

c) What is the ratio of the area of the image square to the area of the pre-image square?

d) How is the ratio of the area related to the ratio of the lengths of the sides?

3. a) Look back at the results of Activity 10. Since it produces an image of a new size, you can think of the change from zoom level 64 to zoom level 32 as a dilation. What is its magnitude?

b) What magnitude dilation would take an image from zoom level 64 to zoom level 16? From zoom level 64 to zoom level 4?

c) How does the area of an object appear to change as its width is halved?

d) How does the area of an object appear to change under a dilation of magnitude 1/4? Of magnitude 1/16?

e) As a source of light moves farther above the ground, its light covers an increased area, although the amount of light being emitted remains constant. By how much will the intensity of light falling on the ground decrease as the source of light doubles in height?

f) Some satellites are sun-synchronous (maintain the same relative position among satellite, earth, and sun) and orbit at a constant height. Other satellites are geo-synchronous (maintaining the same relative location to a point on the earth's surface) and/or orbit at various heights. Write a brief report about how a satellite can best avoid the uncertainty that results from changes in reflectance values as the satellite changes height, or as the angle of the sun changes.

4. a) You should notice a relationship between changes in area and changes in length (i.e., the magnitude of the dilation, or scale factor) as a result of zooming. That relationship was explored in Activity 10 and in Items 2 and 3. What is the relationship?

   b) See **Figure 3.96**. Given the information in the first arrow diagram, complete the second arrow diagram.

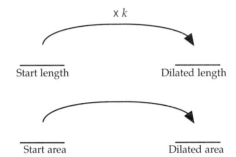

Figure 3.96.
Arrow diagrams for dilated length and area.

5. Two friends enjoy building scale-model replicas of famous aircraft and trains. In fact, each has a fairly extensive collection of hand-painted models. Karen's models are all built to 1/60 scale. Paschal's models are all at 1/120 scale. How should their paint bills compare? Explain.

6. a) Highway departments frequently write instructions to drivers not only on signs posted at the sides of roads, but also directly on the highway surface itself. (If possible and safe, take a car ride to look at signs on the road). The height for one such sign is 8 ft. About how far from the sign do you think the driver is expected to read it?

   b) Sketch a side view of a driver looking at a sign painted on the road. Use your measurements to label all important distances.

   c) Airport runways are labeled with giant numbers, indicating their "names." For example one runway at Atlanta's Hartsfield International Airport bears the label 26R, for runway "26 Right." Assume that pilots approaching for a landing look at these run-way markers at the same angle as drivers on roads, and that pilots need to see the markers while they are still a mile away from the runway (measured along the ground). How tall would the runway markings have to be in order to appear the same to the pilot as road markings do to drivers? How would the amount of paint needed for such a runway letter compare to that for a road letter?

d) Although the runway numbers can be read from as far away as two miles, they do not appear as large to pilots, even at one mile away, as do highway markings to drivers. In fact, according to the FAA, runway markings are "only" 60 feet tall. Assuming that pilots and drivers have the same viewing angle, how far away from the end of the runway does a marking appear to be the same size to the pilot as highway markings are to drivers? Revise your paint approximation from part (c).

e) Research the method airports use for naming their runways. For example, runway 26R in Atlanta is parallel to three other runways: 26L, 27L, and 27R. What do these numbers and letters mean, if anything?

f) Research airplane landing paths. Do pilots have the same angle of vision as the drivers described above?

7. a) Use your calculator to graph the two equations $Y1 = X+1$ and $Y2 = 2-.5*X^2$ in the viewing window having both X and Y between −12 and 12. Tape a small piece of transparency film over your screen and trace the image.

b) Use the "Zoom In" command, centered at $(0, 0)$, and retrace the image. Be sure the transparency film does not move between sketches.

c) Check for similarity and dilation. Justify your answer.

8. **Figure 3.97** shows an irregularly shaped region. The same region is available for exploration on the computer as the file *Region*. Use the image in Figure 3.97 or a computer to determine the area of this region as accurately as you can. Use as many methods as you can. Explain each method carefully. Which method seems most accurate? Why do you think that?

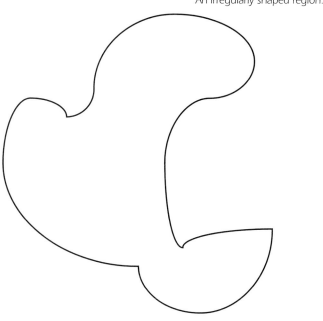

Figure 3.97.
An irregularly shaped region.

NO MATTER WHAT SHAPE

For your final activity in this unit you will complete the Czech deforestation problem. To do that, you need to find the areas of regions covered by conifer trees in 1985 and 1990. Forested regions and lakes generally have irregular boundaries; they rarely occur naturally as rectangles, triangles, or circles.

In this activity, you will use several methods to find the areas of irregularly shaped regions. Compare your results so that you can explain to a friend how to decide when using a particular method makes good sense. Think also about whether each method can be applied to an image that is composed of pixels and to one that is not.

INSCRIBED/CIRCUMSCRIBED RECTANGLE METHOD

This method is a "quick and dirty" method that gives fairly good results, on average, and is very fast and easy to use. It is also easy to explain. First, overestimate the area of a region by circumscribing (enclosing) it in a single rectangle. Then underestimate the area of the region by inscribing (placing inside the region) a smaller rectangle. Average the two areas to give an estimate of the area of the irregularly shaped region.

As an example, find the area of an island near Beverly, Massachusetts. In **Figure 3.98**, a rectangle completely encloses the island.

Note the coordinates displayed in Figure 3.98. Remember, the computer screen is like a large coordinate graph with (0, 0) in the upper-left corner, (512, 0) in the upper-right corner, (0, 512) at the lower left, and (512, 512) at the lower right. In a *MultiSpec* display, lines of pixels run horizontally and can be thought of as rows in a matrix. Columns of pixels run vertically and can be thought of as columns in a matrix.

The coordinates shown in Figure 3.98 are lines 272 to 301 and columns 347 to 370. Since 301 − 272 = 29 lines, the top and bottom lines of the rectangle are 29 line-widths apart, so there are 30 lines

ACTIVITY

NO MATTER WHAT SHAPE

11

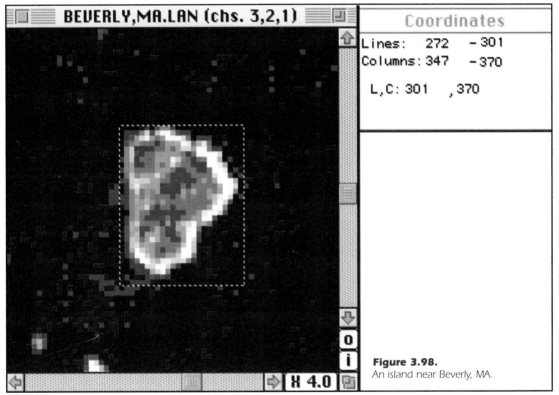

BEVERLY,MA.LAN (chs. 3,2,1)

Coordinates

Lines: 272   – 301
Columns: 347   –370

L,C: 301   ,370

X 4.0

Figure 3.98.
An island near Beverly, MA.

of pixels in the rectangle. Thus the highlighted rectangle is 30 meters per pixel x 30 pixels = 900 meters tall. The difference of the column coordinates is 370 – 347 = 23 pixels, so the width is 24 pixels, or 30 meters per pixel x 24 pixels = 720 meters. The area of the highlighted rectangular region is 900 meters x 720 meters, or 648,000 square meters.

1. a) Follow the same procedure to determine the area of the highlighted region in **Figure 3.99**.

   b) The island is "trapped" between the two rectangles, so its area is, too. Average the areas of the two trapping rectangles to compute an approximate value for the area of the island.

ACTIVITY

NO MATTER WHAT SHAPE

11

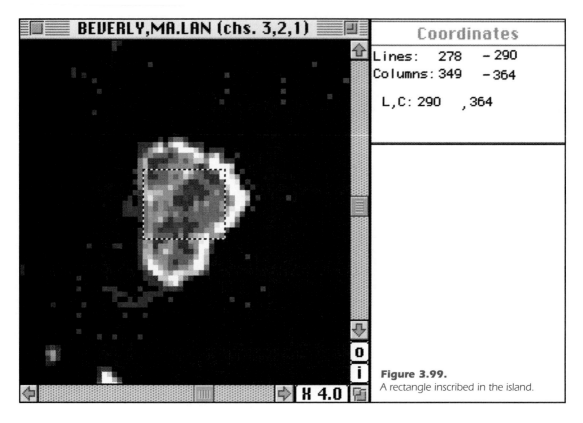

Figure 3.99.
A rectangle inscribed in the island.

Figure 3.100.
Square grid.

SQUARE GRID METHODS

Other methods are based on a similar idea but include some effort to be more precise. Place a grid of squares of known size over the image. Now count the number of squares that fit into the region of interest (the island). Convert the area of the counted squares using the scale of the map.

The length of one side of each small square in the grid in **Figure 3.100** is 0.5 cm.

NO MATTER WHAT SHAPE

2. a) If the scale of a map is based on centimeters and reads 1:100,000, then one cm on the map represents 100,000 cm on Earth. How many centimeters on Earth does 0.5 cm represent on the map?

   b) How many kilometers are represented by your answer to part (a)?

   c) Find the area in square kilometers of one earth-covered square represented by one square in the grid in Figure 3.100.

3. a) Calculate the scale of the printed image in Figure 3.98 in the reading "Inscribed/Circumscribed Rectangle Method."

   b) Place a 0.5 cm square grid (like that in Transparency T3.9) over Figure 3.98, the image of an island near Beverly, MA. Recalculate the area of the island using the 0.5 cm grid.

   c) Explain how you used the grid to find the area.

   d) If you have access to a computer with *MultiSpec* software or other software that will display and zoom the Beverly image, use it to zoom in to pixel level. Apply the logic of your grid method to this image, using pixels themselves as the grid. Comment on the strengths and weaknesses of this method.

INDIVIDUAL WORK 11

Knitty-Griddy

Figure 3.101 is a map of Crater Lake National Park, located in southwestern Oregon's Cascade Mountains. The dominant feature of the park is an enormous caldera, a remnant of the ancient Mt. Mazama. The mountain last erupted about 6,600 years ago, collapsing and creating the caldera that is now Crater Lake. In the past, scientists and map makers have found it very difficult to draw highly accurate maps of mountainous regions because of the steep, changing slope of the land. More recently, map makers have found it useful to use Landsat images and mathematical techniques to map such regions.

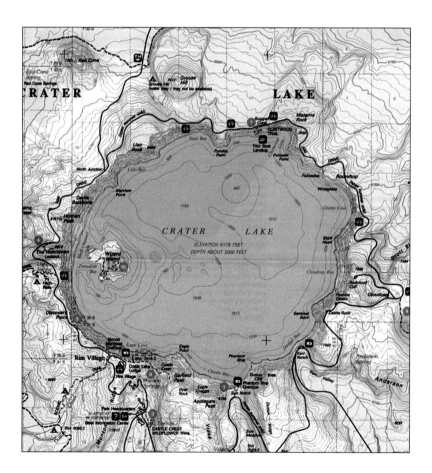

Figure 3.101.
Crater Lake National Park.

1. a) Place a 0.5 cm-square grid (like Transparency T3.9) over the map of Crater Lake. Use the grid to determine the area of the lake. Explain how you determined the area. Be sure to include units.

   b) What decisions did you have to make to find the area?

2. a) When you use the grid method, you often have to decide what to do about squares that include both lake and land. One method is to find the area twice, counting squares each time. To find the maximum area, count every square that contains any portion of the lake, and to find the minimum area, count squares that contain only water and no land. Estimate both the minimum and maximum area of Crater Lake.

   b) Average the minimum and maximum areas to find an area that, in most cases, is more accurate than either one of the preceding estimates.

3. An irregularly shaped region with an overlaid grid is shown in **Figure 3.102**. Trace the figure and lightly shade the squares included in the minimum and maximum area, using a different pattern or colored pencil for each. Explain how you found the average between the minimum and maximum areas.

4. Use both the Inscribed/Circumscribed Rectangle method and some version of a Square Grid method to approximate the areas of each region shown in **Figures 3.103–3.107**. Note the scales given on each image.

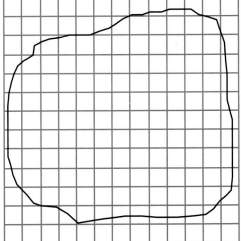

Figure 3.102.
A grid on an irregular region.

5. If you have access to a computer with image software, zoom in to the pixel level to determine the areas of the regions in Item 4. They are available on disk in the files *Zoom1–Zoom5*. If you do not have access to a computer, use **Figures 3.108–3.112**.

6. Use your calculator to graph the two equations $Y1 = X$ and $Y2 = .5 \times X^2$ in the viewing window having both X and Y between 0 and 2.5. Approximate the area trapped between the line and the curve in screen units. Explain your method clearly.

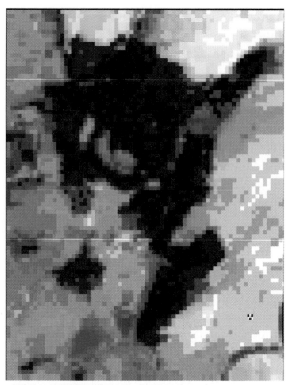

Figure 3.103.
A forest in the Czech Republic. Approximate scale: 1 cm = 270 m.

1 inch

Figure 3.105. A disk.

1 cm

Figure 3.106. A parallelogram.

1 cm

Figure 3.107. A triangle.

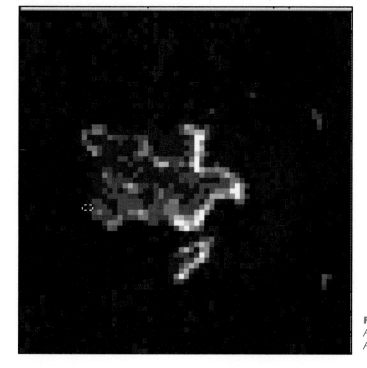

Figure 3.104.
An island near Beverly, MA.
Approximate scale: 1 cm = 200 m.

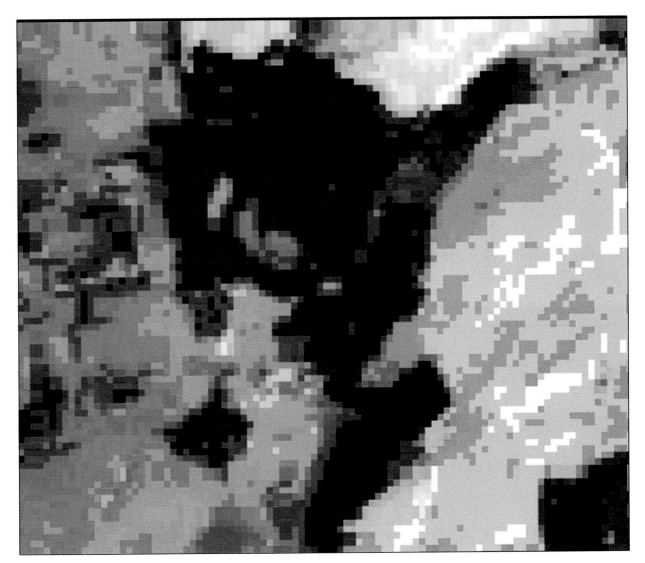

Figure 3.108.
A forest in the Czech Republic, zoomed to pixels.

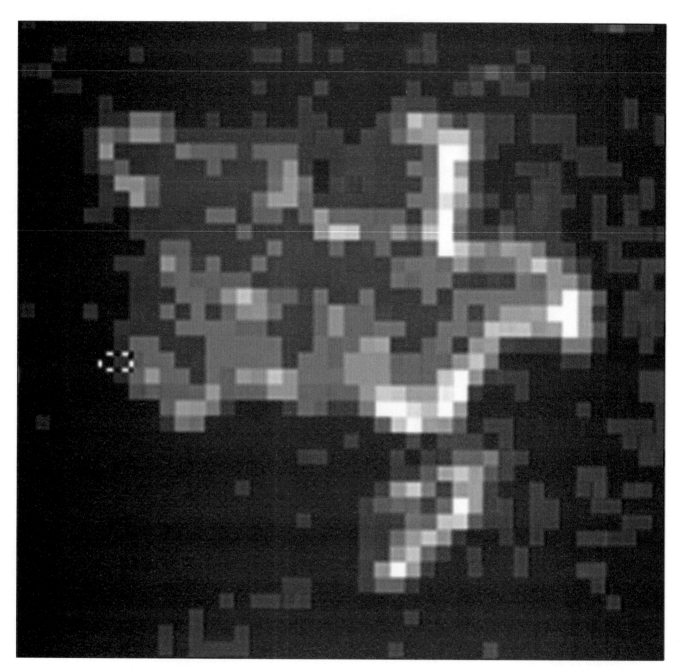

Figure 3.109.
An island near Beverly, MA, zoomed to pixels.

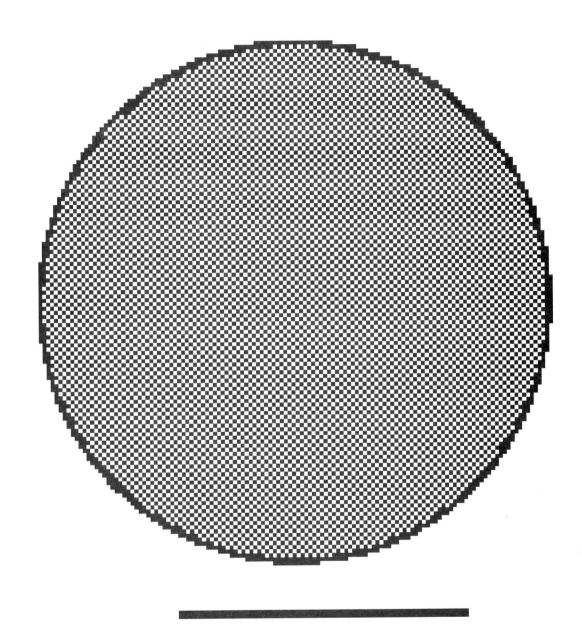

1 inch

Figure 3.110.
A disk, zoomed to pixels.

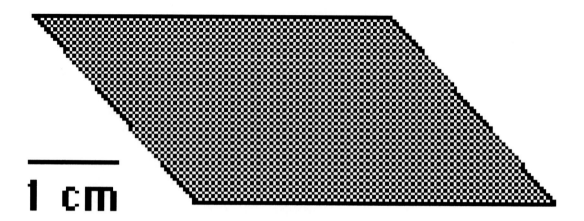

Figure 3.111.
A parallelogram, zoomed to pixels.

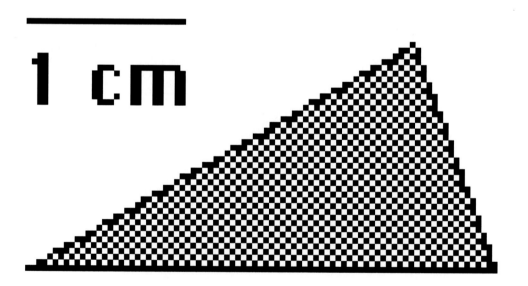

Figure 3.112.
A triangle, zoomed to pixels.

7. a) A number of handcrafts involve units similar to pixels. For example, a decorative fabric art known as "counted cross-stitch" involves sewing lots of tiny "x's" onto a fabric having a square weave. The resulting design can be both intricate and beautiful. One particular cross-stitch pattern has grid squares with sides measuring 1/14 inch. Each cross-stitch has two components: an "x" that runs diagonally across the intersections of the grid and is visible on the front of the fabric; and a pair of strands that connect adjacent squares and appear on the back of the fabric. Explain how an artist can estimate the amount of thread that will be needed for a particular design from a full-size sketch of the design.

   b) Typical cross-stitch charts report thread requirements in yards of six-strand thread, with the understanding that the six strands will be separated into three groups of two strands each before sewing begins. Convert your answer for part (a) to yards of six-strand thread needed.

   c) Adjust your calculations to accommodate a scene that is a scale drawing instead of a full-size drawing.

8. Another hand craft that uses "pixels" is knitting. Again, sweaters and other items to be made by knitting are created one stitch at a time, in regularly-spaced rows and columns. The amount of yarn used per stitch depends on the size of the yarn and needle, and the tension on the yarn as the stitches are made. Research knitting patterns to determine approximate conversion factors relating finished area to amount of yarn.

9. a) Suppose a tiny dot is placed in each square of a 0.5-cm-square grid. What is the conversion factor from squares to dots?

   b) What is the conversion factor from dots to squares?

   c) What is the conversion factor from dots to area?

   d) Explain how you could use these results to modify your square grid methods. Why might that be useful?

WHAT'S UP, DOT?

In the previous two activities you have examined several fairly direct methods for calculating the approximate areas of irregularly shaped regions. Some were quick, but not very precise. Others, though more precise, were somewhat tedious to use. There is no perfect method; almost every real region has an area that cannot be calculated exactly.

Remember, too, that images come in at least two varieties. Some are computer or calculator images. Those are made up of pixels and may lose their sharpness if you enlarge them too much. In addition, since each pixel takes up some space, there is actually area in the boundary. Thus, the tool that seems most promising for precisely finding areas actually leads to somewhat less precision because it blurs edges. Plus, some imprecision is built into the boundary pixels themselves.

The other type of image is the mathematical image. It exists only in your mind, but it has perfectly clear, thin boundaries. Circles, squares, and graphs of equations all fall into this category.

Can you have both precision and ease of use? Look at your work in Item 9 of Individual Work 11. Dots certainly eliminate the need for fractional squares, at least if boundary lines are clearly distinguishable (and have no width!). That means that dots may do well with mathematical figures.

In fact, the dot method also provides another way to find the area of an irregularly shaped region on a map. In this method, a dot is placed in the center of each small square in a grid, as shown in **Figure 3.113**. Each dot represents the entire area of the small square in which the dot appears.

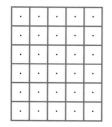

Figure 3.113.
Square grid with dots.

If a dot falls within the region being measured, then the area of the grid square represented by that dot is included in the total area. If the dot is not in the region being measured, the area of the grid square in which the dot is located is excluded from the total area. For the given grid of dots (Figure 3.113), as with the

ACTIVITY

WHAT'S UP, DOT?

12

square grid methods, the length of the side of each grid square is 0.5 cm. If the map scale is 1:100,000, each dot represents 0.25 square kilometers.

In practice, people who use this method do not include the grid lines on the transparency. Transparency T3.10 consists only of the dots. Using that transparency, similar to **Figure 3.114**, each dot represents 0.25 square kilometers.

Figure 3.114.
Grid dots without the grid.

1. a) Open the computer image *Beverly,MA*. If computers are not available, use Figures 3.18, 3.21, and 3.22 from Lesson 2 (pages 183, 186, and 187). Choose an identifiable object or region in the image at a convenient zoom level. Make a careful drawing of the object and describe its position on the image.

   b) Recall that the *MultiSpec* image measures 512 pixels in each direction, and each pixel represents a 30-meter-by-30-meter square. Measure the image on your screen to determine its scale. (You may need to use "Selection Graph" coordinates in this work.) Use that information to find the actual area represented by one square on the 0.5-cm grid (Transparency T3.9). This is equivalent to the area represented by a dot on Transparency T3.10.

   c) Use the dot or grid method to find the area of the region you chose. Describe your measurement procedure and calculations carefully.

   d) Find the area of the same region using the inscribed/circumscribed rectangle method, as described in Activity 11. Again, describe your procedures and how you obtained your results.

   e) Compare the results obtained by your two methods, and state reasons for any similarities or differences. Do you feel your answers are reasonable? Explain.

The Monte Carlo Method

The Monte Carlo approach to problem solving was given its name by physicists working on the Manhattan Project during World War II. The physicists were trying to understand the behavior of neutrons in various materials so that they could construct shields and dampers for nuclear bombs and reactors. Clearly, direct experimentation with nuclear bombs would have been too dangerous, expensive, and time-consuming. So, instead of conducting a direct experiment, the physicists came up with a model that was mathematically the same as the nuclear experiments. Because the model was based on a gambling game, they gave this process the code name "Monte Carlo" after the European city famous for its casinos. Since then, the term "Monte Carlo method" has been applied to the use of chance experiments (either actual or simulated) to estimate values.

One method for estimating areas of irregularly shaped regions (such as forests or lakes) relies on the role of chance. Since chance plays such a significant role in gambling, and because Monte Carlo is one of the world's most famous places to gamble, mathematical methods that use chance, or random processes, are known as **Monte Carlo methods**. You won't be doing any gambling, but you can still take advantage of chance to compute areas.

EXAMPLE:

Suppose you want to determine the area of the irregularly shaped lake shown in **Figure 3.115**. You know that the area of the rectangle surrounding the lake is 200 sq. km.

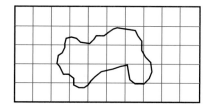

Figure 3.115.
A fictitious lake.

First, randomly toss ten pebbles on the rectangle. "Randomly" means that a pebble is as likely to land on any one point on the rectangle as it is to land on any other point on the rectangle. **Figure 3.116** shows one possible result of randomly tossing ten pebbles.

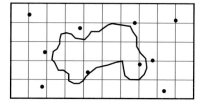

Figure 3.116.
A fictitious lake with ten pebbles.

Two of the ten pebbles that landed on the rectangle also landed on the lake. Using the same logic as with the dot-grid

ACTIVITY

WHAT'S UP DOT?

12

method in Item 1, there are a total of ten pebbles (dots) representing a total of 200 sq. km. That sets up a conversion between scales: pebbles and sq. km (**Figure 3.117**).

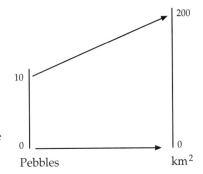

Figure 3.117.
Unit conversion from "pebbles" to square kilometers.

The conversion factor from pebbles to sq. km is

$$\frac{\Delta \text{ sq. km}}{\Delta \text{ pebble}} = 200 \text{ sq. km}/10 \text{ pebbles,}$$

or 20 sq. km/pebble. Thus, for two pebbles, the average associated area is 2 pebbles x 20 sq. km/pebble = 40 sq. km.

2. a) Now toss 50 pebbles onto the rectangle surrounding the lake in Figure 3.116. **Figure 3.118** shows one possible result.

   In this figure, 12 of the 50 pebbles have landed on the lake. Find the new conversion factor, then estimate the area of the lake.

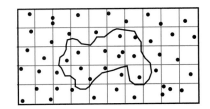

Figure 3.118.
A fictitious lake with 50 pebbles.

   b) Which estimate do you think is more accurate—the one based on ten pebbles or the one based on 50 pebbles? Why?

   c) How could you use the Monte Carlo method to obtain an even better estimate?

INDIVIDUAL WORK 12

Dot's All, Folks

1. Use the dot transparency (T3.10) to recalculate the area of the lake used in the Monte Carlo example in Activity 12. Remember, count only the dots that appear to be in the lake.

2. a) Use the dot transparency (T3.10) to recalculate the area of Crater Lake. Remember, count only the dots that appear to be in the lake.

   b) Describe any difficulties you encountered and how you handled them.

   c) Compare the area you found using the dot method with the area you found using the square-grid method (average of minimum and maximum area). If your results are different, explain the difference.

3. a) To use the Monte Carlo method on the Crater Lake map, first place Transparency T3.9 (the 0.5-cm grid) over the image and label the grid with a coordinate system.

   Now you will need to use your calculator or other randomizing device to produce random numbers for the X- and Y-coordinates of the "pebbles" you will throw. The coordinates need to be selected randomly from the set of all possible coordinates on your grid. If your grid goes from $X = 0$ to $X = 20$, you need to select random X-coordinates between 0 and 20 (not just integers).

Most calculators and spreadsheet programs have a command that generates random numbers between 0 and 1. (On calculators, check under MATH or PROBABILITY menus, for example.) If the command is RAND, then the command N*RAND generates random numbers between (and including) 0 and N. With the example numbers, you would use 20*RAND to generate X-values. Using that command twice, then, gives two coordinates that lie in the 0 to 20 interval. A similar command can be used to get Y-values.

Select N, the interval length, based on your coordinate system covering Crater Lake. Then generate and plot a fairly large number of randomly located "pebbles" on the transparency for your image to get the area approximation. Combine your results with those of other students in your class for a better approximation.

b) Many calculators can store and work with lists of numbers. If yours has that capability, use it or a spreadsheet program on a computer to generate lists of random coordinates within the region containing Crater Lake. Use these points to apply the Monte Carlo method once more.

4. Use at least three methods to find the approximate area of a circle having radius 5 cm. Show your work in each case.

5. Repeat your calculation of the area between the line and curve in the graph of $Y1 = X$ and $Y2 = .5 \times X^2$, this time using a dot method.

WHAT HAPPENED TO THE TREES?

13

You have learned about satellite imagery and how it can help address some global problems. You are now ready to answer a question you first encountered in Lesson 1: What was the extent of forestation in a portion of the Czech Republic in 1985? You may also wish to pursue a second variation of that question: What is the percentage of decrease in the area of forests in a particular region of the Czech Republic between 1985 and 1990?

You have two images of the region: the files *Czech85*, created in 1985, and *Czech90*, created in 1990. If you are using the *MultiSpec* program, open the image using color gun assignments 5-4-3 instead of the usual 3-2-1. This will produce a "false-color" image. With this assignment, the trees in the region (conifers) will appear a dark greenish-blue. All other features will be much lighter, except for water, which will appear almost black. (If you do not have access to computers with image software, use the printed images in **Figures 3.119** and **3.120**, which are also false-color images.) Based on ground truthing, almost all of the trees in this region are Norway spruce, a type of conifer.

Your objective is to prepare a report for EOS (the Institute for the Study of Earth, Oceans, and Space). Use two different methods to estimate the area of conifers, explaining each method carefully and giving the percentage of decrease in forestation between 1985 and 1990. Your report should include the methods you used to determine the area of the forested region in each image and how you determined the percentage of decrease in forested area. Address issues of scale and precision of measurements and calculations explicitly.

Your report will be graded on inclusion of required information, thoroughness of work, clarity of written explanation, neatness, and accuracy.

You may use any of the techniques you have learned in this unit to find the areas of irregularly shaped regions.

WHAT HAPPENED TO THE TREES?

Figure 3.119. A region in the Czech Republic in 1985. Zoom level 1.0.

ACTIVITY

13

WHAT HAPPENED TO THE TREES?

Figure 3.120. A Czech region in 1990 at zoom factor x1.0.

Wrapping Up Unit Three

1. Consider a map on which 2 cm represent 1/2 mile. Measurements may be made on the map (in centimeters) or in the real world (in miles). Here is a table of five corresponding ruler readings for the two measurement systems.

Map(cm)	1.0	2.0	3.0	10.0	20.0
Reality(mi.)	0.25	0.5	0.75	2.5	5.0

   a) Think of this table as a coding rule, with the map units being the plain text and the real locations being the coded message. State the coding rule in words.

   b) Rewrite the coding rule as an arrow diagram.

   c) Use the table or coding rule to represent the process as a graph.

2. Suppose a teacher grades a test using a 0–75 scale. A student's grade using this ruler is called the raw score (R). Grades (G) will be returned to students on a 0–100 scale.

   a) Explain how to convert the raw score to a returned score. Use several representations, including a graph and parallel rulers to illustrate the process.

   b) If two students decide to compare scores, what would you know if $\Delta R = 5$? What would you not know?

3. First check your estimation skills by writing down your estimate before computing.

   a) How fast is 20 mi./hr. in in./sec.?

   b) Give your answer in appropriately significant digits.

4. **Figure 3.121** shows a gray-scale chart and a sailboat scene.

   a) Using graph paper or a blank grid, assign each square a reflectance value (digit) that corresponds to a gray-scale level on the chart.

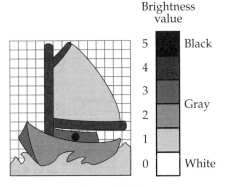

Figure 3.121.

b) How did you decide which reflectance value to assign to pixels that contain more than one shade of gray?

c) Use a pencil to shade each pixel on blank graph paper or on a carefully-drawn grid with the intensity of gray associated with the number in that pixel (its reflectance value).

d Your new picture will probably look different from the original. Why?

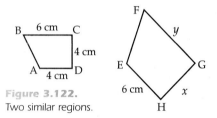

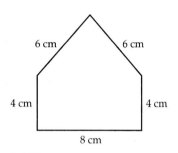

Figure 3.122.
Two similar regions.

5. **Figure 3.122** shows two regions that are intended to have the same shape. Making that assumption, find the lengths of the sides labeled *x* and *y*, and explain how you got your answers.

6. Find the area of the pentagon in **Figure 3.123**.

7. a) Find the center of the dilation in **Figure 3.124**.

b) Estimate the magnitude of the dilation.

For Items 8 and 9 use **Figure 3.125** (x1) and **Figure 3.126** (x.75), two images of Cape Canaveral, Florida. Note that the dark rectangle near the center is the launch pad with light glaring from the gantry and the polygon outlined in green at top left is not a geographic feature.

8. a) Choose some feature of the region and use a transparency pen to outline it as carefully as you can in each image.

b) Remove your transparency from the page and determine whether the shape of the feature has changed from one image to the other. Justify your conclusion using the definition of similarity.

9. a) Estimate the area of the body of water contained within the polygon outlined in green in Figure 3.125.

b) Explain how you could make this estimate more precise.

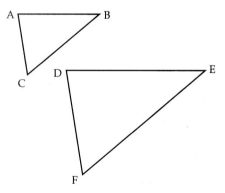

Figure 3.123.
A pentagon.

Figure 3.124.
A dilation.

Figure 3.125.
Cape Canaveral, Florida (x1).

Figure 3.126.
Cape Canaveral, Florida (x.75).

Mathematical Summary

SCALES

Unit conversions and scaling measurements are essentially identical processes, both similar to the coding process for using a secret code (**Figure 3.127**). The idea that two "rulers," or scales, can be matched up in a very precise way lies behind all three processes.

Figure 3.127
Three side-by-side scales: (clockwise) A coding table, a Fahrenheit/Celsius thermometer, and a map scale showing centimeters/kilometers.

PLAIN TEXT	1	2			. . .	2 6
CODED VALUES	3	6			. . .	7 8

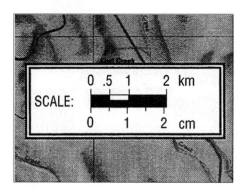

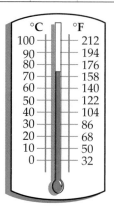

Words, tables, equations, arrow diagrams, and graphs all are useful representations for these processes.

Distance is the difference between two different "ruler" readings. For scaling situations, comparing distances within different measurement systems identifies a multiplier—the scale factor, or conversion factor. To turn image X into image Y, multiply by:

$$\text{Scale factor} = \frac{\text{distance in final image}}{\text{distance in initial image}} = \frac{\text{distance in } Y}{\text{distance in } X} = \frac{\Delta Y}{\Delta X}.$$

Scale factors and conversion factors always have units determined by the "direction" of the conversion. In going from X units to Y units, the conversion factor will have Y/X units. In **Figure 3.128** an X-distance has been converted to a Y-distance, using an arrow diagram. The corresponding graph will have the starting units (X) along the horizontal axis and the ending units (Y) along the vertical axis.

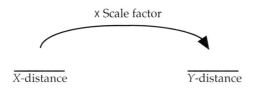

Figure 3.128.
Arrow diagram representation of scaling.

Conversions may be made using the graph by "reflecting" horizontal segments using the graph as the "mirror." In **Figure 3.129**, an *X*-distance (in inches) has been scaled to a *Y*-distance (in miles).

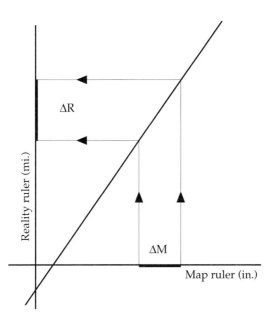

Figure 3.129.
Graph representing scaling from map distances to real distances.

PRECISION

Measurements are never perfectly precise, and calculations cannot increase precision. Always round your final answers to any computation to the same level of significance as the least certain measurement in the problem.

DIGITIZATION

Each individual area whose brightness gets recorded as a separate numerical value is called a pixel. Because its pixels represent regions that are 30 meters square, Landsat 5 is said to have 30-meter spatial resolution.

DILATION

Zooming in on an image results in a dilation. The size changes, but the shape and orientation remain constant. A dilation has a center and a magnitude. The center is the common point of intersection of the lines joining corresponding points in the two images.

By definition, the magnitude of a dilation is the ratio of the distances from the center of the dilation, computed as "final over initial." The magnitude is exactly the scale factor relating the pre-image to the image.

SIMILARITY

Ratios of distances within any figure can be used to describe the figure's shape. If every corresponding pair of shape ratios within two figures is the same, the two figures have the same shape.

Figures with the same shape are said to be similar. Dilations always produce similar images. Their corresponding angles are equal, their corresponding sides have equal scale factors, and their shape ratios are equal. These relationships are summarized in the arrow diagram in **Figure 3.130**.

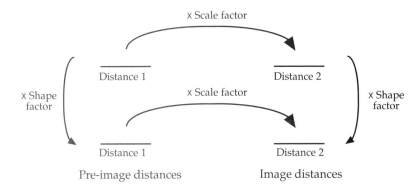

Figure 3.130.
An arrow diagram representing relationships among parts of similar images.

AREA

Circumscribed and inscribed rectangles may be averaged to approximate the area of an irregular region. A grid overlay can be used for a more accurate measurement; the smaller the grid, the more accurate the estimation of the area. On a computer display, pixels can be used as a grid to determine area.

Dot methods are variations of the grid method, using a conversion factor to move from dots to squares to area. A second dot method for approximating area, known as Monte Carlo simulation, relies on chance.

With any method, you must be careful that you are using the map's scale accurately. In addition, no matter which method you use, it is necessary to determine the edges of the region whose area you seek. That may prove to be as difficult as any other aspect of the problem!

Glossary

CONSTANT OF PROPORTIONALITY:
The value of the constant ratio between two proportional quantities.

CONVERSION:
The process of changing a measurement from one system of units to another.

CONVERSION FACTOR:
The factor by which measurements are multiplied during the conversion process.

CORRESPONDING PARTS:
Features that are "matched" in similar figures.

Δ (DELTA):
The symbol for a difference between two measurements; a distance.

DIGITIZATION:
The representation of information with a single number.

DILATION:
A scaling that takes each point to a point "M times its original distance" from a fixed point O. O is the center of the dilation; M is the magnitude.

GROUND TRUTHING:
Visiting a physical site to compare it to its appearance in a satellite image.

IMAGE:
A figure resulting from applying some transformation to another figure.

MAGNITUDE:
The magnitude of a dilation is the factor by which distances from the center of the dilation are multiplied.

MONTE CARLO METHODS:
Mathematical calculations based on repeated random selection.

PIXEL:
One "picture element"; the smallest building block of a digital image.

PRE-IMAGE:
A figure to which some transformation is applied.

PROPORTIONAL:
Two variable quantities having a constant ratio.

REMOTE SENSOR:
Any device that collects measurements without being in physical contact with the measured object.

SCALE:
n. A "ruler" for making measurements.

n. The ratio of distances in a map to the distances they represent in reality.

v. The process of changing one image into another by changing its size.

SCALE FACTOR:
The factor by which distances are multiplied during the scaling process.

SCALING:
The process of converting distances from one system to another when the corresponding distances are not really equal.

SIGNIFICANT FIGURES:
The number of meaningful digits in a reported measurement.

SIMILAR:
The property of having the same shape; corresponding distances within each figure have exactly the same ratios.

UNIT

4

Prediction

Have you ever made a prediction? Was your prediction correct? If so, and it wasn't luck, your prediction was probably correct because you saw a pattern or recognized a special relationship. Mathematics is all about identifying patterns and describing relationships. In this unit, you will work with some of the tools mathematicians have developed to make reasonable predictions based on relationships. In the process, you will extend your understanding of lines and the equations that describe them. You will learn to "fit a line" to data and use the equation of your line to make predictions.

A PREDICTION

Would you like to be able to predict the future or a solve a past mystery? Equipped with a crystal ball or a time machine, what information would you seek? Next week's winning lottery number? Your first job after finishing school? Who's buried in the tomb of the unknown soldier?

Alas, you don't have a crystal ball or a time machine. That doesn't mean, however, that you can't try to predict future outcomes or solutions to past mysteries. Weather forecasters predict tomorrow's weather, and scientists worry that long-term climate predictions show continued global warming. Economists predict unemployment; the government predicts its tax revenues; and businesses predict sales. Anthropologists "predict" features of early humans, and forensic scientists make predictions about assailants and murderers.

Instead of crystal balls or time machines, all of the people mentioned above use real data and build mathematical models to help them make useful predictions. In this unit, you too will use real data and employ mathematical models to make predictions about real-world situations—from analyzing archeological data to helping to save a gentle giant of the sea.

LESSON ONE

The Hip Bone's Connected

KEY CONCEPTS

Explanatory and response variables

Scaling

Graphs of lines

Prediction error

Linear patterns in data with two variables

Linear equations

Ratios

The Lost Dutchman Mine

Legend has it that over 100 years ago, somewhere in Arizona's Superstition Mountains, a Dutchman by the name of Jacob Walz murdered a group of gold miners in order to claim their mine for himself. Over the years, he was periodically seen in Phoenix with saddlebags filled with rich ore. After squandering his money, he would return to the mine for more gold. Attempts were made to follow Walz, but he always managed to lose trackers in the rugged wilderness. In 1890, after another round of digging, followed by a year of spending, Walz again set out for the mine. This time, however, even he was unable to locate it! Walz eventually straggled back to Phoenix and died in 1891.

For over a century people have searched without success for the Lost Dutchman Mine. Some have lost not only time and money, but their lives. At least two searchers are known to have been murdered during their quest. Others, unable to meet the physical challenges of the rugged area, never returned from their trek and remain missing.

PREPARATION READING

If These Bones Could Speak

A contemporary gold digger searching for the Lost Dutchman Mine might come across the bones of earlier explorers. Whenever human remains are found, experts (archaeologists or forensic scientists, depending on the circumstances) are called upon to investigate and explain their findings.

For example, suppose a skull and eight long bones are found fully intact. In addition, suppose numerous bone fragments are found close to the intact bones. After first documenting the exact location and position of the bones and fragments at the site, the scientists record information about the bones, such as their size and general condition.

From their research, the scientists are able to classify the intact bones. Their findings appear in **Figure 4.2**.

After studying the data, scientists conclude that the bones belonged to at least two different people. Assume there were only two people, Bones 1 and Bones 2. Can you figure out which bones belong to which person? Later, if you find evidence that the bones belonged to more than two people, you can change your assumption.

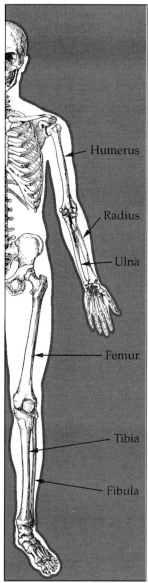

Figure 4.1.
A human skeleton.

Bone type	Number found	Length (mm)
Femur	3	413 414 508
Tibia	1	416
Ulna	2	228 290
Radius	1	215
Humerus	1	357
Skull	1	230
Fragments	More than 10	From 30 to 50

Figure 4.2.
Sample record of bones found at a site.

CONSIDER:

1. How did the scientists decide that the bones belonged to at least two people?

2. Assume Bones 1 is the taller of the two. Which bones do you think belong to Bones 1, and which to Bones 2? Why?

3. About which bones are you uncertain? Why?

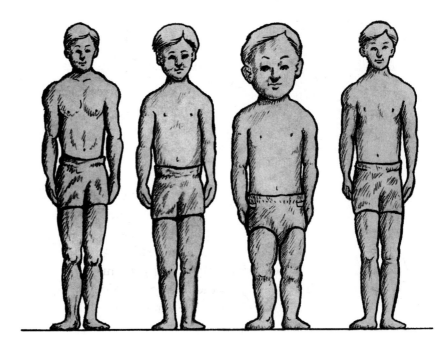

ACTIVITY

1

USING YOUR HEAD

When artists sketch people, they use their understanding of relationships between lengths of different body segments to draw figures that appear normal. In this activity, you will examine one such relationship. At the end of the activity, this relationship will allow you to make a prediction about the bones discussed in the preparation reading.

Figure 4.3.
Sketches of four people of different ages.

1. **Figure 4.3** shows sketches of four people ranging in age from one to 16 years old. The sketches are the same *size*, not the same *scale*. Arrange the sketches by age from youngest to oldest. Describe all the features you found useful in ordering the sketches.

   In arranging the people by age, you may have noticed that heads get smaller in relation to the length of the body as a person grows. You have to keep this in mind when you are

ACTIVITY

USING YOUR HEAD

1

Head length (cm)	Height (cm)
5	
6	
7	
10	
20	
	175
	210

sketching people. To help, artists have rules of thumb or guidelines for drawing people of different ages. For example, if you want to draw a 14-year-old, the sketch should be about 7 head-lengths tall.

2. Suppose an artist sketches a head that is 3 cm long and then completes the figure according to the artists' rule of thumb for 14-year-olds. How tall will the completed figure be?

3. Make a copy of the table in **Figure 4.4**. Complete your table using the artists' guideline for drawing 14-year-olds.

Figure 4.4.
Head length and height for sketches of 14-year-olds.

One major principle in examining data is to look at the numbers for patterns. A second principle is to use a graph to help spot patterns. In the next item, you will plot the data from your table. However, first, you need to choose which axis to label as head length and which as height (the horizontal or vertical). This decision customarily depends on the roles played by the two variables. The artists' guideline can be represented by an arrow diagram (**Figure 4.5**).

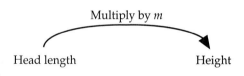

Figure 4.5.
Arrow diagram for head length and height.

From the arrow diagram, you can see that the head-length measurement "explains" the height measurement; height is seven times head length. So, head length is called the **explanatory variable**. In turn, the height measurement "responds" to changes in the head-length measurement. So, it makes sense to call height the **response variable**. According to mathematical convention, you should use the horizontal axis for the explanatory variable and the vertical axis for the response variable.

ACTIVITY

1

USING YOUR HEAD

> Mathematical Convention:
> The explanatory variable is displayed on the horizontal axis, and the
> response variable is displayed on the vertical axis.

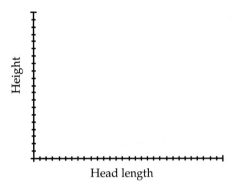

Figure 4.6.
Axes with head length as the explanatory variable and
height as the response variable.

4.a) On graph paper, draw a horizontal axis to represent head length and a vertical axis to represent height (similar to the set of axes shown in **Figure 4.6**).

b) After drawing the axes, you need to decide on appropriate scalings for them. In other words, for each axis you need to decide how many centimeters are represented by the distance between each pair of consecutive tick marks. (For example, if you decide that the distance between two tick marks represents 2 cm, then you should write 2, 4, 6, and so forth next to the tick marks.) The largest value on your scale for the horizontal axis should be somewhat larger than the maximum head length recorded in your table. For the vertical axis, the largest value on your scale should be somewhat larger than the maximum height. Choose appropriate scalings for the two axes, and label the axes accordingly.

c) Plot the data from your table. Describe the pattern made by the points.

d) Draw a smooth line through the points in your plot. Then, write an equation that describes the relationship between height, H, and head length, L, shown by your graph.

e) How does your equation compare to the rule you used to make the original table of values?

ACTIVITY

USING YOUR HEAD

5. Juan decides to draw a picture of a friend standing by a window. He follows the artists' guidelines for drawing 14-year-olds. He makes a preliminary sketch, but he decides that the figure is too small. So, for his final sketch, he draws the head of his figure 1 cm longer than in his preliminary sketch and continues to follow the artists' guidelines.

   a) Suppose the head length of the preliminary sketch measured 8 cm. How much taller than his preliminary sketch is Juan's final sketch? Justify your answer.

   b) Suppose you don't know the head length of the preliminary sketch and decide to represent it by x. Write algebraic expressions involving x that indicate how the head length of the final sketch, the height of the figure in the preliminary sketch, and the height of the figure in the final sketch are related to x.

   c) The change in head length between Juan's preliminary and final sketches, expressed in mathematical shorthand as Δhead length, is 1 cm. What is Δheight, the corresponding change in height?

6. Artists have found that the rule of thumb, "draw a 14-year-old 7 head lengths tall," helps them draw teenagers with heads correctly proportioned to their bodies. But how closely do the dimensions of real teenagers match the ideal relationship suggested by artists?

   a) Within your group, measure each person's head length (from the chin to the top of the head). Record your data in a table similar to the one in **Figure 4.7**. Be sure to specify your units of measure at the top of the last four columns.

Name	Head length	Predicted height	Actual height	Error: Actual − Predicted

Figure 4.7.
Teenagers' head length and height data.

b) Use the relationship "height = 7 head lengths" to predict each person's height. Record the results in your table.

c) Next, measure and record each person's actual height.

In almost every situation in which predictions are made from data, it is useful to examine the **prediction errors**. Prediction errors are defined as the difference between the actual value and the predicted value for each point in your data. That is, prediction error = $y_{actual} - y_{predicted}$.

d) Calculate the prediction errors for your data, and record the results in your table.

e) How well did the relationship "height = 7 head lengths" do in predicting the actual heights of members of your group?

f) Would a multiplier different from 7 do a better job? If so, what multiplier would you choose? What process did you use to determine this multiplier? Why do you think it does a better job than the multiplier 7?

7. The relationship between height and head length changes with age. Therefore, artists adjust their guidelines based on the age of the person they are drawing.

a) When drawing sketches of adults (ages 18 to 50) artists follow this guideline: Draw the figure of an adult approximately $7\frac{1}{2}$ head lengths tall.

Write a formula that describes the relationship between height, H, and head length, L, according to the artists' guideline for drawing an adult.

b) One of the most noticeable features of babies is their large heads. In drawing newborn babies, artists use this guideline: A baby should be approximately 4 head lengths long.

Write a formula that describes the relationship between

ACTIVITY

USING YOUR HEAD

1

the baby's height (which is really body length since babies don't stand), H, and head length, L, according to the artists' guideline for drawing newborns.

Using the artists' guidelines for drawing teenagers, adults, and newborn babies, you can calculate the figure's height by multiplying its head length by a number. Therefore, in each case, the equation describing the relationship between height and head length could be written in the form

$H = mL$, where m is a number.

This is equivalent to saying that the ratio $H/L = m$, a constant. That is, H and L are proportional, and m is the constant of proportionality.

However, in order to use your calculator to graph such a relationship, you will have to rename variables H and L as y and x, respectively. As the arrow diagram in **Figure 4.8** shows,

$H = mL$ and $y = mx$

represent the same relationship. Only the names of the variables have been changed.

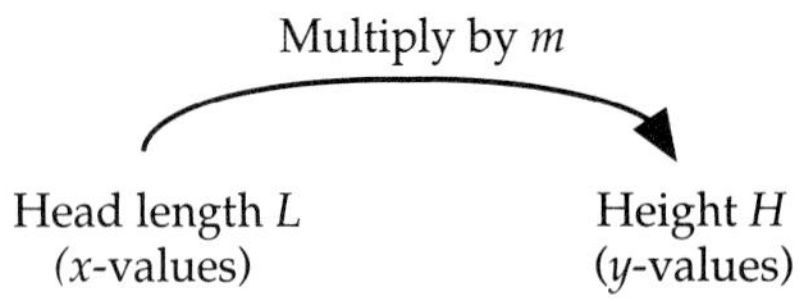

Multiply by m

Head length L Height H
(x-values)   (y-values)

Figure 4.8.
Arrow diagram of the relationship between height and head length.

8. a) Using a graphing calculator, on the same set of axes, graph the equations describing the relationships between height, H, and head length, L, for teenagers (age 14), adults (ages 18–50), and newborn babies. Set your window to [–5, 30]x[–50, 200] (the minimum x-value is –5, maximum x-value is 30, minimum y-value is –50, and maximum y-value is 200). Then make a careful sketch of your three graphs. Be sure that you label each graph with its equation and indicate the scale on each axis.

  b) How are the three graphs the same, and how are they different? What effect does changing the value of the number m have on the graph?

ACTIVITY

1

USING YOUR HEAD

c) Only a portion of each graph describes the relationship between height and head length. Assuming that the head length and height are in centimeters, shade the portion of each graph that you think makes sense in the given context. (Recall that 1 in. = 2.54 cm.) Explain how you decided on the shading for each graph.

9. After completing Item 8, you can use one of the artists' guidelines to make a rough prediction of the height of the person whose skull was found in the preparation reading (either Bones 1 or Bones 2).

a) What assumptions might you make in order to make the prediction?

b) Recall that the skull measured 230 mm in length. Predict the height of the person in cm. Describe the process you used in making your prediction.

c) Convert your prediction to inches. Does your prediction result in a height that is reasonable for a person?

d) Do you think your prediction is likely to be close to the actual height of the person. Why, or why not?

You may not have realized that artists' guidelines for drawing figures can be used to make predictions. In this activity, you have examined relationships in which one variable is a multiple of another variable. Though the variable names may differ, these relationships all belong to the $y = mx$ family. You indicate individual members of the family by stating the choice for m. The remaining problems in this activity will help you to become more familiar with members of this family and their properties.

ACTIVITY

USING YOUR HEAD

1

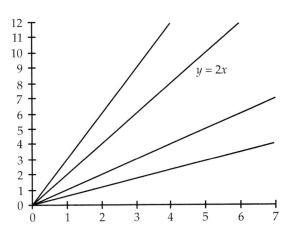

Figure 4.9.
Graphs of four members of
the $y = mx$ family.

10. **Figure 4.9** shows graphs of four equations from the $y = mx$ family. The line corresponding to the choice $m = 2$ has already been labeled with its equation, $y = 2x$.

   a) Look at the line $y = 2x$. What is the value of y when x has value 3? What is the value of y when x has value 4? What effect did changing the value of x from 3 to 4 have on the value of y?

   b) What are the equations for the other three lines? Explain how you determined their formulas.

   c) The mathematical name for the number m in equations of the form $y = mx$ is **slope**. Do you think this is a reasonable name? Explain.

   d) Finally, sketch the graphs of two more members of the $y = mx$ family, $m = -1$ and $m = 0$.

ACTIVITY

1

USING YOUR HEAD

11. Answer the following questions without graphing.

   a) Does the point (3, 4) lie on the line $y = 3x$?
   How can you tell?

   b) Does the line that passes through the points (2, 8) and (5, 20) belong to the $y = mx$ family? If so, what is its equation? If not, why not?

   c) Does the line that passes through the points (3, 9) and (6, 21) belong to the $y = mx$ family? If so, what is its equation? If not, why not?

12. a) Esther bought a present for Michele that cost $40 before tax. If the sales tax in her state was 6%, how much tax did she have to pay?

   b) If the sales tax is 6%, specify an equation from the $y = mx$ family that describes the tax you will pay on an item given its price.

   c) If the tax on an item is $9, how much did the item cost before tax?

By now you should be aware that the graph of any equation from the $y = mx$ family is a line that passes through the origin, the point at which the two axes intersect. The appearance of each line is controlled by the number m.

In the next activity, you work with relationships that belong to the larger $y = mx + b$ family. You can indicate individual members of this family by stating choices for m and b. As you might guess, m and b control the appearance of lines in that family.

INDIVIDUAL WORK 1

Leg Work

The artists' guidelines in Activity 1 allowed you to make predictions, but these predictions were not very precise. In this assignment you work with equations that are a bit more complicated than the artists' guidelines for drawing figures. The added complexity allows you to improve your predictions. You also learn how you can determine whether two different equations lead to the same predictions.

Figure 4.10.
The femur (thighbone).

Recall from the preparation reading and Activity 1 that you need to know two things in order to approximate a person's height based on the length of his skull. First, you must know how much larger a person's head length is than his skull length. Second, you need to know how well the multiplier from the artists' guideline, $H = 7.5L$, works for the relationship between head length and height of real people. How can you test this prediction method to see how well it works? Can bones other than the skull also be used to predict the height of the person from whom they came?

Dr. Mildred Trotter (1899–1991), a physical anthropologist, was well known for her work in the area of height prediction based on the length of the long bones in the arms and legs.

Here is one of the relationships proposed by Dr. Trotter. $H = 2.38F + 61.41$ where H is the person's height (in cm) and F is the length of their femur (in cm).

Note that your bones data are in millimeters, not centimeters.

1. a) Draw an arrow diagram for Dr. Trotter's formula.

   b) Which is the explanatory variable? Which is the response variable? How can you tell?

Dr. Mildred Trotter had a long, distinguished career as a physical anthropologist that included working as a special consultant to the U. S. government during World War II. Her task during the war involved the identification of skeletal remains of servicemen. At the time, she realized that bone size and proportions vary based on age, sex, and racial/ethnic background. Forensic scientists and law enforcement agencies are still using Trotter's formulas for estimating a person's stature based on the lengths of their bones.

(Conroy, Glenn, et. al. 1992)

2. Suppose, for most adults, femurs range in size from about 38 cm to 55 cm. According to Dr. Trotter's formula, how tall is a person with a 38-cm femur? How tall is a person with a 55-cm femur?

3. On graph paper, draw a set of axes similar to that shown in **Figure 4.11**.

   Notice that the horizontal axis is scaled from around 35 cm to 60 cm (slightly wider than the minimum and maximum femur lengths) with tick marks every 5 units. A zigzag has been added to indicate that there is a break in this scale between 0 and 35.

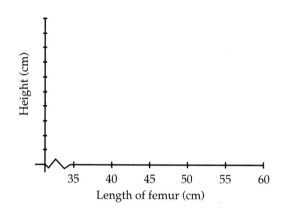

Figure 4.11.
Axes for height and femur length.

   a) Why has the vertical axis been used for height and the horizontal axis for femur-length? Would it matter if you switched them?

   b) Draw a scale on the vertical axis that would be appropriate for data on adult heights (in cm).

   c) Choose three possible femur lengths, and then use Dr. Trotter's equation to calculate "length of femur and height" pairs to complete the entries in the table in **Figure 4.12**.

   d) Next, use the data from your table to graph Dr. Trotter's relationship on the set of axes you have drawn. (Save your graph for use in Activity 2.)

Length of Femur (cm)	Height (cm)
35	
60	

Figure 4.12.
Table of values for Dr. Trotter's equation.

4. Jason's femur measures 40 cm. His brother's measures 41 cm. Predict the difference in the brothers' heights.

5. The femurs of two men differ by one centimeter. Predict the difference in their heights. Explain how you are able to determine your answer even though the lengths of the two men's femurs are not given. In addition, explain how you can read your answer directly from Dr. Trotter's equation.

6. Suppose a woman is 172.7 cm (about 5 ft. 8 in.) tall. Explain how you could use your graph to estimate the length of her femur. What is your estimate?

7. a) The arrow diagram that you drew for Item 1 tells you how to predict a person's height from the length of the femur. Now, reverse the process (similar to decoding in Unit 2, *Secret Codes and the Power of Algebra*) and draw an arrow diagram that tells you how to use a person's height to predict femur length.

   b) Write an equation that describes how you can predict the length of the femur from a person's height. For this equation, which is the explanatory variable and which the response variable?

   c) Now, use your equation from part (b) to predict the length of a woman's femur if the woman is 172.7 cm tall. Compare your answer to the one from Item 6.

As you have just observed, you can use Dr. Trotter's formula to predict height from the length of a person's femur (use $H = 2.38F + 61.41$), and you can also work backwards and estimate the length of the femur using a person's height (use $F = (H - 61.41)/2.38$). The two equations are algebraically **equivalent**. This means they have the same solution: any height and femur-length pair that makes $H = 2.38F + 61.41$ true will also make $F = (H - 61.41)/2.38$ true.

8. Decide if the following equations are algebraically equivalent. How did you decide?

   a) $y + 3 = 2x + 4$ and $y = 2x + 1$

   b) $H + 10 = 5L + 6$ and $H = 5L + 4$

   c) $2y = 4x$ and $y = 2x$

   d) $3y = 9x + 9$ and $y = 3x + 9$

9. The display in **Figure 4.13** represents the equation $2y + 6 = 6x + 12$.

   Imagine the two large circles are pans on a balance scale. The variable y is represented by a square weight, x by a triangular weight, and the units by circular weights. Because $2y + 6$ and $6x + 12$ are equal, the scale is balanced.

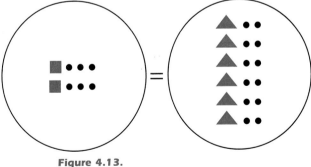

Figure 4.13.
Representation of $2y + 6 = 6x + 12$.

   a) If you took two circular weights from each pan would the scale still balance? If so, what equation would be represented by the weights remaining on the scale? If not, explain why not.

b) If you doubled the weights by putting two square and six circular weights in the left pan and six triangular and 12 circular weights in the right pan, would the scale still balance? If so, what equation would be represented by the weights presently on the scale? If not, explain why not.

c) Use the analogy of the scale to explain why adding the same amounts to both sides of an equation will produce an equivalent equation. Give at least two example equations equivalent to the equation in Figure 4.13.

d) If three squares were added to the left pan and three triangles to the right pan, would the scale still be balance? If so, what equation would be represented by the weights presently on the scale? If not, explain why not.

e) Being careful to keep the scale balanced, add weights or remove weights so that you are left with a single square weight in the left pan. What equation is represented by the weights remaining on the scale? In other words, how is y related to x?

After completing Item 9, you should not be surprised that you can transform one equation into an equivalent equation by

(1) adding or subtracting the same amount to both sides of the equation, or

(2) multiplying or dividing both sides of the equation by the same non-zero amount.

Instead of transforming Dr. Trotter's formula with an arrow diagram, you can apply (1) and (2) to solve Dr. Trotter's equation for F. That means transforming her equation into an equivalent equation of the form F = (expression involving H but not F). Here's how.

$H = 2.38F + 61.41$

Step 1: Subtract 61.41 from both sides of the equation.

$H - 61.41 = 2.23F + 61.41 - 61.41 = 2.38F$

Step 2: Divide the expressions on both sides of the equation by 2.38.

$(H - 61.41)/2.38 = 2.38F/2.38 = F$

Final result: $F = (H - 61.41)/2.38$

10. Another of Dr. Trotter's equations predicts height from the person's tibia: $H = 2.52T + 78.62$.

    a) Write a set of algebraic steps to solve $H = 2.52T + 78.62$ for T.

    b) If a man is 172.7 cm tall, predict the length of his tibia.

11. In yet another equation, Dr. Trotter used both the tibia and the femur to predict height: $H = 1.30(F + T) + 63.29$.

    a) Suppose you want height and tibia length to predict the length of a person's femur. Solve for F symbolically. (In other words, find an equivalent equation of the form $F =$ (expression involving T and H, but not F). Show how to use an arrow diagram in this situation.

    b) If a person is 178 cm tall and has a tibia that is 42 cm long, predict the length of the person's femur.

12. a) Use one of Dr. Trotter's equations to estimate the heights from the femur lengths of Bones 1 and Bones 2.

    b) Use one of Dr. Trotter's equations to help you decide if the tibia described in the preparation reading belongs to Bones 1 or Bones 2.

In Activity 1 and Individual Work 1, you examined and interpreted equations established by artists and by a scientist. You used equations from a famous scientist to estimate the heights of Bones 1 and Bones 2 from the preparation reading. However, the formula you used is accurate only if the bones came from males. Dr. Trotter adjusted her formulas for age, gender, and racial/ethnic background, but you don't know the gender or racial/ethnic background of Bones 1 and Bones 2.

In addition to working with Dr. Trotter's equation, you have learned how to transform an equation into different but equivalent equations. In lessons that follow, you will determine your own equations from data and use your equations to make predictions. It will be helpful to know when two different equations produce the same predictions, that is, when the two equations are algebraically equivalent.

UNDER INVESTIGATION

Unlike the artists' guidelines for drawing figures, Dr. Trotter's equation, $H = 2.38F + 61.41$ (where height, H, and femur length, F, are in cm), is not a member of the $y = mx$ family, but instead belongs to the larger $y = mx + b$ family. You indicate members of this family by choosing values for m and b. (What were Dr. Trotter's choices for m and b?)

You already have seen many members of the $y = mx + b$ family in Unit 2, *Secret Codes and the Power of Algebra*. In Unit 2, you were generally given the choices for m and b or given a situation that allowed you to determine values for m and b. As part of the process of making predictions, you may have a preliminary model (equation) and find that you need to tinker with your model in order to improve your predictions. An understanding of how changes in the values of m and b affect the graph can help you make the needed adjustments to your model. By the end of this activity, you should be familiar with the graphs described by members of the $y = mx + b$ family and understand how your choices for m and b affect the appearance of the graphs.

Recall that Dr. Trotter's equation $H = 2.38F + 61.41$ was designed to work well for a particular population, white males. She later modified her formula to adjust for racial/ethnic background and gender. In order to make appropriate adjustments in models used for prediction, you will need to know how changes in m and b affect the graph. What happens when you make changes to m and b? Complete the following investigation and find out!

Because there are two quantities to change, m and b, it may help to divide the investigation into two parts, as described below.

ACTIVITY

UNDER INVESTIGATION

2

PART I: KEEP m THE SAME AND CHANGE b

(1) Choose a value for m and one for b. What is your equation?

(2) Graph your equation.

(3) Choose several other values for b. What equations correspond to these choices?

(4) Graph several of the equations from (3) and the equation from (2) in the same window.

PART II. KEEP b THE SAME AND CHANGE m

Repeat Part I, reversing the roles of m and b.

1. Use your graphing calculator to investigate how changing the values of m and b affect the graph of a member of the $y = mx + b$ family.

   a) How does changing the value of b affect the graph of a member of the $y = mx + b$ family? Illustrate using several examples. Continue experimenting with choices for b until you know what b controls on the graph.

   b) How does changing the value of m affect the graph of a member of the $y = mx + b$ family? Illustrate using several examples. Continue experimenting with choices for m until you until you know what m controls on the graph.

   c) The numbers m and b are called the slope and **y-intercept**, respectively. Do you think that slope and y-intercept are descriptive names for m and b? Why or why not?

By changing your window settings, you can affect the appearance of a line described by a member of the $y = mx + b$ family without changing the values of m or b. At times, you may want to adjust your window settings to display your graph more

ACTIVITY

UNDER INVESTIGATION

2

effectively. However, you should also be aware that some people, driven by an interest in distorting the truth, will tinker with their window settings until they produce a graph that achieves their purpose. Your understanding of how scale change affects the appearance of the line will help you interpret graphs correctly and avoid being misled by their distortions. The next investigation will help you learn the effects on a graph of changing the maximum settings for the horizontal or vertical axis.

2. In Individual Work 1, you drew a graph of Dr. Trotter's equation by hand. Now you will reproduce your hand-drawn graph using a graphing calculator.

a) Set the viewing window on your calculator to match the scalings on the axes of your hand-drawn graph. (For example, set Xmin = 35, Xmax = 60, Xscl = 5. The *y*-settings will depend on your choice of scale for the vertical axis.) Enter Dr. Trotter's equation into your calculator and then graph the equation. How does your calculator-produced graph compare with your hand-drawn graph?

b) Experiment with changing the scale on the vertical axis by first increasing the value of Ymax and then decreasing the value of Ymax. How would you change the value of Ymax to make the graph of Dr. Trotter's equation appear very steep? How would you change the value of Ymax to make the graph appear much flatter?

c) Without actually changing the scaling on the horizontal axis, predict what would happen to the appearance of the graph if you changed the value of Xmax from 60 to 120. Why do you think your graph will change as you predicted? Finally, check your prediction by changing the Xmax setting from 60 to 120.

ACTIVITY

UNDER INVESTIGATION

2

3. Answer the following questions without graphing the equations.

   a) Which graph is steeper, the graph of $y = 3.48x + 20$ or $y = 5.78x + 5$? How do you know?

   b) Which graph crosses the y-axis at 30, the graph of $y = 30x + 15$ or $y = 15x + 30$? How do you know?

   c) Which graph slants downward as the x-values increase, $y = \frac{1}{2}x + 15$ or $y = -2x + 5$?

In this activity you discovered how modifying a member of the $y = mx + b$ family by changing the value of m or b affects its graph. You also discovered that rescaling can change your perception of how steeply a line rises or falls, even though you are graphing the same equation. This understanding will come in handy when you want to select members of the $y = mx + b$ family to describe patterns in data.

INDIVIDUAL WORK 2

Line Up

This activity gives you an opportunity to practice graphing a line from its equation and determining an equation from its graph. In order to be successful with the material in this and later units, you must be able to switch from one type of description to another.

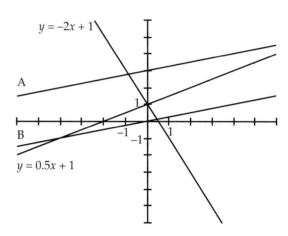

Figure 4.14.
Graphs of four lines.

1. The line corresponding to $y = \frac{1}{2}x + 1$ has already been labeled with its equation. Recall that the value multiplying x, in this case $\frac{1}{2}$, is called the slope of the line.

   a) For this line, what is the value of y when x has value 0? How can you read this information from the equation?

   b) Suppose you change the value of x by 2 units, $\Delta x = 2$. What is the value of Δy?

   c) What is the value of $\Delta y / \Delta x$? How is this ratio related to the equation of this line?

2. Next, look at the line corresponding to $y = -2x + 1$.

   a) What is the slope of this line?

   b) Suppose you change the value of x by 3 units so that $\Delta x = 3$. What is the value of Δy?

   c) What is the value of $\Delta y / \Delta x$? How is this ratio related to the equation of this line?

3. What is the slope of line B? What is its equation?

4. Find an equation describing line A.

5. How are lines A and B alike? How are they different? How are the equations describing lines A and B alike? How are they different?

In Unit 2, *Secret Codes and the Power of Algebra*, you learned how to draw graphs of equations from the $y = mx + b$ family quickly. Below are some practice problems to help you recall the method.

6. Draw the graphs of $y = 2 + 3x$ and $y = 3 - x$ using the following hints.

Hint 1: The first graph starts at $y = 2$ on the y-axis, then goes up 3 units and across 1 unit, then up 3 units and across 1 unit, and so on. The slope of the line is $3/1 = 3$.

Hint 2: The second graph starts at $y = 3$ on the y-axis , then goes down 1 unit and across 1 unit, then down 1 and across 1, and so on. The slope is $-1/1 = -1$.

7. Copy the two grids in **Figure 4.15**, including the letters. Draw the graphs of the following lines on one of the two grids. If you choose the correct grid, then the line will go through one of the letters positioned around the sides of the grid. Then create and complete a table like the one in **Figure 4.16** by entering the letter that corresponds to the Roman numeral of the item. Read the secret message.

I. $y = -2 + x$

II. $y = -2 - x$

III. $y = 3 + 2x$

IV. $y = 2 - 2x$

V. $y = -3 + 2x$

VI. $y = -3 - 2x$

VII. $y = 2 - x$

VIII. $y = 1 + (1/2)x$

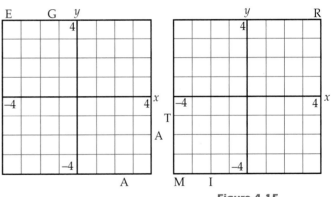

Figure 4.15.
Graphical code.

Item number	I	II	III	IV	V	VI	VII	VIII
Letter								

Figure 4.16.
Decoding table.

8. "Understory" trees are the short trees among much taller trees in a forest or jungle. Their growth is stunted because of the thick vegetation above them. Although understory trees are shorter than other trees, their crowns can be very wide.

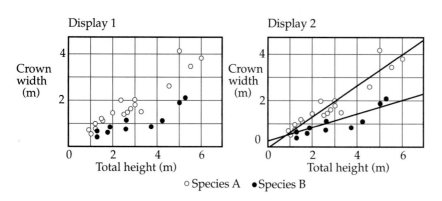

Figure 4.17.
Understory trees in a forest.

Biologists studied two species of understory trees and recorded their measurements in the scatter plot shown in **Figure 4.17**, Display 1. To sharpen the relationship between height and width, they drew lines that they thought described the general pattern of the data for each species of tree. (See Figure 4.17, Display 2.)

a) For each species, predict the crown width when the tree height is 4 meters.

b) For each species, predict the tree height when the crown width is 2 meters.

The two lines in Display 2 are examples of straight-line relationships between two variables. In this case the variables are tree height and crown width. The official name for such relationships are **linear relationships,** and the equations that describe these relationships are called **linear equations.**

c) Which of the two lines in Display 2 can be described by a linear equation from the $y = mx$ family? How can you tell? What is the value for m (approximately)? How did you determine m's value?

d) The other line can be described by a linear equation from the $y = mx + b$ family. (The value of b will not be 0 for this line.) Determine an equation for this line.

e) In your equation for Species B, what does m mean in this context? What does b mean?

Up to this point, most of the linear equations you have worked with have been written in **slope-intercept form**, meaning as members of the $y = mx + b$ family. Suppose you choose $b = 3$. Then, all members of the $y = mx + 3$ family will pass through the point $(0, 3)$. (Why?) Or suppose that you wanted your lines to pass through the point $(4, 3)$? What equations would you use to describe these lines? The key to the answer is contained in the next item.

9. Check that the following lines all pass through the point (4, 3). Two of the equations specify the same line. Which two?

a) $y - 3 = 2(x - 4)$

b) $y = 5(x - 4) + 3$

c) $y - 3 = -2(x - 4)$

d) $y = 2(x - 4) + 3$

e) $y = m(x - 4) + 3$ (Even though you don't know the value of m, you can still check that y has value 3 when x has value 4.)

10. A scatter plot is shown in **Figure 4.18**. Its center is marked with an "X" at (10, 20), and a line is drawn through the X in the general direction of the pattern of dots.

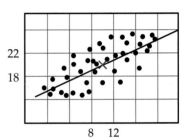

Figure 4.18.
Data with X at center.

a) Explain why the following statement is true: The graph of any linear equation from the $y = m(x - 10) + 20$ family will pass through the X in this scatter plot.

b) Approximately what is the slope of the line? Explain how you arrived at your answer.

c) Determine an equation for the line.

d) Find an equation from the $y = mx + b$ family that is algebraically equivalent to your equation in part (c).

11. Here is a general equation for a line: $y - k = m(x - h)$. x and y are variables; m, h, and k represent constants.

a) What letter matches the slope of this line?

b) What two letters tell you the coordinates of a point on this line?

c) This form of linear equation is called the **point-slope form**. Why is this a good name?

As you have seen in this assignment and those preceding it, linear relationships can be described by words, tables of values, equations, and graphs. In the next activity, you will determine a line that describes the pattern of data in a table. In order to do this, you will need to apply your knowledge about the slope-intercept or the point-slope forms of equations for lines.

ACTIVITY

FROM HEAD TO TOES

3

Length (in.)	Head circumference (in.)
$18\frac{1}{4}$	$12\frac{3}{4}$
$18\frac{3}{4}$	13
$19\frac{1}{4}$	$13\frac{1}{4}$
20	$13\frac{3}{4}$
$20\frac{1}{2}$	14
21	$14\frac{1}{2}$
$21\frac{1}{2}$	$14\frac{3}{4}$

Figure 4.19.
Boys at birth.

Length (in.)	Head circumference (in.)
$17\frac{3}{4}$	$12\frac{3}{4}$
$18\frac{1}{4}$	13
19	$13\frac{1}{4}$
$19\frac{3}{4}$	$13\frac{1}{2}$
20	$13\frac{3}{4}$
$20\frac{1}{2}$	14
$20\frac{3}{4}$	$14\frac{1}{4}$

Figure 4.20.
Girls at birth.

Up to this point, you have worked with linear relationships described either by equations or by graphs. In this activity, you will apply what you have learned about linear equations to develop models based on real data. In addition, you will begin to develop criteria for selecting a good model.

Doctors look at many factors when assessing the health of a newborn. They observe the baby's color, listen to its heartbeat, measure the length of the baby, and check the size of the baby's head relative to the length of its body. Abnormalities in any of these areas might indicate future health problems for the baby. For example, a large head relative to the body might indicate fluid in the brain, while a small head relative to the body might indicate an underdeveloped brain.

As you might expect, body length and head circumference measurements of newborns vary from baby to baby. So, how do doctors determine what is normal? The tables in **Figures 4.19** and **4.20** contain body-length and head-circumference data on boy and girl babies. (The data, taken from a reference book for pediatricians, have been selected from baby measurements ranging from the comparatively small to the comparatively large.) One of your tasks will be to develop models (equations) for predicting a baby's head circumference given its body length. Perhaps your models could be used to help doctors determine if a baby has developed normally.

ACTIVITY

FROM HEAD TO TOES

3

1. Why do you think the data are separated by gender?

2. a) If you want to determine a model that predicts a baby's head circumference from its body length, which would be the explanatory variable, and which the response variable? Justify your answer.

   b) If your model turns out to be a line, what does its slope tell you about babies?

In Items 3–13, for each gender, you will determine two models that describe the relationship between body length and head circumference, one from the $y = mx + b$ family and the other from the $y = mx$ family. Then you will have a contest to decide which of your models is best. Divide the work among the members of your group; half of your group should work on the models for boys, the other half on the models for girls.

3. On graph paper, draw a set of axes, and label each axis with the appropriate variable. Include the units of measurement and a reasonable scale for each variable. Remember to insert a zigzag if you break the scale near zero. Then plot the data from Figures 4.19 and 4.20. A graph of ordered pairs of data is called a **scatter plot**.

4. On your scatter plot, use a ruler to draw a straight line that you think describes the pattern in your data reasonably well.

   a) What is the approximate slope of your line? How did you determine the slope? What does that slope mean about babies?

   b) On your line, determine the coordinates of one point that lies somewhere near the center of your scatter plot.

   c) Now, use your answers to parts (a) and (b) to find an equation in point-slope form that describes your line.

ACTIVITY

3

FROM HEAD TO TOES

d) What member of the $y = mx + b$ family is equivalent to your equation from part (c)?

e) Does the value of b make sense in this context? Explain.

5. a) Enter the data (either for the boys or the girls) into your calculator.

   b) Set your window to match the scaling on the graph that you drew by hand. Make a graph of the line and a scatter plot of the data. Compare the display on your screen with the one that you drew by hand. (If the line is not where you intended, check to see that you have determined the equation of your hand-drawn line correctly.)

Length (in.) Boys	Girls	Predicted head circumference (in.)	Actual head circumference (in.) Boys	Girls	Errors in prediction (in.)
$18\frac{1}{4}$	$17\frac{3}{4}$		$12\frac{3}{4}$	$12\frac{3}{4}$	
$18\frac{3}{4}$	$18\frac{1}{4}$		13	13	
$19\frac{1}{4}$	19		$13\frac{1}{4}$	$13\frac{1}{4}$	
20	$19\frac{3}{4}$		$13\frac{3}{4}$	$13\frac{1}{2}$	
$20\frac{1}{2}$	20		14	$13\frac{3}{4}$	
21	$20\frac{1}{2}$		$14\frac{1}{2}$	14	
$21\frac{1}{2}$	$20\frac{3}{4}$		$14\frac{3}{4}$	$14\frac{1}{4}$	

Figure 4.21.
Predicted head circumference and prediction errors from $y = mx + b$ model.

6. Copy the table in **Figure 4.21** (for either the boys' or girls' data, not both).

   a) Use your model from Item 4 to complete the entries for the Predicted head circumference column. (If you apply your model to the list containing the length data in a calculator list or spreadsheet, you can calculate the entries for the Predicted head circumference column very quickly.)

ACTIVITY

FROM HEAD TO TOES

3

b) Compute the errors—the difference between the actual head circumference and the predicted head circumference—as "actual minus predicted."

7. If an error is positive, what does that tell you about your prediction? What if an error is negative? What if an error is 0?

8. a) Are the errors fairly evenly divided between positive and negative values?

b) Are the errors small in comparison to the head circumferences you are trying to predict?

c) Does your model do a good job of predicting head circumference?

9. a) What is the absolute value of the worst error?

b) In general, if you knew that the absolute value of the worst error was small relative to the values you were predicting, what would that tell you about your model?

10. a) What is the average of the errors?

b) In another situation, would it be possible to have a worst error that is very large but an average of the errors that is close to 0? If so, how could that happen?

Doctors frequently prefer a simpler model, one from the $y = mx$ family. This type of model allows them to look at ratios of head circumference to body length (y/x) and check to see that it is close to the value of m.

11. Determine a model from the $y = mx$ family that you think describes the relationship between a baby's head circumference and body length. Explain how you determined your model.

FROM HEAD TO TOES

12. Complete a table similar to the one in Figure 4.21 for your model from the $y = mx$ family.

13. Set up criteria for deciding what makes one model better than another. Then decide which models describe the pattern in the data better, your models from the $y = mx + b$ family or your models from the $y = mx$ family. Choose one model for the girls and one for the boys. Justify your answer according to your criteria.

14. Bring the two halves of your group together and compare the model you chose to describe the boys' data with the one for the girls' data. Do the relationships between head circumference and body length appear to be different for boy babies and girl babies?

Now that you have completed this activity, you have an idea of what "fitting" a model (an equation) to data is all about. You determined lines that described the patterns of dots in two scatter plots. Your models, the equations of these lines, are called "linear models."

What makes fitting a model difficult is that real data rarely fall exactly on a line. In later lessons, you will learn about methods used by statisticians to specify a "best-fitting line," which will be particularly useful when your data do not fall as close to a line as the data in this activity.

INDIVIDUAL WORK 3

Find the Line

This assignment will give you more practice in finding equations of lines from their graphs.

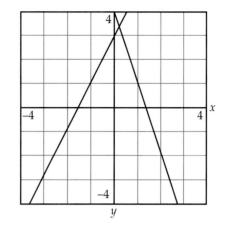

Figure 4.22.
Two lines.

1. Using the point-slope form, write two possible equations for each of the lines in **Figure 4.22**.

2. On a single set of axes, draw these four lines:

    a) $y - 4 = -2(x - 4)$

    b) $y - 4 = \frac{1}{2}(x - 4)$

    c) $y = -2(x - 3) + 1$

    d) $y = \frac{1}{2}(x - 3) + 1$

    What shape do the lines enclose?

3. Make up, and answer, a problem similar to the one you answered in Item 2. (Make sure the equations that you specify enclose some shape.)

4. Play a game using the calculator program *Target* (or another program provided by your teacher). The program shows you a graph of several dots. (For example, the graph in **Figure 4.23** contains 13 dots.) Your goal is to input equations of lines that will intersect as many dots as possible. There are a variety of ways to find a score in this game. For example, you could base your score on a table like that in **Figure 4.24**. The *Target* program uses its own scoring system and calculates your score automatically.

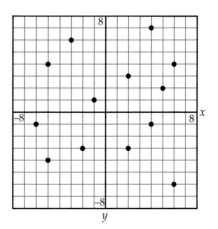

Figure 4.23.
Targets.

Dots on the line	1	2	3	4	5	6	7	8	9	...
Score	1	2	4	8	16	32	64	128	256	...

Figure 4.24.
Table assigning scores to the number of dots on the line.

The rules of the game are:

(a) Each line must be described by its equation.

(b) The line need not go through the center of the dot, it can go through any part of the dot.

(c) Each dot must lie on at least one line.

(d) Your total score is the sum of the scores for each line.

LESSON TWO

You're So Predictable

KEY CONCEPTS

Linear model

Variability

Precision of predictions

Assessing the fit of a model

Mean

Prediction error

Hulton Deutsch Collection/Corbis

PREPARATION READING

It's All Relative

Measurements can be used to make predictions for the past, present, or future. Often the subjects that scientists study cannot be directly observed. As a result, scientists have to take measurements and apply them to rules in order to draw conclusions.

Archaeologists are scientists who study ancient human life. They learn about early humans by studying their bones, fossilized footprints, clothes, and other artifacts. Usually they dig in the earth to uncover bones that have been preserved. Often unable to find a complete skeleton intact, archaeologists must gather information about ancient humans from only a few bones and artifacts.

One example of how an archaeologist might collect information about a specimen is by using a general rule about proportions. For instance, a typical teenage female might have forearms that are 16% of her total height. If you knew the length of the teenager's forearm, you could calculate her height.

CONSIDER:

1. If a teenager's forearm is 16% of her height, does that mean that every teenager has a forearm length that is 16% of their height?

2. Do you think that 16% will be the same for boys and girls?

3. Do you think that adults' forearms are 16% of their total height?

4. Do you think that the relationship between a person's height and forearm length was the same long ago as it is today?

MEASURING UP

So far in this text, your analysis has included parts of a process known as mathematical modeling (this will be covered more formally in Unit 6, *Wildlife*). The modeling process begins when you identify a problem for which you need an answer or a situation that requires further understanding.

In Lesson 1, you learned that during World War II, the armed services sometimes had problems identifying the remains of dead soldiers. Dr. Mildred Trotter was asked to help them. She wondered if there were relationships between the height of a person and the lengths of his long bones.

Having posed this question, Dr. Trotter's next step was to collect relevant data. She needed measurements of people's heights and the lengths of their long bones. Her model $H = 2.38F + 61.41$ expresses the relationship she observed between the height and femur-length measurements from her data.

Because it depends on data, the model is only as good as the quality of the data on which it is based. Dr. Trotter took special care to check that her data were collected by individuals who followed detailed instructions for taking the measurements. In this way, she was able minimize the variability in her data that was due to the measurement process.

Dr. Trotter used lengths of long skeletal bones to predict height. You can't directly measure the bones in your body. Instead, in this activity, you will design methods for collecting data on students' heights and the lengths of their forearms. Later, you will develop a model to predict classmates' heights using the length of their forearms. Then you can compare your equation to the 16% relationship discussed in the preparation reading.

ACTIVITY

MEASURING UP

4

Before you collect your data (height and forearm length from each student in your class), you need to establish a method for taking the measurements. Remember, the worth of your model will depend on the quality of the data that you collect. Everyone who will be doing the measuring must use the same method and then record their data to the same degree of precision (for example, to the nearest eighth of an inch, or to the nearest millimeter).

1. With members of your group, discuss methods for measuring
   (1) the heights of students and
   (2) the lengths of their forearms.

2. Test your methods as follows:

   a) Have two different students measure the height of the same individual using your method. Are both height measurements roughly the same? Are they recorded to the same degree of precision? If not, modify your method and test it again. Keep modifying your method until there is only a reasonably small amount of variation in the measurements taken.

   b) Repeat part (a), but this time measure forearm length.

3. Discuss the methods for measuring height and forearm length. Then select one method. Write a brief description of the method that the class uses to collect the data.

4. Take measurements of classmates' forearms and heights. Record your results on Handout H4.8, *Class Data Recording Sheet*. Leave the last column blank. (You will collect more data from your classmates later.) Be sure to record the units you used for height and forearm length at the top of the appropriate columns.

Note: Save your data for use later in this lesson and in Lesson 3.

INDIVIDUAL WORK 4

Follow in My Footsteps

The length of a person's stride is also related to a person's height. Now you will develop a method for measuring a person's stride. Later you will use your measurements in a model to predict a person's height.

To collect reliable data, you need to carefully plan the method that you will use to collect the data. Remember, your model will only be as good as the data on which it is based.

Design a method for measuring the length of a person's stride.

Here are some items to consider.

How will the person walk? Do you plan to measure from heel to heel or heel to toe? Since step lengths for the same person can vary, does it makes sense to have the person take more than one step and average the results? If so, how many steps should she take?

Determine the measuring instrument (ruler, tape measure, meter stick, etc.) you will use to make the measurement.

Specify the precision of the measurement.

After you have decided on your method, test it as you did the methods for measuring height and forearm length.

When you are satisfied with your method, describe it with a set of written instructions. Give your instructions to a friend to see if someone else understands what you mean. If necessary, revise your instructions. Save them until your class is ready to collect the stride-length data needed for Lesson 3, Activity 10.

ACTIVITY

I PREDICT THAT

5

The data in **Figure 4.25** show the forearm lengths of a tenth-grade class.

FEMALE			MALE		
Name	Forearm length (cm)	Height (cm)	Name	Forearm length (cm)	Height (cm)
Angela	24	157	Ahmed	26.5	173
Bia	24.5	166	Brian	27	177
Carmen	27	164	Daniel	27	174
Chantalle	24	164	Davis	31	192
Emily	23	161	Hiroshi	28	172
Jennifer	27.5	164	Kurt	29	180
Ji-Hyun	27	167	Leon	27	174
Kim	26	162	Luis	28	175
Kirsten	26	175	Max	32	185
Miriam	28.5	166	Nathan	30	185
Tanner	26.5	172	Roger	30	178
Teresa	25.5	176			

Figure 4.25.
Height-forearm data from Class A.

1. Look over the data from Class A. By how much do the heights vary from the shortest student to the tallest?

Recall the general principle that data can be examined graphically. A graphic representation of the height data might help you to assess the amount of variation in student height. Follow the instructions in Item 2 to construct your own **dot plot** for the height data.

ACTIVITY

I PREDICT THAT

5

2. On a piece of graph paper, draw a number line similar to the one in **Figure 4.26**. To make a dot plot, place a dot above each number that corresponds to a student's height. If there are two heights that are the same, place one of the dots directly above the other. Dots representing Angela, Bia, Carmen, and Chantalle's heights have already been marked. Complete the dot plot for the remaining students.

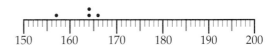

Figure 4.26.
Partial dot plot of height data from Class A.

3. Suppose another tenth-grader joined Class A. Would it be reasonable to predict that the tenth-grader would be between 164 cm and 180 cm tall? Explain your answer.

4. One way to predict a new student's height is to take the average of all the heights.

   a) Find the average of all the student's heights. Mark the average with an "X" on your dot plot.

   b) Suppose the new student is as short as the shortest student in Class A. How far off was the prediction in part (a)?

   c) Suppose the new student is as tall as the tallest student in Class A. How far off was the prediction in part (a)?

   d) Do you think that taking the average helped to make a good prediction? Explain. Can you suggest a better one?

ACTIVITY

5

I PREDICT THAT

5. Notice that the data separate into two groups, one to the left of 170 and the other to the right. Do you think that this separation shows the split in height by gender? **Figure 4.27** shows number lines for use with two dot plots, one showing only girls' heights and the other showing only boys' heights. They use the same scale.

a) On your own paper, draw these number lines and include the data for the two dot plots.

b) What do these dot plots tell you about the heights of the Class-A tenth-grade girls and boys? Do girls or boys tend to be shorter in Class A?

Female

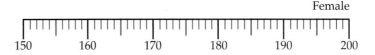

150   160   170   180   190   200

Figure 4.27.
Number lines for
comparative dot plots.

Male

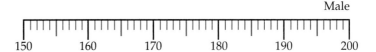

150   160   170   180   190   200

6. Suppose the new student's name is Malisa. Realizing that the new student is a girl may change your prediction.

a) Find the average for girls' heights in Class A to predict Malisa's height. If Malisa is as short as the shortest girl in Class A, how far off is your prediction? What if Malisa is as tall as the tallest girl?

b) Do you think this is a better prediction than the prediction made in Item 4? Explain.

7. Suppose the new student turns out to be Martin (a boy), not Malisa.

a) Choose a method for predicting Martin's height. Give your prediction and describe your method.

I PREDICT THAT

5

b) If the new student's height is somewhere between that of the shortest boy and the tallest boy, what is the largest possible error that could have resulted from your prediction?

Using the height data from Class A, you have computed at least two and possibly three different averages: an average of all the data, an average for the girls, and an average for the boys.

The term **mean** is another name for average. For the remaining items, when you are asked to calculate the mean, just find the sum of the data and then divide by the number of data points. It's no different from calculating an average.

Shorthand notation	Meaning
Σx	The sum of the data.
n	The number of data.
$\bar{x}$	The mean: the sum of the data divided by the number of data.

Figure 4.28.
Table of shorthand.

If your data have been entered into one of your calculator's lists, you can use a built-in calculator command to compute the mean. However, you will need to know some mathematical shorthand in order to understand what your calculator is telling you. (See the table in **Figure 4.28**.)

8. For example, suppose you want to find the mean height of only a small group of students in Class A. After entering the data into your calculator, and pressing a few keys, the screen in **Figure 4.29** appears on your calculator.

a) What is the sum of these data?

b) How many people are in this small group?

c) What is the mean height for the people in this group?

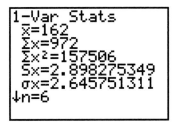

Figure 4.29.
One-variable statistics screen.

9. Now compare the heights of students in your class to the heights of students from Class A.

a) Make two dot plots for your class data, similar to the ones that you made for Item 5(a), one for the boys' heights and one for the girls' heights.

b) Enter the boys' heights and girls' heights into separate lists in your calculator. What is the mean height for the boys? What is the mean height for the girls?

c) Based on your dot plots and the means of boys' heights and girls' heights, do the boys in your class tend to be shorter or taller than the girls?

d) Compare the data from your class to the data from Class A. Describe the similarities and differences between the two data sets.

If you find it helpful, you may use your calculator's built-in statistical capabilities to calculate the means in the remaining items.

10. A researcher gathered data on the number of gray hairs on the heads of 25-year-olds. These are the data she found.

0	23	45	6	8	9	33	15	0	2	4	10
12	13	34	67	40	38	27	25	0	13	34	23
56	34	7	789	44	6	4	0	31	22	5	16
17	11	2	1								

a) Represent these data in a dot plot. (How do you plan to deal with the largest data point?) Then use your dot plot to help you list your data from smallest to largest.

b) Take the smallest ten numbers and calculate the mean (the average) of these ten data. Then take the largest ten numbers and calculate the mean of these ten data. Which of the two means is a better predictor for the number of gray hairs on the head of a random 25-year-old? Justify your answer.

c) Next, calculate the mean using all of the data.

I PREDICT THAT

d) Statisticians use the term **outlier** when referring to data much larger or smaller than the rest of the data. How does the outlier in this data set, 789, affect the mean of the data? To find out, calculate the mean again, this time leaving out the outlier.

e) You have calculated four means, two for part (b) and one each in parts (c) and (d). Which one of these means do you think is the best predictor of the number of gray hairs on 25-year-olds? Why?

11. When the researcher (from Item 10) gathered data on the number of gray hairs on the heads of 20-year-olds, the data looked quite different than for the 25-year-olds. Her data are displayed in the dot plot in **Figure 4.30**.

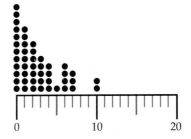

Figure 4.30.
Number of gray hairs on the heads of 20-year-olds.

a) Suppose a 20-year-old student teacher will be visiting your class tomorrow. Predict the number of gray hairs on the student teacher's head.

b) If you had bet money on your prediction, would you prefer to predict the number of gray hairs on the head of a 25-year-old or on a 20-year-old? Why?

If the data have a lot of variability (in other words, the data are very spread out) then it is difficult to make precise predictions. If, instead, the variability in the data is small, so that the data are very concentrated, it is much easier to make fairly precise predictions.

INDIVIDUAL WORK 5

Exercising Judgment

*E*ach of the items in this assignment provides an opportunity to compare data from two groups. When you make comparisons to analyze data, use what you have learned from Activity 5 as well as common sense.

1. The table in **Figure 4.31** lists the weights of babies at birth for two groups of babies. The first group of babies had mothers who never smoked. The second group of babies had mothers who smoked at least ten cigarettes per day. From these data, does it appear that smoking has an influence on a baby's birthweight? Explain your answer.

Figure 4.31.
Babies' birthweights.

Never smoked	6.3	7.3	8.2	7.1	7.8	9.7	6.1	9.6	7.4	7.8	9.4	7.6
Smoked ten or more cigarettes per day	6.3	6.4	4.2	9.4	7.1	5.9	6.8	8.2	7.8	5.9	5.4	6.3

2. Two groups of high school students were asked how much they typically spend on a date. The first group consisted of 12 students who did not exercise; students in the second group exercised at least twice a week. The results of this survey are displayed in **Figure 4.32**.

Figure 4.32.
Cost of a date (dollars).

Does not exercise	10	5	20	4	20	20	15	0	8	40	8	15
Does exercise	15	15	15	5	10	5	5	6	30	25	30	60

a) From the data in Figure 4.32, make two dot plots using the same scaling on each. Place one dot plot directly above the other.

b) What can you learn from your dot plots?

c) Predict the amount spent on a date by a person who exercises. Explain why the average amount spent by the "exercise" group might not be a good choice for your prediction.

d) Complete the following sentences: I predict that a person from the "does not exercise" group will spend _____ on their next date. However, given what this group has spent on dates in the past, this person might spend as little as _____ or as much as _____. So my prediction might be as far off as _____ . (Add any additional comments that you think shed light on your prediction.)

e) Now compare your predictions. Which of the two groups of students spends more on a date? Does your dot plot support the same conclusion?

f) Make a scatter plot of these data. Label the vertical axis "Does not exercise" and the horizontal axis, "Does exercise."

g) Is it valid to claim that there is a direct connection between the exercise and non-exercise groups? Could you use the average amount spent by one in the "Does exercise" category and predict how much the person in the "Does not exercise" category would spend? Explain.

ACTIVITY

6

FOREARMED IS FOREWARNED

In Activity 5, *I Predict That*, you predicted the heights of tenth-grade students based on the heights of the students in Class A. Later, you used your knowledge of whether the student was male or female to improve the precision of your predictions. In this activity, you will return to a question raised in the preparation reading: Can you use the relationship "a person's forearm measures 16% of his or her height" to make accurate predictions of heights of tenth-grade students? You will also use two different methods to develop alternative models for predicting height. Then, you will decide which of the models is the best.

Take a minute to study the data in Figure 4.25 (in Activity 5) and then try to answer the following Consider questions.

CONSIDER:

1. If someone in Class A has a forearm length of 27 cm, what can you predict about their gender and height?

2. If a girl of the same age as the students in Class A has a forearm that measures between 25 and 27 cm, what would you predict for her height? How accurate do you think your prediction would be?

You might have had trouble answering these questions. Even though the data are neatly arranged in four columns, they have not been organized in a way that is helpful for answering the Consider questions. One of your tasks in this activity is to organize the data in a way that shows the relationship between a person's height and the length of his or her forearm.

ACTIVITY

FOREARMED IS FOREWARNED

1. a) On graph paper, represent the data with a scatter plot. Remember to label each axis with its variable and an appropriate scale for that variable. To differentiate the data for the boys from the data for the girls, use two colors, one to represent the girls' data, and the other the boys'.

   b) Now return to the Consider questions and try to answer them using your scatter plot.

2. Use your scatter plot to make the following predictions.

   a) Predict the height of a tenth-grade boy with a 28.5-cm forearm. Explain how you determined your answer.

   b) Predict the height for a tenth-grade student who has a forearm length of 33 cm. How did you do this?

3. The forearm length of a teenager is 16% of her height.

   a) Translate this relationship into an equation that relates forearm length, x, and height, y. Then test a few points to be sure your equation makes sense.

   For the remainder of this activity this equation will be referred to as the "16% model."

   b) Sketch a graph of the 16% model on the same set of axes as your scatter plot.

   c) Is the 16% model true for all the people in Class A? Justify your answer based on your graph.

4. Do you think that the 16% model fits the boys' data or the girls' data better? Justify your answer, based on the model's prediction errors.

ACTIVITY

FOREARMED IS FOREWARNED

6

In Lesson 1, Activity 3 the scatter plot for the baby data (head circumference and body length) fell almost perfectly on a line. That made it fairly easy to draw a line that closely matched the data. However, for the height-forearm data from Class A, the points are much more spread out. Picking a line that describes the data, or makes good predictions for height given forearm lengths, is more difficult in this situation because of the increased variability.

5. Below are two methods to help you select a line that describes the pattern of the height-forearm data from Class A. Divide your group in half. Half of your group should use Method #1 and the other half should use Method #2.

METHOD #1:

Pick a point that appears to lie in the middle of the points displayed in your scatter plot. What are the coordinates of this point? Now anchor your line to this point and adjust the slope until you find a line that you think best describes the pattern of the data. What is the equation of the line you have selected? How did you decide which line was best?

METHOD #2:

Draw two lines in such a way that the points on your hand-drawn scatter plot are bounded as tightly as possible between these lines. (The lines don't have to be parallel.) Now draw one line halfway between the two lines that you have drawn. What is the equation of this line? How did you decide which line was closest to the middle of the two outer lines?

6. In Item 5, your group used two methods to determine a model (equation) that described the data in your scatter plot.

a) Find members of the $y = mx + b$ family that are equivalent to your models determined by your graphs created by

FOREARMED IS FOREWARNED

Methods #1 and #2. How different are the $y = mx + b$ forms of the models determined by the graphs from the two types of methods?

b) Which model, the one from Method #1 or the one from Method #2, appears to describe the pattern of the data better?

c) What were your criteria for choosing which model was better?

d) Using your criteria, does your selected model from part (b) appear to fit the data better than the 16% model? Explain.

7. Use your model from Item 6 (b) or the 16% model (whichever you think is better) to make the following predictions:

a) Predict the height of a student whose forearm is 27 cm. Use the data from Class A to assess the precision of your prediction.

b) Predict the height of a student whose forearm is 33 cm. Do the data provide any clues to suggest how precise this prediction might be? Explain.

c) The forearm lengths of two students differ by 1 cm. Predict how much their heights differ. What if their forearm lengths differed by 2 cm? Justify your answers.

d) What does the value of the slope in your model tell you about people?

The height-forearm data from Class A were fairly scattered about the line that you chose for your model. The amount of scatter (variability in the data) made determining a model that describes the data somewhat difficult. In the next lesson, you will develop criteria that will help you select a single line that "fits" or describes the data as well as possible.

LESSON THREE

Save the Manatee

KEY CONCEPTS

Positive and negative relationships

Linear and non-linear forms

Criteria for model selection

Least-squares line (linear regression line)

Residual plots

The Image Bank

PREPARATION READING

The Endangered Manatee

On the coast of Florida lives the manatee, a large, endearingly ugly, friendly marine mammal. Unfortunately, the gentle Florida manatee is one of the most endangered marine mammals in the United States.

One long-term threat to the survival of the manatee is loss of habitat due to coastal development. Many of the manatees' natural feeding, resting, mating, calving, and nursing areas have been eliminated by construction and water pollution.

The deaths of manatees have steadily increased over the last 20 years. About one-third of all manatee deaths are from human-related causes. Regulations have been put into action to help protect the manatee. Agencies, such as the U. S. Fish and Wildlife Service and the Florida Department of Environmental Protection, study data on manatee deaths annually. Based on trends in the data, these agencies assess the effectiveness of regulations designed to preserve the manatee and make new recommendations to protect the manatee better.

In this lesson you will imagine working for the Florida Department of Environmental Protection. After reviewing data about the incidence of manatee deaths due to accidents with powerboats, you will decide whether to recommend limiting the number of powerboat registrations. Imposing such a limit would not be popular with many boaters, and failing to impose the limit may anger environmentalists. So, you will need to present a strong argument for your recommendation. To make your argument more compelling, you need to include predictions of the effect of imposing or neglecting to impose a limit on powerboat use on the rate of manatee deaths.

CONSIDER:

1. As the number of powerboats in Florida coastal waters increases, would you expect the number of manatee deaths to increase, decrease, or stay the same?

2. If slower speed limits on powerboats are enforced in regions inhabited by manatees, would you expect the number of manatee deaths to increase, decrease, or stay the same?

3. As the human population along the Florida coast increases, would you expect the manatee population to increase, decrease, or stay the same?

ACTIVITY

THE NATURE OF OUR RELATIONSHIP

7

When writing a formal report, authors agree to use certain words and phrases to make sure their intentions are understood. In this

assignment, you will learn to use some standard phrases to make sure that your words are also precise and easily understood. The standard descriptive phrases that you learn will be highlighted in bold throughout the text. These terms are used to describe the direction, form, and strength of the data.

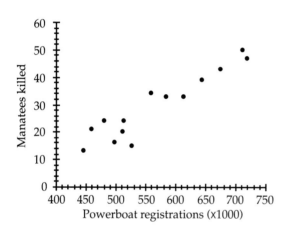

Figure 4.33.
Scatter plot of manatees killed versus powerboat registrations.

One relationship that you will analyze is the connection between the number of registered power-boats and the number of manatee deaths caused by powerboats. The Florida Department of Environmental Protection keeps track of this relationship. **Figure 4.33** shows a scatter plot of the number of manatees killed **versus** the number of powerboat registrations. (The placement of the word "versus" means that the number of manatees killed is the response variable and the number of powerboat registrations is the explanatory variable.)

DIRECTION

1. Describe the relationship between the number of manatees killed and the number of powerboat registrations.

In your description, you probably noted that as the number of powerboats increased, the number of manatees killed also increases. When the response variable and the explanatory variable increase together, the two variables are **positively related**. If instead, one variable decreases while the other increases, then the variables are **negatively related**.

THE NATURE OF OUR RELATIONSHIP

2. a) Suppose the pattern in a scatter plot moves from the upper left to the lower right. Are the two variables in the scatter plot positively or negatively related? Explain.

   b) What if the pattern in a scatter plot moves from the lower right to the upper left? Are the variables positively or negatively related? Explain.

3. Reread the last two Consider questions at the end of the preparation reading. Which of the variables are positively related, according to your response? Which are negatively related, according to your response? Explain your answers.

4. Read each of the scenarios (a and b) below and then make a scatter plot to fit each one. For each scenario, specify which variable is the explanatory variable and which is the response variable. Explain your answer. Also, state whether you suspect that the relationship between these two variables is positive or negative. Explain your answer.

   a) For nine weeks the number of umbrellas sold and the hours that it rained were measured.

   b) For nine weeks in the summer, the amount of ice cream sold and the amount of rainfall were measured.

FORM

So far, most, if not all, of the scatter plots you have drawn have a **linear form**. When a scatter plot has a linear form, it is possible to draw a line that describes the general flow of the data. However, sometimes the data do not fall along a straight line. In that case, the scatter plot has a **nonlinear form**.

How can you tell if your scatter plot has linear or nonlinear form? Here are two analogies that might help:

Linear form:

Imagine a flea walking along a straight line, scattering its eggs on either side of the line. The pattern made by the eggs resembles a scatter plot that has a linear form.

ACTIVITY

10

THE PLOT THICKENS

6. Determine a relationship between height and forearm length for your actual data, first using the entire data set, then using the girls' data, and finally using the boys' data.

   a) Compare the three models. Is there much difference between them? Explain.

   b) Assess the strength of the linear relationship for the model based on the class data and the models based on the single-gender data. (What numeric measure will you use to assess the strength?)

   c) Which model would give more precise predictions if, in fact, the thief were male? What if the thief were female? Justify your answer based on your data.

   d) Make a residual plot for each of the models. Do the dots in the plot appear to be randomly scattered or is a clear pattern apparent? Are there any unusually large residuals?

7. Repeat Item 6 for the relationship between height and stride length.

8. Select the best model for the job of predicting the height of the thief. Support your selection. Finally, use this model as the basis for completing the following conclusion.

   I predict that the thief is _____ cm tall. But the thief might be as short as _____ or as tall as _____.

LESSON FOUR

Remodeling

KEY CONCEPTS

Outliers

Assessing the model

Revising the model

The Image Bank

PREPARATION READING

If Noah Knew What NOAA Knows

Some models that describe relationships are very complicated. However, a good model can be well worth the effort required to create it.

For example, the National Oceanic and Atmospheric Administration (NOAA) has helped hundreds of participants involved in the annual Chesapeake Bay Swim for the March of Dimes. NOAA provides predictions of the velocity of tidal currents along with up-to-date weather forecasts to aid race organizers in setting the optimum start time for the race. In 1992, before the use of NOAA's predictions, only 164 out of 884 entrants successfully completed the race. However, in 1993 after NOAA's involvement, 504 out of 521 swimmers finished the race.

The model that NOAA uses to predict the velocity of tidal currents is considerably more complicated than the linear regression models that you have worked with in this unit.

Although at times there is a need for more complicated models, the linear regression model continues to be very popular and useful.

The linear regression model, however, is not adequate to describe the patterns in all two-variable data sets. In this lesson, you will observe how something as simple as a single **outlier** can greatly affect the equation of the least-squares line. In a collection of data, an outlier is an individual data point that falls outside the general pattern of the other data. In the examples posed in this lesson, determining a good model relies on identification of outliers followed by refitting the regression line to the data that remain. Evaluating the fit of a model and then tinkering with the model to compensate for aberrations in the pattern of the data is part of the model building-process. This is true for simple models such as the linear regression model and for more complicated models such as the one used by NOAA.

CONSIDER:

1. NOAA's model for predicting the velocity of tidal currents depends on more than one explanatory variable. What variables might be useful in predicting the velocity of tidal currents?

ACTIVITY

YOU ARE WHAT YOU EAT

11

Selecting a line according to the least-squares criterion often produces a line with good properties. That's why selecting a line using the least-squares criterion is so popular. However, sometimes this line does a terrible job in describing the pattern of the data.

In this activity, you will explore what happens when you add an outlier to your data. How bad could a single outlier be? Explore and find out!

1. Begin with the data set in **Figure 4.56**.

   a) What is the least-squares line for these data? Make a display that shows a scatter plot of the data and a graph of the line. Then make a residual plot. Based on these plots, does the line appear to fit the data reasonably well?

   b) Next, change the y-value associated with $x = 1$ from 8 to 10. Predict the effect this change will have on the slope of the regression line. (Do you think the slope will increase or decrease?) Explain.

   c) Now, what is the least-squares line? Again, make a display that shows a scatter plot of the data and a graph of the line. Then plot the residuals v. x-values. Based on these displays, does the line appear to fit the data reasonably well?

   d) Compare the slopes of the least-squares lines from parts (a) and (c). Was your prediction in part (b) correct?

   e) Experiment with changing the y-value associated with $x = 1$. How large would this value need to be before the least-squares line turns out to have a negative slope?

   f) Does the least-squares line that you found for part (e) (the one with a negative slope) still appear to fit the data reasonably well? Explain why or why not.

x	y
1	8.0
2	9.0
3	7.7
4	8.6
5	11.4
6	9.7
7	9.2
8	10.9
9	11.6
10	13.4
11	14.6
12	14.5

Figure 4.56.
A data set.

ACTIVITY

YOU ARE WHAT YOU EAT

11

g) What do you think would have happened if the outlier had an *x*-value of 6 (a value in the middle of the domain of the data) instead of 1 (at the edge of the domain of the data)? Experiment and then summarize your findings.

2. What is the relationship between the number of calories a food actually has and the number of calories people think it has? A food industry group surveyed 3368 people, asking them to guess the number of calories in several common foods.

Figure 4.57.
Guessed calories and
actual calories.
(USA TODAY, October 12, 1983)

Food	Guessed calories	Actual calories
8 oz. whole milk	196	159
5 oz. spaghetti with tomato sauce	394	163
5 oz. macaroni with cheese	350	269
One slice of wheat bread	117	61
One slice white bread	136	76
2-oz candy bar	364	260
Saltine cracker	74	12
Medium-size apple	107	80
Medium-size potato	160	88
Cream-filled snack cake	419	160

ACTIVITY

YOU ARE WHAT YOU EAT

11

Figure 4.57 shows a table of the average guessed calories and the actual calorie counts.

a) The goal is to predict the guessed calories from the actual calories. Enter the data into your calculator and make a scatter plot with this in mind.

b) Describe in words the most important features of the scatter plot.

c) Find the regression line for predicting guessed calories from actual calories. Then make a residual plot. Does the regression line adequately describe these data?

d) Would you classify any of the data as outliers? If so, identify them. What do they tell you?

e) If you found outliers, remove them and re-calculate the regression line. Compare your new equation to the one from part (c).

f) Do the actual calories in a food item enable you to predict accurately what people will guess? Explain.

g) Interpret the meaning of the slope of your model for predicting guessed calories from actual calories.

INDIVIDUAL WORK 9

It's All in the Timing

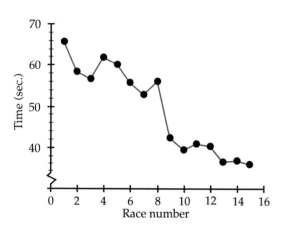

Figure 4.58.
Swim times v. race number.

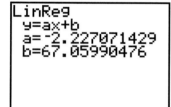

Figure 4.59.
Output from LinReg.

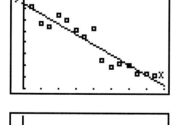

Figure 4.60.
Scatter plot and least-squares line.

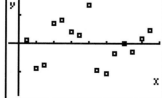

Figure 4.61.
Residual plot.

You will need to use a graphing calculator for the last half of this assignment.

A swimmer who has been competing for many years has a variety of experiences at swim meets. Sometimes she does very well at swim meets. She swims the race faster than she has ever done before and betters her times. But in other competitions, she swims more slowly, adding seconds to her previous time. **Figure 4.58** shows a scatter plot of her times v. the race number. (In other words, "time" is on the vertical axis and "race number" is on the horizontal axis.) To help you see the trend from race to race, dots that corresponded to consecutive races are connected.

1. Describe the general trend for the swimmer's times.

2. Identify three races in which she would have been disappointed with her times.

3. The swimmer's coach gives "Personal Best" ribbons each time a swimmer beats a previously held best time. Identify the races for which she received a "Personal Best" ribbon.

   Figures 4.59–4.61 show three screens of calculator output for the least-squares line and a residual plot.

4. Does the residual plot indicate that the linear regression equation is adequate to describe these data? How can you tell?

5. What is the equation of the least-squares line? (Round the numbers to two decimal places.)

6. What does the slope tell you about the swimmer's racing history?

7. Predict the time for her 16th race. For her 17th race. For her 30th race. How reasonable are your answers?

   The swimmer's favorite stroke is the butterfly. Her times are listed in **Figure 4.62**.

8. Use your calculator to make a scatter plot of the data. Describe the nature of the relationship between time and race number. Are any outliers apparent? If so, describe their general location relative to the non-outliers in the scatter plot.

9. What is the least-squares line for these data? Use your calculator to make a residual plot. Based on the residual plot, does this linear model appear to describe the data adequately? Explain.

10. In the scatter plot that you observed in Item 8, you should have noted two outliers. These correspond to the times for the first two races. (The swimmer had just started swimming butterfly, so her times were unusually slow.) What effect do these points have on the least-squares line? How can the least-squares criterion be used to explain why these points had this effect?

Race number	25-yard butterfly time (sec.)
1	60.81
2	66.11
3	47.32
4	42.69
5	43.40
6	44.82
7	42.67
8	45.17
9	41.20
10	43.68
11	42.47
12	41.74
13	40.40
14	42.90

Figure 4.62.
Butterfly times.

11. How do you think the least-squares line would change if the two outliers were removed from the data? Explain your reasoning.

12. Remove the outliers from the data. What is the equation for the least-squares line now? Again, use your calculator to make a residual plot. Based on the residual plot, does this linear model appear to describe the data adequately (with the exception of the outliers)? Explain.

13. Based on the model from Item 12, predict the swimmer's time for her 15th race. For her 16th race. On average, by how much are her times decreasing from race to race? Can this pattern continue indefinitely? Explain.

14. Which model would the swimmer prefer her times to follow, the model from Item 9 or 13? Explain. Which model is more reasonable, based on the data? Explain.

As you have just seen for the butterfly data, the least-squares line is very sensitive to outliers. One or more outliers have the effect of pulling the line in their direction, thus diverting it from the general pattern of the rest of the data. In these situations, the least-squares line does a poor job of describing the pattern of the majority of the data. In those cases where you can determine that the outliers are "unusual points" that are not representative of the relationship, remove these points and recalculate the equation of the least-squares line using the remaining data. In the situation of the butterfly data, a good argument could be made that the first two times were not "typical" because the swimmer was still learning the butterfly. In this case, it seems reasonable to remove the outliers and refit the model.

One way to adjust your model is to remove outliers that you have identified as unusual points that are not representative of the relationship. Another way is to add another variable to the mix. For example, Dr. Trotter determined equations to predict a person's height from both the femur and the tibia. Her equations for predicting the heights of African-American females from the lengths of their femurs and tibias are given below. (All measurements are in cm.)

height = 2.28(femur) + 59.76

height = 2.45(tibia) + 72.65

15. Suppose the femur of an African-American woman measures 47.4 cm, and the tibia 40 cm.

Use Dr. Trotter's equations to predict the height of the woman. How far apart are your two predictions?

16. Dr. Trotter also proposed more complicated equations for predicting height. For example, in the following equation she combined two explanatory variables:

height = 1.26(femur + tibia) + 59.72

a) Explain how to calculate a least-squares equation similar to the one above if you were given data on height, femur length, and tibia length.

b) Use this more complicated equation to predict the height of the woman.

17. The actual height of this woman is 170 cm. Which of your three predictions comes closest to her actual height?

Now you are able to analyze a problem like identifying Bones 1 and Bones 2. First, you need data similar to the data that Dr. Trotter used in creating her models. Once you have the data, you know how to determine linear regression models based on different explanatory variables. You also have learned a method for selecting the more reliable model from models based on different explanatory variables. You've discovered the effect of outliers and when it is appropriate to remove them and, in this last example, you have been given an idea of how to create a model based on more than one explanatory variable. You have all the tools that you need.

Wrapping Up Unit Four

The data in **Figure 4.63** provide information on people's heights as children and again as adults. Use the data in this table to answer Items 1–4.

1. Suppose one of the people from this study planned to visit your school.

   a) If you find out that the visitor will be a woman, predict her height. How did you decide on your prediction?

   b) What if the visitor is a man?

Girls' height at 1.5 years (cm)	Girls' adult height (cm)	Boys' height at 2 years (cm)	Boys' adult height (cm)
78.0	157.0	89.0	178.0
79.4	158.4	89.9	177.1
80.4	161.4	90.3	179.6
81.3	164.7	90.8	181.8
81.3	160.4	90.9	184.0
82.1	163.7	91.0	180.5
83.2	164.4	91.1	182.0
83.2	170.2	91.2	183.1
83.9	170.5	91.4	180.1
84.9	166.5	91.9	185.1
86.2	171.3	92.9	182.0
87.9	170.7	93.3	186.3
88.2	179.7	94.7	187.4
89.4	176.9	95.4	187.9
90.1	176.9	96.1	189.4

Figure 4.63.
Height data.

2. Create a display that compares the men's heights to the women's heights. Describe in words the information that your display conveys.

3. Suppose you wanted to predict how tall a $1\frac{1}{2}$-year-old girl would be when she reached adulthood.

   a) Which is the explanatory variable and which is the response variable?

   b) Make a scatter plot of the relationship between women's adult heights and their heights when they were $1\frac{1}{2}$ years old.

   c) Would you describe the relationship between women's heights and girls' heights as linear or nonlinear? Positive or negative? Explain.

   d) Fit a least-squares line to the data in your scatter plot. Write its equation, and sketch its graph on your scatter plot.

   e) Make a residual plot. Based on your residual plot, does the least-squares line appear to describe the relationship between women's adult height and childhood height adequately? Explain.

   f) Use your equation to predict the adult height of a $1\frac{1}{2}$-year-old girl who is 82.5 cm tall.

4. a) Determine the least-squares line for predicting men's heights from their heights when they were 2 years old.

   b) If two 2-year-old boys differ in height by 1 cm, predict how much their heights will differ when they are adults.

   c) What if their heights as 2-year-olds differ by 2 cm?

   d) Does the y-intercept of the least-squares line have any meaning in this context? Explain.

   e) Does the slope of the least-squares line have any meaning in this context? Explain?

5. For each description that follows, determine an equation of a line that satisfies it.

   a) The line passes through the point (1, 4) and has slope $\frac{1}{2}$.

   b) The line passes through the points (2, 5) and (3, 2).

   c) The line has y-intercept 5 and is parallel to the line $y = 2x + 6$.

Mathematical Summary

SCATTER PLOTS

In many situations, one is confronted with questions such as "Are values of quantity 1 related to values of quantity 2?" For example, a forensic scientist might ask, "Is height related to femur length?" In general, such questions suggest the use of graphs called scatter plots.

Since the question implies that one quantity might help explain, or predict, values of the other quantity, it is common to refer to the quantities as the explanatory and response variables, respectively. A scatter plot is a graph in which the response variable's values are represented on the vertical axis and the explanatory variable's values are represented on the horizontal axis. This is also referred to as a graph of the response variable versus the explanatory variable.

A scatter plot is an ideal tool in looking for patterns in a relationship between two quantities.

LINEAR RELATIONSHIPS

Linear relationships between two variables can be described by graphs, equations, tables, and arrow diagrams.

- Graphs of linear relationships are lines.

- The amount of "tilt" in the graph of a line is measured as the slope of the line. A line with slope of 0 is horizontal; the further from 0 the slope, the steeper the graph of the line.

- Two common forms of linear equations are:

  the slope-intercept form, $y = mx + b$, and

  the point-slope form, $y - k = m(x - h)$, or equivalently,
  $y = m(x - h) + k$.

Given any two points on a line, you can determine the value of m, the slope of the line, by computing the ratio $\Delta y / \Delta x$ between the two points. Note that in each of these forms, the slope appears as the number multiplying the explanatory variable, x.

- A table represents a linear relationship if a plot of the (x, y)-values lies on a straight line or if $\Delta y / \Delta x$ has the same value for every pair of (x, y)-values in the table.

- Arrow diagrams that consist of a combination of rules of the form "multiply by a number" or "add a number" represent linear relationships. For example, the equation $y = 3x + 2$ can be represented by the

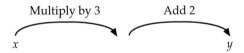

Multiply by 3   Add 2

x y

arrow diagram

EQUIVALENCE

Two equations are equivalent if they have the same solution(s). Two operations, each of which transforms one equation into an equivalent equation, are:

- adding the same quantity to both sides of an equation

- multiplying both sides by of an equation by the same non-zero amount.

For example, the graph of the equation $y - 5 = 3(x - 1)$ is a line that passes through the point (1, 5) and has slope 3. The equation $y = 3x + 2$ is an equivalent equation because using the distributive law on $3(x - 1)$, then adding 5 to both sides transforms the first equation into the second one. Its graph is a line with slope 3 and y-intercept 2. Because graphs are plots of the solutions of an equation and equivalent equations have identical solutions, equivalent equations have the same graphs. So the graph of $y = 3x + 2$ also passes through the point (1, 5). You could verify that fact by substituting the pair (1, 5) for x and y into each equation.

FITTING AND EVALUATING EQUATIONS

The main question of this unit is, "How can you identify and describe a relationship between two variables so that you can predict values of one variable from values of the other?"

First, collect data on the two variables. As noted above, a scatter plot is a useful display for gaining insight into possible relationships. From the scatter plot check the direction (positive, negative, or neither) and the form (linear or nonlinear) of the relationship.

If a scatter plot has a linear form, then you can "fit" a line to the data and use the equation of your line to make predictions. The principal tool in evaluating the fit of your line is the set of residual errors—the differences between the actual and predicted values of the response variable. Different criteria based on the residual errors can be used to determine the "best-fitting" line. Unfortunately, the "best-fitting" line according to one criterion is not always the best according to another. However, a good fit should always have residuals that are randomly scattered around the horizontal axis.

One of the most commonly used criteria for determining the "best-fitting" line is called the least-squares criterion. The least-squares line has the smallest sum of the squared errors (residuals). Also referred to as the regression line, it is popular because it generally does a good job of describing data that have a linear form. However, when outliers are present or when the scatter plot does not have a linear form, the least-squares line can do a very poor job of describing the pattern of a scatter plot.

A plot of the residuals versus the explanatory variable can be very helpful in spotting outliers or nonlinear data. Such plots can display outliers more prominently than a scatter plot of the original data. Also, if the data have a nonlinear form, a residual plot will show a strong pattern.

When outliers are present, removing the outliers and refitting a linear model to the remaining data may produce a better prediction model. However, when data have a nonlinear form, no line will adequately describe the pattern of the data. In this situation, look for a different kind of model.

THE PRECISION OF A PREDICTION

The precision of a prediction is linked to the variability inherent in the data. For example, suppose you had the following data on student heights (in cm): 150, 152, 154, 156, 158. If you were asked to predict the height of a student in this group, you might decide to chose the mean height of 154 cm for your prediction. In this case, the actual height could be as short as 150 cm or as tall as 158 cm; so you could be as far off as 4 cm. You can use a similar approach when dealing with relationships between two variables by examining the variability in the residuals.

CHOOSING BETWEEN TWO LINEAR MODELS

In some situations, you may have two explanatory variables that are linearly related to the same response variable. In this case, it is generally best to base your predictions on the explanatory variable that has the strongest linear relationship with the response variable. Strong relationships have low variability, so one way of determining the strength of the linear relationship is to use the sum of the squared errors. For example, you could select the least-squares line associated with the explanatory variable that has the smaller sum of squared residuals. If the data on the two explanatory variables contain different numbers of observations, then select the least-squares line associated with the explanatory variable that has the smaller average squared error.

Glossary

ABSOLUTE REFERENCE:
In a spreadsheet formula, referring to a cell by its name.

DOT PLOT:
Display in which dots are placed above a number line to represent the values of data for a single variable.

EQUIVALENT EQUATIONS:
Two equations that have exactly the same solutions.

EXPLANATORY VARIABLE:
The variable on which a prediction is based; the variable that "explains" the response variable. Mathematicians frequently use the letter "x" to represent this in noncontextual situations.

LEAST-SQUARES CRITERION:
Choose the line with the smallest sum of squared errors (SSE).

LEAST-SQUARES LINE:
The line that satisfies the least-squares criterion.

LINEAR EQUATION:
An equation relating to x and y that can be put in the form $y = mx + b$.

LINEAR FORM:
When a scatter plot has a linear form, it is possible to draw a line that describes the general flow of the data.

LINEAR REGRESSION:
Fitting a line to data using least-squares criterion.

LINEAR RELATIONSHIPS:
Relationships that can be described using linear equations.

NEGATIVE RELATIONSHIP:
A relationship between two variables in which one variable tends to decrease while the other increases.

NONLINEAR FORM:
On a scatter plot, the general flow of the data is not well described by a straight line.

OUTLIER:
In a collection of data, an outlier is an individual data point that falls outside the general pattern of the other data.

PIECEWISE DEFINED FUNCTION:
A function defined by one equation for part of its domain and by another equation for another part of its domain.

POINT-SLOPE FORM:
$y - k = m(x - h)$; a form for a linear equation where (h, k) is a point on the line and m is the slope of the line.

POSITIVE RELATIONSHIP:
A relationship between two variables in which both variables tend to increase together.

PREDICTION ERROR:
The difference between the actual value and the predicted value (see also residual).

REGRESSION:
Fitting lines or curves to data.

RELATIVE REFERENCE:
In a spreadsheet formula, referring to a spreadsheet cell by its location.

RESIDUAL ERRORS:
Actual value of the response variable minus the predicted value.

RESIDUAL PLOT:
A scatter plot of the residuals versus the explanatory variable.

RESPONSE VARIABLE:
The variable that is to be predicted; the variable that "responds" to changes in the explanatory variable; it changes as a result of a change in the explanatory variable. Mathematicians frequently use the letter "y" to represent this in noncontextual situations.

SCATTER PLOT:
A plot of ordered pairs of data.

SLOPE:
$\Delta y / \Delta x$; the ratio of the change in the response variable for a given change in the explanatory variable, the value of m in the form $y = mx + b$.

SLOPE-INTERCEPT FORM:
$y = mx + b$; a form for a linear equation where m is the slope of the line and b is its y-intercept.

SSE:
The sum of the squared errors.

STRONG RELATIONSHIP:
A scatter plot of the data lies in a narrow band.

WEAK RELATIONSHIP:
A scatter plot of the data does not lie in a narrow band; they are more scattered.

Y-INTERCEPT:
The value of y that corresponds to the intersection of a graph with the y-axis; the value of b in the form $y = mx + b$.

VERSUS:
When used in the phrase y versus x, it describes a scatter plot of y and x in which y is the response variable and x is the explanatory variable.

UNIT

5

Animation/ Special Effects

Animation has always entertained and fascinated people, from the first television cartoons made from artists' plates to the latest computer-generated movies. Virtual reality, a type of animation, is a passport to the world. You can visit faraway and exotic places without leaving your room. Put on the special headset and fly a space shuttle, battle an enemy, design your room, or learn how to drive a car.

The mathematics of geometry and algebra are fundamental to animation. Graphs, equations, and matrices are central to computer animation programs that simulate motion. Coordinates are used to identify the location of an object. Parametric equations represent locations that change with time. Closed-form parametric equations describe a new location in reference to the beginning location, and recursive parametric equations describe location in reference to the previous location. Both forms of parametric equations rely on rate of change and velocity to represent motion. Matrices help organize the data that contain information about a figure and its movements. Matrix operations may be used to slide, rotate, and change the size and shape of animated objects.

Computer programmers use mathematics to write software that allows children, adults, and professional animators to create their own animated cartoons. In this unit, you will use mathematics and program language to create your own simple animation. You will design the instructions that create motion on a calculator screen and, in the process, understand the basics of how computer animation really works.

ANIMATION/SPECIAL EFFECTS

Animation is fun. Cartoons, animated movies, video games and virtual reality engage your attention and captivate your imagination. You laugh at the misfortune of cartoon characters, and you gasp as a beautiful face is transformed into something grotesque. A virtual ride at an amusement park takes you on a terrifying trip through the universe without actually going anywhere.

Animation is also useful. Students learn how to drive cars and pilots learn how to fly planes using simulators. There is no risk of someone getting hurt. Engineers can design race cars, buyers can tour houses, and designers can decorate rooms using virtual reality. Software programs give students and professional animators the tools to make their own cartoons.

As you watch animated movies and play video games, the central question must be, "How do they do that?" How do animators make objects appear to move and change on a picture screen?

Early cartoons used hundreds of hand-drawn frames, or cels. Incredible advances in animation have been made since those early days. Imagination, mathematics, and advanced computer technology have brought to life special effects such as morphing (the gradual transformation from one shape to another) and virtual reality (the creation of the illusion that you see from within the animator's world). Animation is so rich in mathematics and visual stimulation that it makes a wonderful context for investigating a wide range of mathematical ideas.

Engineers and software designers spend years acquiring the knowledge and skill to design animation and virtual reality programs. Use your imagination and math as you enter the world of the animator and create your own simple animation.

Get Moving

KEY CONCEPTS

Coordinate systems

Continuous and discrete representations

The Iamge Bank

PREPARATION READING

A Photo Finish

*I*n a photo finish, two racers arrive at the finish line at the same time. Observers cannot tell who was first. Fortunately, many races are now videotaped. The videotape can be replayed and advanced, frame by frame, until the exact moment the first racer reached the finish line. Each frame is a snapshot, preserving the critical moment.

Videotape captures life in a series of frames or snapshots. When the video is played again at the same speed at which it was taped and you watch the series of frames, you are convinced you are seeing the event again. You are not. The videotape simulates motion by showing the series of frames faster than your eye can detect the changes.

Traditional animation is like video. Hundreds of cartoon frames or drawings, each slightly different from the one before, create the illusion of motion when viewed in rapid succession. Advances in technology have changed traditional animation, bringing new tools and capabilities to the animator.

How does the animator use the computer to create the illusion of motion? What role does mathematics play in the design of the complex computer software used by animators? The key is communication. Computer programs are the set of instructions that tell the computer what to do and what to display. Math is critical to the programming language.

In the process of animation, or simulating motion, you begin with the essential elements and create a simple model. The purpose of this first lesson is to identify the basic elements of animation. In the lessons that follow, you start with a simple model and add complexity. Eventually you will design your own simple animation. The final product will depend on how well you apply the language of mathematics.

CONSIDER:

1. In previous units you have used mathematics to model real-life contexts. You began the modeling process by building a simple model based on the essential elements of the context. "Simplify and conquer" is a characteristic of mathematical modeling. In Unit 1, *Pick a Winner*, the model you used to begin the process was simple majority. What are the simple models you used to begin the modeling process in Units 2, 3, and 4, *Secret Codes and the Power of Algebra, Landsat*, and *Prediction*?

2. Coaches and teachers often focus on basic skills and fundamentals. The soccer coach emphasizes passing and conditioning. The piano instructor teaches students how to read music and place fingers properly. Choose an area of life familiar to you that requires you to learn new skills. What are the three most basic or fundamental skills for the area chosen?

3. Animation simulates movement of objects. What are the basics of simulating motion? What should you learn how to do first?

4. Animation involves creating images on a television or computer monitor. You studied images taken by a satellite in Unit 3, *Landsat*. The images taken by satellite cameras have different resolutions. Resolution is important to animation, too. How does the resolution of the images affect what you see?

THE BIG STRETCH

In your role as animator, you need to know the most effective ways to communicate the specific changes involved in animation to a graphing calculator or computer.

Do this activity with a partner. You and your partner take turns giving and following directions that explain specific motions.

In your notebook or journal, write what you or your partner says, what you do, and how things turn out. Be as accurate in your record-keeping as possible.

Sit back-to-back with your partner.

Using a number 2 pencil, copy one of the images in **Figure 5.1** onto a piece of paper. Then press Silly® Putty on to your pencil drawing. The Silly Putty image can now be stretched or changed.

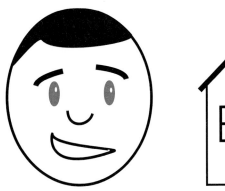 

Figure 5.1.
Smiley face and house.

THE BIG STRETCH

The student with the clay or putty is the "Animator." The other student is the "Drawer." Only the Animator is allowed to talk during this activity. The Animator should begin with the figure placed in the center of a piece of paper. The Drawer may check to see that the correct starting placement is being used, but from that point on neither student may look to see what the other is doing.

ANIMATOR: Make some change in the figure, but be sure that it ends up back on the paper. After the change is made, describe the change to the Drawer. Remember, neither person can look at the other's work, and the Drawer may not talk.

DRAWER: Draw on your paper the results of the Animator's description.

After the drawing is complete, look at each other's work, and record how successful you were. Make special notes of exactly how your figures differed (if they did). Keep a list of descriptions and directions that were most effective. Make another list of directions that were confusing or ineffective.

Repeat the entire activity several times, swapping roles. Save your records for discussion with the rest of the class and refer to your notes as you answer the following items.

THE BIG STRETCH

1. Describe any difficulties you had re-creating the figure changes. Be specific.

2. Select the three or four most useful descriptions from your activity notes. (Example: When the figure was rotated, it helped to know the approximate number of degrees and the direction—clockwise or counterclockwise—of the rotation.)

3. It might have helped to limit the kinds of changes that could be made. Write two or three restrictions that might have made the drawing task easier.

4. Describe how a common grid or numbering system, worked out in advance, would have increased your success.

5. Each frame of a cartoon is slightly different from the previous frame. Explain how a series of small changes is easier to describe than one large or complicated change. If you disagree, explain your reasons.

6. In computer animation, thousands of points change from one frame to the next. Mathematical modeling begins with a simplified version of a problem. How can you simplify the process of changing several points?

7. Real-life motion involves several types of movement. In the process of mathematical modeling you begin with the simplest type of movement. Identify the type of movement you think would be simplest to model. (Examples of movement include rotation, following a curved path, and shrinking in size.)

INDIVIDUAL WORK 1

The Next Frame

1. Describe how the object changes from one frame to the next in **Figures 5.2–5.5**. Draw the next frame in the sequence.

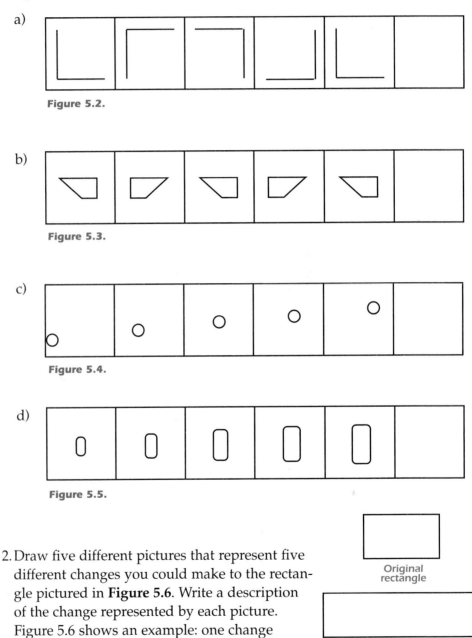

a)

Figure 5.2.

b)

Figure 5.3.

c)

Figure 5.4.

d)

Figure 5.5.

2. Draw five different pictures that represent five different changes you could make to the rectangle pictured in **Figure 5.6**. Write a description of the change represented by each picture. Figure 5.6 shows an example: one change might be a horizontal stretch from the center.

Original
rectangle

Original rectangle with a
horizontal stretch from the center

Figure 5.6.
Original rectangle and stretched rectangle.

3. On your paper, complete five frames of an animation (as shown in **Figure 5.7**) that might show a ball bouncing. Write a description for how to draw the sixth frame.

Frame 1	Frame 2	Frame 3	Frame 4	Frame 5	Frame 6
					?

Draw in the first five frames only.

Figure 5.7.
Blank frames for bouncing ball sequence.

4. The five frames pictured in **Figure 5.8** represent the word "HEAR" on a moving marquee.

| R | AR | EAR | HEA | HE | |

Figure 5.8.
Moving marquee for HEAR.

a) Draw the sixth frame.

b) Describe what is wrong with the marquee.

5. A photocopy machine can be used to enlarge or reduce the size of an object drawn on a page. Describe how, without such a machine, you would draw a house that looks exactly like the one in **Figure 5.9**, only twice as big.

Figure 5.9.
House to enlarge.

6. When you see the reflection of an object in a flat mirror, the reflection usually looks identical to the original object except it is backwards (reversed). When the mirror is rounded, the reflection of the object looks different from the original object. Describe or draw how the reflection of the house in Figure 5.9 would appear if you were viewing it in a rounded mirror. Rounded or spherical mirrors are often

found in hospitals. They are placed in the ceiling where hallways meet so a person walking toward the mirror can see if someone is approaching from a side hallway. You get a similar effect by looking at the reflection of an object in the back of a spoon.

7. Describe how to move the bold line to the location indicated by the fine line in **Figure 5.10**. Assume that the person to whom you are giving the directions does not know the location of the fine line.

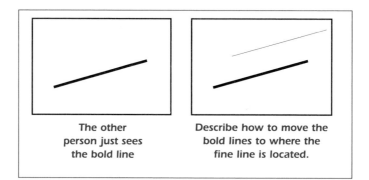

Figure 5.10.
Before-and-after location for line.

The other person just sees the bold line

Describe how to move the bold lines to where the fine line is located.

To describe changes in a geometric object, you need a language that allows you to compare "before" and "after" situations. Geometric ideas related to your work thus far include location and distance. In order to make an understandable description using either of these ideas (location and distance), start with a **reference point** and build a **reference system**. In a reference system, a **starting location** is chosen and a consistent measuring scale is used to indicate another point's distance or displacement from that starting point. The starting location is the location of an object when $t = 0$ in the chosen reference system for time. The reference point is a point that both you and your audience can find easily; it is the zero point in the reference system. For example, you might refer to the lower-left corner of your paper as a reference point. Another name for a reference point is "origin," which you first encountered in Unit 2, *Secret Codes and the Power of Algebra*. When distance and location are described using a fixed point, a reference system is created. Mathematicians frequently use grids and number lines as reference systems. In such grid systems, it is customary to designate the origin as (0, 0).

Another geometric idea related to animation is direction. Examples of words that describe direction are: up, down, left, right, north, southwest, and so forth. Think of the descriptions used during Activity 1. Which did the best job of describing a particular change? Words like "two inches below and one inch to the left of the top-right corner" do a good job of conveying precise meaning. Notice this description uses *distances in a particular direction from a specific point*. Mathematicians agree that location, distance, and direction are key ingredients of a reference system.

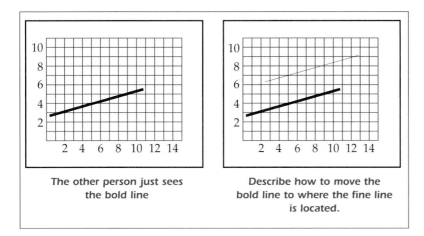

The other person just sees the bold line

Describe how to move the bold line to where the fine line is located.

Figure 5.11.
Move the line using a grid.

8. Repeat Item 7 using a grid reference system (see **Figure 5.11**).

9. Explain why a description that uses a reference system is better than a description that does not use a reference system. If you disagree, explain why.

10. a) Numbers are used in reference systems to determine how far you have moved from a starting location or reference point. A ruler uses numbers to tell how many centimeters or inches you are from one end. Innings are used in baseball to tell how far you have come from the beginning of the game. Often highways have markers that tell how far you have come from the beginning of the highway (or the state line). Number lines are used in mathematics. In such number line systems, left is negative and right is positive. There is nothing special about these designations, but in math and science there is general agreement that these directions are standard.

Describe three other situations in which numbers are used to determine how far you have come from a starting location.

b) Some reference systems use a pair of numbers or letters to describe a location. Street maps use a reference system that identifies a location using a letter and a number. The reference point is usually the upper-left corner of the map. The letter determines how far you have moved down from the reference point in a vertical direction and the number identifies how far you have moved to the right of the reference point in a horizontal direction.

Describe one other situation in which a reference system uses a pair of numbers or letters to determine a location. Identify the reference point.

11. Motion is continuous. Your eyes view motion continuously; there are no gaps or spaces in your viewing (except when you blink). Video is not continuous. Animation and video capture motion in a series of discrete frames.

a) Simulate the way a video camera records motion by using a strobe light and watching someone walk across a room. If you don't have a strobe light you can move a folder back and forth in front of your eyes or blink in a regular pattern as you watch the person walk. Describe the difference between viewing a moving object continuously and viewing an object with a strobe.

b) How does the rate of the strobe affect the way you view motion? That is, what differences do you notice when you view motion with a strobe flashing slowly and a strobe flashing rapidly?

12. You performed Activity 1 in class. Try the same activity at home with someone in your family or a neighbor. Discuss similarities and differences between your classroom experience and your home experience.

LESSON TWO

Get to the Point

KEY CONCEPTS

Coordinate systems

Continuous and discrete representations

Rates of change

Variables and constants

Linear functions

Recursive and closed-form representations

The Image Bank

PREPARATION READING

Simplify and Conquer

A s you discovered in Unit 3, *Landsat*, if you take a close look at a television picture or computer screen, you will see thousands of small pixels. The higher the resolution, the greater the number of pixels. Most computer monitors are 1024 pixels across and 768 pixels high; a total of 786,432 pixels. A motion picture changes at a rate of 30 frames per second (a **frame** is one picture in a series of pictures that change): 23 1/2 million pixels could change every second! That is a lot of changes for a computer animator to make each second.

The process seems incredibly complex. Do what mathematicians do: simplify! "Simplify and conquer" is at the heart of modeling. Whether you animate a message across an electronic sign or simulate a journey through deepest space, all animation begins with a single point. Start by animating one object or even one dot the size of a pixel. Simplify the motion by moving the object or dot in a horizontal direction only. Marquee lights are an example of this type of motion.

How do you tell a computer or calculator to make a dot move to the right on the screen? The activities in this lesson are designed to help you understand and answer this question. You must learn how to move one dot in one direction before you can move entire figures in many directions. Once you have mastered this challenge, you will be ready to tackle more complex movements. Eventually, you will be asked to create a simple animation of your own using the tools and concepts of this unit.

This lesson also introduces the use of a motion "control panel." A control panel summarizes the position and movements of an object. It is a table that includes information about where an object begins, in what direction it is going, and how fast it is moving. Like arrow diagrams, equations, tables, and graphs, control panels are a way to represent information. You will use the control panel when you develop spreadsheets to track moving objects.

CONSIDER:

1. Videos, motion pictures, and animations simulate motion. A motion picture may show you 30 frames in one second, but it cannot show you the movement between frames. How are the scales of a piano like animation and the string on a violin like actual motion; how is a trumpet like animation and a trombone like actual motion? How can something else from your own experience illustrate the difference between something that is continuous and something that is separated?

ACTIVITY

THE LIVING MARQUEE

2

Marquee lights look like a parade of dots all moving in the same direction. Each dot seems to move along in a line. In this activity you simulate the movement of a single dot or point in a marquee.

The purpose of this activity is to determine the mathematical language needed to tell a computer how to move a single point in a horizontal direction.

PART 1:
Simple Marquee

1. Each person in your group needs to stand in a line, shoulder-to-shoulder, facing the same direction (as shown below). For example, you could all stand in a line at the front of your classroom, facing the rear of the classroom. Each person in your line should be holding a "motion card" that is one color on one side and another color on the other. Your task is to make a point appear to move from one end of the line to the other at a predetermined rate. Groups should be at least ten people.

2. Demonstrate your "living marquee" for the rest of the class (or videotape it and play it back). Try doing the simulation with your eyes closed.

3. Prepare a table, graph, equation, or arrow diagram to represent your living marquee. (Hint: Be sure to identify the explanatory variable and the response variable.)

THE LIVING MARQUEE

PART 2:

Creative Marquee

4. Be creative with your living marquee. Write the design for your own unique marquee motion. Here are some examples:

   - Simulate two points that cross or collide.

   - Simulate a line that grows longer as it moves from left to right.

   - Change the rate at which the point moves from one end of the line to the other.

   Demonstrate your creative marquee for the class.

PART 3:

Reflect and Discuss

5. How did you know when to flip *your* card?

6. How do you make the motion look smooth? What could you change to make it smoother?

7. What is the most challenging part of this activity?

INDIVIDUAL WORK 2

Next in Line

1. The members of your class stand in a line and form a living marquee. To make references less confusing, consider the first person in line to be location 1. The second person in line is location 2. Each change in the cards is considered a frame for the animation.

   a) Suppose at $t = 0$ seconds the moving dot is at location 0, which means no person in the line displays the moving dot. If frames change every half-second, where is the dot at $t = 3$ seconds?

   b) If frames change every half-second and the dot begins at location 0 for $t = 0$, when does the dot reach location 16?

   c) Suppose the dot begins at location 5 when $t = 0$ seconds and frames change 10 times each second. Where is the dot at $t = 4$ seconds?

   d) Suppose the dot begins at location 12 when $t = 0$ seconds and changes 2 times each second. Express the relationship between location and time with an arrow diagram that begins with time and concludes with the location.

   e) Express the relationship between location and time as described in part (d) using a symbolic equation. Use L for location (person) and t for time (seconds).

   f) Express the relationship between location and time as described in part (d) by completing a table of sample pairs (see **Figure 5.12**) matching location to time.

Time (sec.)	0	3	5	8	10	15
Location (person)						

   Figure 5.12.
   Table for time and location.

   g) In parts (d–f), you used three different representations to communicate the location of a moving dot. What is an advantage of using equations rather than an arrow diagram or a table?

In Unit 4, *Prediction*, you worked with the concept of slope. In that unit, slope is defined as the ratio of the change in the response variable to the corresponding change in the explanatory variable. Thus, slope is the rate of change of y with respect to x. In Activity 2 the words **rate of change** are used (instead of slope) in reference to how fast the dot appears to move along the line of people. It is a ratio of the change in location to the change in time. If the dot moves from location 15 to location 27 in 3 seconds, then the rate of change is 12/3 people per second or 4 people per second.

Thus, the rate of change is a fraction or ratio that compares the change in one quantity to the change in another. Rates of change are always calculated as:

change in the first quantity / change in the second quantity.

Rates are expressed as units of the first quantity per units of the second quantity.

Every rate of change can be seen as a slope by graphing the first quantity as y on the vertical axis and the second quantity as x on the horizontal axis.

2. Sometimes rates are negative.

   a) What does a negative rate mean?

   b) What does it mean if a company is earning –$1000 per month?

   c) What does it mean if water level is rising at a rate of –3 feet per hour?

   d) What does a negative rate of change mean in the context of animating a point along a horizontal line?

3. Calculate the rates of change.

   a) Suppose you are charged $3.89 for 5 pens. What is the cost per pen?

   b) Reynaldo is traveling at a steady speed. He passes mile marker 78 at 2 minutes after the hour. He passes mile marker 92 at 30 minutes past the hour. Calculate how fast Reynaldo is traveling in miles per minute, then convert your answer to miles per hour.

   c) The Animator's Club had 153 members in April and 216 members by August. Calculate the average rate of increase in members per month.

   d) Angie started the day with 500 figurines to sell. After 6 hours she has 371 figurines remaining. Calculate the average change of her inventory in figurines per hour.

   e) A person orders 12 (cubic) yards of compost to be delivered to his home. The total bill is $200, which includes the cost for delivery. A neighbor pays $300 to have 20 yards of compost delivered. What is the cost per (cubic) yard? What is the cost for delivery?

Animation involves objects that move from one location to another. Recall from Unit 3, *Landsat*, that locations in a digital image (such as a computer monitor or television screen) are called pixels, for "picture elements." Three different rates of change are commonly used in animation:

You can increase or decrease the number of times the animation changes frames in a second. (Most movies are changing at a rate of 30 frames per second.)

$$\frac{\text{change in frames}}{\text{change in seconds}} \quad \text{or frames per second}$$

You can change how far a dot moves (in pixels) from one frame to the next.

$$\frac{\text{change in pixels}}{\text{change in frames}} \quad \text{or pixels per frame}$$

The combination of pixels per frame and frames per second results in pixels per second, or the speed at which the dot appears to move.

$$\frac{\text{change in pixels}}{\text{change in seconds}} \quad \text{or pixels per second}$$

Thus, you have rates involving the change of the location of the object on the screen, how often the screen changes, and the combination of those two rates.

4. Each person in a living marquee is like a pixel on a computer screen. If a dot moves 2 people with each frame (skips over 1 person), that is the same as 2 pixels per frame.

   a) If the dot moves 2 pixels per frame and 30 frames per second, how fast is the object moving in pixels per second?

   b) How far (how many pixels) does the dot move in 3 seconds?

   c) How many frames would it take to travel 94 pixels?

   d) How many seconds would it take to travel 94 pixels?

5. Suppose you build a living marquee with 500 people all standing in a line and holding a card they can flip to simulate the movement of a point or dot.

   a) To find the location of the point after 6 seconds, what do you need to know?

b) Suppose the dot moves two people with each frame (which means it skips a person each frame). Where is the dot after 10 frames?

c) The animation begins with person 7 and advances two people with each frame. Where is the dot after 5 frames? After 8 seconds?

d) Suppose you want each person in a living marquee to know exactly when to flip his or her card. Which representation would you use to communicate the exact time to each person: a word description, a table, a graph, an equation, or an arrow diagram? Explain your answer.

6. Assume a dot moves 3 people each frame (skips over 2 people) and the frames change 2 times each second. The marquee simulation begins with person 9.

a) Where is the dot after 6 seconds? After 15 seconds?

b) Write a symbolic equation to represent the location based on the number of seconds. Let L represent location and t the number of seconds.

c) Prepare an arrow diagram.

d) Complete a sample table of values (see **Figure 5.13**).

Figure 5.13.
Table of sample values.

Time (sec.)	1	2	3	4	5
Location (person)					

e) Use graph paper and the table to prepare a graph of location versus time. The graph will not be a solid line connecting the sample points because you are not graphing real motion, you are graphing an animated simulation.

f) Which representation would you use—word description, arrow diagram, table, graph, or equation—to determine when person P should flip his or her card? Give reasons for your answer.

7. A dot begins at pixel 5 and moves to the right at a rate of 2 pixels per frame in **Figure 5.14**.

Frame 1

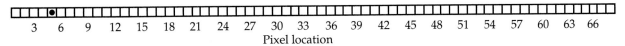

Pixel location

Frame 2

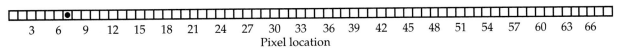

Pixel location

Frame 3

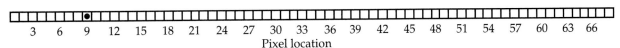

Pixel location

Figure 5.14.
Three frames for moving dot.

a) Write an equation to represent the location L after f frames.

b) Where is the object located at the 12th frame?

c) When does the dot reach location 37? What equation do you solve?

d) If you were to graph location versus time, would the graph be a continuous line or a group of discrete points that form a linear pattern? Explain your answer.

e) Suppose frames change at a rate of 10 frames per second. The point is located at position 5 when $t = 0$ seconds. Write an equation to represent location L at time t in seconds.

f) Where is the point located after 30 seconds?

g) When does the point reach location 173? What equation do you solve?

h) If you graph location versus time in seconds, would the graph be a continuous line or a group of discrete points that form a linear pattern? Explain your answer.

8. An object moves continuously in a horizontal direction from left to right at a rate of 5 feet per second. At $t = 0$ seconds the object is located 36 feet to the right of the origin or reference point.

   a) Use an arrow diagram to show how to find the location of the object at any given time.

   b) Write a symbolic equation for location in terms of time.

   c) Will the graph of location versus time be a continuous line?

   d) Graph location versus time on graph paper.

9. The equation $L = 5d + 100$ represents the location in miles from a known landmark for a group of hikers after d days.

   a) The explanatory variable is d. The response variable is L. One of the constants or control numbers in the equation is the number 5. What is the meaning of the number 5 in the context of hikers?

   b) What is the meaning of the constant 100?

   c) The graph of $L = 5d + 100$ is linear. How does the graph change if you change the constant 100 to 60?

   d) How does the graph change if you change the constant 5 to 8?

   e) Suppose you replace the constant 5 with –3. Explain the meaning of the negative in this context?

10. Old cartoons used flip books or flip frames. Each page of the flip book is slightly different than the previous frame, so when viewed in rapid succession, the objects appear to move.

    a) On your paper, draw five frames (as pictured in **Figure 5.15**) to show an arrow moving horizontally.

Frame 1	Frame 2	Frame 3	Frame 4	Frame 5

Figure 5.15.
Pattern for arrow sketch.

    b) Is it possible to write an equation or make a table to represent the horizontal movement of the arrow? Explain your answer.

11. Suppose the equation L = 3t + 27 represents a living marquee with t measured in seconds and L in people.

    a) How does the person at location 42 know when to flip her card?

    b) Some members of another living marquee take a different approach. They know to watch the person before them in the line and flip the card exactly 1 second after that person. Write a word equation to represent this approach.

12. Describe how a two-step coding process is similar to a living marquee.

Recall equations of the form $y = mx + b$ (from Unit 4, *Prediction*) and $c = 2p + 4$ (from Unit 2, *Secret Codes and the Power of Algebra*). They are examples of **closed-form equations**, which are equations that allow you to calculate the location at any time in just two steps.

This two-step process can be represented using an arrow diagram (see **Figure 5.16**).

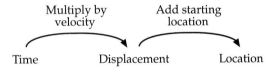

Figure 5.16.
Arrow diagram to find current location.

Notice the roles played by **velocity** and starting location. "Velocity" is the speed at which an object moves in a specific direction. Multiplying the velocity by the time gives you **displacement**. Displacement is distance in a particular direction.

    displacement = velocity × time

Adding displacement to the starting location gives the current location.

    current location = starting location + displacement

In describing motion using formal terminology, velocity is used instead of "speed" and displacement is used instead of "distance."

The essential difference in these terms is that speed and distance are generally thought of as always being positive, while velocity and displacement may be either positive or negative, with the signs indicating a direction.

Suppose you want to write a closed-form equation to represent the location of an object at a particular time. The starting location of the object is 15 miles from the reference point. The object moves in a horizontal direction at 30 miles per hour.

The word equation representing displacement is:

displacement (miles) = 30 (mph) × time (hours)

The closed-form word equation for location is:

current location (miles) = 15 (miles) + 30 (mph) × time (hours)

If the letter x represents current location in miles and the letter t represents time in hours, then the symbolic equation for location is:

$x = 15 + 30t$

You are encouraged to use both word equations and symbolic equations. When you use symbolic equations make sure you understand the meaning of the letters used as variables. In the equation $x = 15 + 30t$, you should be able to state that t represents time in hours and x represents location in a horizontal reference system that uses miles.

13. Suppose $x = 3 + 2.5t$ represents the horizontal position at t seconds for an object moving from left to right on a grid measured in pixels.

a) What is the starting location for the point?

b) The point (8, 23) is a point on the graph. Explain the meaning of the point (8, 23).

c) What equation would you solve if you wanted to know when the point reached the location $x = 53$? Solve the equation to find the time.

d) In Unit 4, *Prediction*, you studied the concept of slope. What is the slope of the line representing the equation $x = 3 + 2.5t$? What is the meaning of the slope?

You can graph the closed-form equation $x = 3 + 2.5t$ on the graphing calculator, but you must be careful. Most calculators require that you use x and y as the variables, where x is the explanatory variable and y is the response variable. In the equation $x = 3 + 2.5t$, t is the explanatory variable instead of x and x is the response variable instead of y (see **Figure 5.17**):

Figure 5.17.
Diagram comparing time-equation variables to calculator-equation variables.

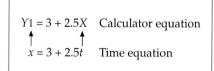

e) Graph the equation $x = 3 + 2.5t$ on the graphing calculator.

f) The object is moving horizontally, but the graph of $x = 3 + 2.5t$ is not horizontal. Why not?

CONSIDER:

1. In Unit 2, *Secret Codes and the Power of Algebra*, you used variables to represent the position number and coded value. In Unit 4, *Prediction*, you used variables to represent several different quantities including arm length, time in years, and powerboat registrations. In the context of animation, what are some concepts that you might represent with variables?

2. In Unit 2, you used word descriptions, arrow diagrams, tables, graphs, and equations to communicate a coding process. What are the advantages of using equations to communicate a coding process? Explain how equations may be more useful than other representations in the context of secret codes.

3. How might equations be useful in the context of computer animation?

ACTIVITY

3

PIXILATION

How do you tell a computer to change thousands of points and produce animation? The task is complex. Answer a simpler question first: How can you make one dot move in one direction on a graphing calculator? Once you have accomplished the simpler task of moving a single point, you can work you way up to more complicated animations involving thousands of points.

This activity is an important transition from the physical world of the living marquee to the symbolic world of the graphing calculator. You need to develop equations to simulate the motion you acted out in Activity 2. To do that, you need to understand the mathematical concepts and skills required to create a simple animation using a graphing calculator or computer.

Imagine you are viewing the animation of two points moving across a calculator screen. You are able to freeze the motion and examine the position of both points frame by frame. The movement of the two points has been summarized in the tables in **Figure 5.18** and recorded in the pictures or frames shown in **Figure 5.19**. A calculator program was used to create this simple animation. Can you figure out how to design instructions for a graphing calculator to animate a single point?

Figure 5.18.
Frame-by-frame locations for two moving points organized in a table.

Time (frame)	2	3	4	5	6	7
First point location (pixels)	(6, 4)	(7, 4)	(8, 4)	(9, 4)	(10, 4)	(11, 4)

Time (frame)	2	3	4	5	6	7
Second point location (pixels)	(8, 4)	(10, 4)	(12, 4)	(14, 4)	(16, 4)	(18, 4)

PIXILATION

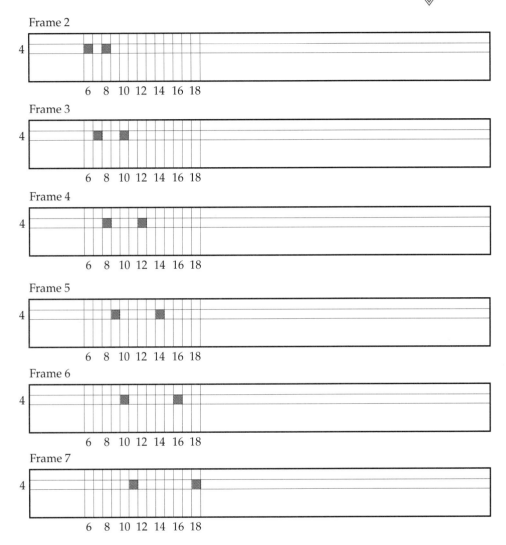

Figure 5.19.
Frame-by-frame views of two moving points.

Each picture represents a magnified view of a portion of a
calculator screen. In Frame 2, one point is located at (6, 4) and a
second point is located at (8, 4). In Frame 3, the first point is now
located at (7, 4) and the second point is at (10, 4). Frame 1 has
been omitted so you can reconstruct it later.

ACTIVITY

3

PIXILATION

1. Equations use constants and variables to represent a relationship between two quantities. Examine the movement of the first point in Figures 5.18 and 5.19.

   a) What are the quantities that vary?

   b) What relationships remain constant?

2. a) Use constants and variables to describe the motion of each point.

   b) Describe how to find the location of each point in frame 8.

   c) Describe how to find the location of each point in frame 30.

   d) When you simulated the motion of a point in Activity 2, you needed to know exactly when to flip your card. The graphing calculator must know exactly when to turn a pixel on and off to simulate the motion of a point. Translate your descriptions from Activity 2 and your description in part (a) into word equations. For each point in Figure 5.18, write a word equation that describes exactly when to turn the correct pixel on and off on the graphing calculator screen.

   e) Use your word equations in part (d) to write symbolic equations that represent the location for each point with respect to the frame. State the meaning of each letter used as a variable.

In Activity 2, you used at least two kinds of descriptions to determine the location of the moving point (or when a person turned his card around). You have practiced translating your descriptions into equations so the motion may be simulated. Compare your descriptions and equations with the following reading.

MOTION EQUATIONS

Two forms of equations are used to represent the current location of a point. The first is the closed-form equation, which you reviewed at the end of Individual Work 2. You may have used a closed-form description in Activity 2.

PIXILATION

The closed-form pattern for the motion of a point is

current location = starting location + velocity × time.

Another kind of description that may have come up during discussion of Activity 2 is called a recursive description. "Recursive" means that you describe the value of a variable in terms of the previous value of the same variable.

The second type of equation, the **recursive equation**, allows you to calculate the current location by adding the displacement during one time period to the previous location. The recursive form is:

current location = previous location + displacement

If the displacement in one time period is +3 units, the recursive equation is:

current location = previous location + 3

You must indicate the starting location when you use a recursive form.

For example, consider an object for which the starting location is 15 miles and the velocity is 30 miles per hour. The recursive word equation is:

current location (miles) = previous location (1 hour ago) + 30 miles (the displacement in 1 hour). The initial location (0 hour) is 15 miles.

The variables used for symbolic recursive equations are different from the variables used for closed-form equations.

Let $x_{current}$ represent current location and $x_{previous}$ represent the previous location, then the symbolic recursive equation for location is:

$$x_{current} = x_{previous} + 30, \, x_{initial} = 15.$$

Notice that x is used to represent location, but a different word or subscript is used to distinguish the current location from the previous location.

> The step or time increment from previous time to current time is 1 hour.
> $x_{initial}$ represents the location at time 0.

ACTIVITY

PIXILATION

3

Mathematics often uses subscripted variables in recursive equations. x_t is the symbolic form for current location at current time and x_{t-1} is the symbolic form for previous location 1 unit of time before the current time. x_0 is the symbol for initial, or location at 0 time. The most common symbolic form is:

$$x_t = x_{t-1} + 30, x_0 = 15.$$

Remember, when you use symbolic equations, make sure you understand the meaning of the letters used as variables. In the equation $x_t = x_{t-1} + 30$, you should be able to state that x_t represents current location in miles and time in hours (if miles and hours are the units of measure) and x_{t-1} represents previous location exactly 1 hour before the current time and location.

Closed-form and recursive equations are used to model motion and identify location. Both are used in computer and calculator programs to model other situations including total distance, number of people, total time, cost, revenue, and profit. You are expected to use both forms throughout this unit. You will discover when it is better to use closed-form equations, when it is better to use recursive equations, and when it may simply be a matter of personal preference. You need to understand both forms for your work in other contexts in this book; they represent two more tools in your mathematical toolbox.

3. Refer to the table in Figure 5.18.

   a) Write a closed-form equation you could use to find the horizontal location of the second point when time is measured in frames. Let t represent time, the explanatory variable, and let x represent horizontal position, the response variable.

   b) Write a recursive equation you could use to find the horizontal location of the second point when time is measured in frames. Write the recursive equation in words or let x_t represent the current position and x_{t-1} represent the position one frame earlier. Use x_0 to identify the starting location.

ACTIVITY

PIXILATION

3

c) Would you use a recursive or closed-form equation to find the location of the second point in frame 50? Give reasons for your answer.

The next five items focus on using equations to represent the motion of points. Work with your group to solve and explain at least two of these. Share your results with the rest of the class.

4. Write a closed-form equation to represent a point that moves vertically 2 units per second and begins at location (3, 5).

5. Write a recursive equation and a closed-form equation in which the second point in Figure 5.19 begins at the same starting location and travels in the opposite direction. Assume that frame 1 is the starting point at time = 0 seconds.

6. Suppose you have two points moving at the same time. One point is moving at a rate of 1 pixel per frame, and the second is moving twice as fast. One point moves down in the vertical direction. The second point moves to the right. Both points start at (6, 8) at time = 0 or frame 0. Write equations or prepare tables to represent the motion of the two points.

7. Model a chase situation. The moving target begins at (13, 8) and moves horizontally at 1.5 meters per second. The chaser begins at (5, 8) and moves horizontally at 2.0 meters per second. Write closed-form or recursive equations to represent the horizontal motion for both points. When and where will the chaser catch the target?

8. In every example so far a particular point has moved at a constant speed. The table in **Figure 5.20** represents the horizontal location of a point measured in meters. Identify explanatory and response variables, and determine the best-fitting line for these data. Use the line of best fit to approximate the horizontal location of the point at $t = 6$ seconds.

Time (sec.)	0	2	3	5	8	10
Horizontal location (m)	8	10.5	12	15	20	22

Figure 5.20.
Table to represent location of a moving point.

INDIVIDUAL WORK 3

Lateral Movement

1. How is measuring the motion of an object in seconds different from measuring the motion of an object in frames as in Figure 5.19?

2. Refer to Figures 5.18 and 5.19. Suppose frames change every second, beginning with frame 1 at 0 seconds.

   a) Describe how to determine the location of both points at 7 seconds.

   b) Where are both points located at 2.5 seconds? How do you know?

   c) Suppose frames change at a rate of 2 frames per second, beginning with frame 1 at time 0. Where are both points located at 2.5 seconds? How do you know?

   d) Using a reference system with animation can lead to confusion. If you measure time in seconds, then you should be able to determine the beginning location of an object at time = 0 seconds. The confusion comes when you measure time in frames of an animation. You must decide if the movement of the object begins with frame 1 or frame 0. Many reference systems use the number 0 as the starting point (also known as the reference point or origin). What are some reference systems for which the reference number is not 0, or 0 is not used because it doesn't make sense?

3. The table in **Figure 5.21** shows sample locations for a point moving steadily horizontally.

Figure 5.21.
Table for point moving
steadily horizontally.

Time (sec.)	2	3	4	6	10	13
location (pixels)	19	23	27	35	51	63

   a) Where was the point when it started moving at $t = 0$ seconds? How do you know?

   b) Where is the point at $t = 8$ seconds?

   c) Where is the point at $t = 4.5$ seconds? How do you know?

   d) Write a closed-form equation to show location with respect to time.

e) The graph representing location versus time for this moving point passes through the point (9, 47). Explain the meaning of the point (9, 47) in this context.

f) What equation do you solve to find the time when the object is located at pixel 67? Solve the equation to determine the time.

g) Write a recursive equation using the format $x_{current} = x_{previous} + b$. Remember to designate the starting location using $x_{initial}$. Count time in intervals of 1 second.

h) Which equation would you use, closed-form or recursive, to find the location of the point at $t = 30$ seconds?

i) Suppose the rate of the point changed to 6 pixels per second. Write the closed-form equation, and describe how the graph of location versus time would change.

4. The horizontal location of an object with respect to time is represented by the arrow diagram in **Figure 5.22**. Use x for location in meters and t for time in seconds.

Figure 5.22.
Arrow diagram to represent location of a moving object.

a) Where is the object when the motion begins?

b) What is the rate of change of location with respect to time?

c) Where is the object located after 12 seconds? How do you know?

d) Write the closed-form equation to represent horizontal location with respect to time.

e) When does the object reach the location 200 meters from the reference point? What equation do you solve?

f) Draw an arrow diagram to show the process used to solve the equation in part (e).

From your work so far, you know that when a point moves in a horizontal direction at a constant velocity, the general equation that represents horizontal location x with respect to time t is $x = a + bt$. The letters x and t are the variables representing location and time—quantities that change throughout the motion. The letters a and b are the constants or control numbers because they represent starting location and velocity and do not change during the motion. If you change the value of a, then you change the starting location of the point at $t = 0$. Change the value of b, and the point moves at a different rate.

Starting location (m)	Rate (mps)
24	8

Figure 5.23.
Motion control panel.

Starting location (pixel)	Rate (pixels/sec.)
52	6

Figure 5.24.
Control panel for 5(a).

Starting location	Rate

Figure 5.25.
Blank motion control panel for 5(b).

Figure 5.26.
Table comparing distance to cab fare.

The equation $x = 8t + 24$ represents a point that starts at location 24 meters and moves at a rate of 8 meters per second. The constants or control numbers, 24 and 8, can be inserted in a control panel like the one pictured in **Figure 5.23**. The control panel is another way to identify the key numbers in a process. It may be incorporated into a spreadsheet program or a computer program (see Item 20). In motion, starting location and velocity play controlling roles.

5.a) Write a closed-form equation with starting location and rate as identified in the control panel in **Figure 5.24**.

b) Complete a motion control panel (see **Figure 5.25**) to match the equation $x = 1.2t + 17$. The units of measurement are miles and minutes.

c) Write a symbolic equation and complete a control panel to match the table in **Figure 5.26**. Use d for distance and f for fare.

Distance (mi.)	0	0.1	0.2	0.3	0.4	0.5
Cab fare ($)	$1.50	$1.85	$2.10	$2.45	$2.80	$3.15

6. A point moves in a horizontal direction at a rate of 12 pixels per second. At $t = 2$ seconds, the point is located at pixel 35.

a) Where was the point located at $t = 0$ seconds, before the motion began? How do you know?

b) Prepare an arrow diagram, and write a symbolic equation for location in terms of time.

c) Copy and complete the control panel in **Figure 5.27**.

d) Complete the arrow diagram from part (b) to show the inverse process used to find the time given the location.

e) When does the point reach pixel 167? What equation do you solve?

Starting location	Rate

Figure 5.27.
Control panel for 6(c).

f) There is a problem with finding the time when the point is at pixel 88. Explain the problem.

7. The equation $x_t = x_{t-1} + 8$ is a recursive equation to represent location x_t at time t hours based on the location the previous hour, x_{t-1}. At $t = 0$ hour the point is located at mile marker 24, that is, $x_0 = 24$. The moving object is a bicyclist traveling at a constant velocity.

a) Copy and complete a table of sample values (**Figure 5.28**) representing location with respect to time.

Time (hr.)	0	1	2	5
Location (mi.)				

Figure 5.28.
Table of sample values for traveling bicyclist.

b) In case you have not worked with subscripts before (the 0 in x_0 is a subscript), when you find the location at $t = 1$, your answer is the value for x_1. The value for x_2 is the location when $t = 2$. What is the value of x_6?

c) How can you find the location at 50 hours? That is, what is the value of x_{50}? Explain why the answer does not make sense.

d) Write a closed-form equation to represent the location of the bicyclist.

e) Suppose you start at the same location but double the velocity. Write the recursive and closed-form equations that incorporate this change.

8. The closed-form equation $p = 21 + 2.5m$ represents the number p of a page a student is beginning in a book she is reading, based on the number of minutes m she has read.

a) On what page is the student starting?

b) What page is the student on after 1 minute?

c) What is her reading rate (with correct units).

d) How long will it take the student to reach the end of the book (page 495)? What equation do you solve?

e) Write a recursive equation using 1-minute intervals.

9. Practice converting from closed-form equations to recursive and recursive to closed-form.

a) Write a closed-form equation for: $x_t = x_{t-1} + 12$. $x_0 = 15$.

b) Write a closed-form equation for: $x_t = x_{t-1} - 28$. $x_0 = 278$.

c) Write a closed-form equation for: $x_t = x_{t-1} + 3.8$. $x_0 = -11.4$.

d) Write a recursive equation for: $x = 4t + 72$.

e) Write a recursive equation for: $y = 198 - 2t$.

You may recall the discussion of velocity, speed, and displacement in Individual Work 2. Duration is the length of time for a process or animation. Displacement is the distance a point moves in a particular direction. Velocity is the displacement divided by the duration. Speed, distance, and duration are always non-negative (positive or zero) numbers. Displacement and velocity may be positive or negative with the signs indicating a direction.

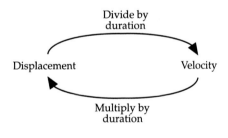

Calculate velocity by dividing displacement by duration. Calculate displacement by multiplying velocity by duration (see **Figure 5.29**).

If the velocity of an object is –75 mph, then the object is moving at a speed of 75 mph in a negative direction. If the displacement of a moving object is –8 km, then the object has traveled a distance of 8 km in a negative direction.

Figure 5.29.
Arrow diagram for calculating velocity and displacement.

10. A point moves 10 pixels each frame and the frames change at a rate of 20 frames per second.

a) What is the rate of the point in pixels per second?

b) Find the displacement of the point when duration is 9 seconds.

c) Explain why the answer to part (b) may not make sense.

d) Find the displacement of a point that begins at pixel 28 and moves 6 pixels to the right with each frame; the frames change 15 times each second, and the animation lasts 2.3 seconds.

e) A point moving at a rate of 8 pixels per second has a displacement of 164 pixels. Calculate the duration.

11. "Current" and "next" are descriptive words sometimes used in a recursive process instead of current and previous.

$x_{next} = x_{current} + 2$. The object is at location 10 by the fourth frame ($x_4 = 10$).

a) Where is the object by the eighth frame?

b) Where was the object when it started moving?

c) What is the current/next equation representing a point that moves 92 pixels in 4 seconds?

d) How is the use of "current" and "next" different from "current" and "previous" as used in earlier questions?

12. The table in **Figure 5.30** represents the progress of a moving object:

Time (sec.)	0	1	3	5	8	10
Location (mi.)	8	11	17	23	32	38

Figure 5.30.
Sample locations for moving object.

a) Write a closed-form equation to represent the motion of this object.

b) Write a recursive equation to represent the motion of this object. Write a word equation using $x_{current}$, $x_{previous}$, and $x_{initial}$ or write a symbolic equation using x_t, x_{t-1}, and x_0.

13. The table in **Figure 5.31** represents sighted locations of a moving object.

Time (min.)	2	4	10	15
Location (km.	7	11.5	24	31

Figure 5.31.
Sighted locations of a moving object.

a) Determine the best-fit line to represent these data.

b) Use your best-fit equation to determine the location after 12 minutes.

c) Describe another way, other than using a best-fit equation, to estimate the object's location at $t = 12$ minutes.

d) Use your best-fit equation to project where the object might be after 30 minutes.

14. Suppose $h = 8 + 1.5t$ represents the level of water in a reservoir t hours after a steady storm causes the water level to rise. The reference system for location measures water level in feet.

   a) What is the water level when the storm begins ($t = 0$ hours)?

   b) The point (6, 17) is a point on the graph of water level versus time. Explain the meaning of the point (6, 17) for this context.

   c) The graph of water level versus time is a straight line. What is the slope of the line? What is the meaning of the slope for this context?

   d) What would it mean in this context if the slope were negative?

   e) Although the equation will accept any numbers in place of h and t, not all numbers make sense in the context. What values would not make sense for the domain (time)? Range (water level)?

   f) Suppose the height of the reservoir at flood stage is 36 feet. What equation would you solve to determine when the reservoir is projected to reach flood stage? Solve the equation, and project when the reservoir reaches flood stage.

15. The recursive equation $x_t = x_{t-1} + 5$, $x_0 = 35$ and the closed-form equation $x = 5t + 35$ both represent the location of an object.

   a) Which equation would you use to find the location when $t = 30$? Explain why.

   b) Which equation would you use to find the time when the object reaches location 160? Explain why.

   c) Discuss when you would prefer a recursive equation to a closed-form equation.

16. A two-step (stretch, then shift) coding process was used to code the word *animation* as

   5 31 21 29 5 43 21 33 31.

   Prepare a control panel similar to **Figure 5.32** to represent this coding process.

Figure 5.32.
Blank control panel for Item 16.

Stretch	Shift

17. Suppose you use a recursive process to code the alphabet as numbers. The coding process is: next = current + 3. Start with the letter *A* coded as 7.

a) Code the letter *F*.

b) Decode 52.

c) Code the word *velocity*.

d) Write a closed-form equation to represent this coding process.

Time (min.)	Balloon ht. (mi.)	Car distance (mi.)
0	0.5	12
1	0.6	13.2
2	0.7	14.4

Figure 5.33.
Table representing balloon and car locations.

18. One object, a car, moves in a horizontal direction at a constant rate. A second object, a balloon, rises at a constant rate. The location of the car is measured in miles. The height of the balloon is measured in miles. Time is measured in minutes. Sample locations for both objects are listed in the table in **Figure 5.33**.

a) Write a closed-form equation to represent the height of the balloon.

b) Write a closed-form equation to represent the horizontal location of the car.

c) Where is the car when the balloon reaches a height of 1 mile? How do you know?

d) How would you write an equation to represent the motion of an object that is rising with the balloon at the same time it is keeping pace with the car? (Imagine a car traveling 30 mph and a balloon rising while it is being blown by a 30-mph wind.)

19. Two cars are moving along the same highway at the same time in the same direction. One car passed the 20-mile marker at $t = 0$ minutes. This car is traveling at a rate of 0.9 miles per minute. The second car passes the 20-mile marker at $t = 5$ minutes. This car is traveling at a rate of 1.2 miles per minute.

   a) When does the first car reach the 40-mile marker? What equation do you solve?

   b) When does the second car reach the 40-mile marker? What equation do you solve?

   c) When does the second car catch up to the first car? Explain how you determine the answer.

20. Model the previous problem using a spreadsheet. The following instructions will help you set up the spreadsheet for the first car. You will need to create the columns and formulas for the second car.

   (1) Name each column (see **Figure 5.34**). Column A is reserved for the Control Panel Information. Column B will keep track of time. Column C is for the location of Car 1.

   (2) The starting location for Car 1 is stored in location A5. The rate is stored in A7 (see Figure 5.32). You can make changes easily when you arrange important numbers in a control panel.

   (3) The number 0 is entered in B2. Time begins counting at 0.

   (4) Enter the formula to increment time in cell B3 (see **Figure 5.35**). In *Excel*, the formula is:

   =B2+$A$16

   The $ sign preceding the column letter or row number is an "absolute" reference. When you copy the formula to the other cells in the column, you will see that only the B2 changes in the formula. Start at B3 and copy the formula through row 32 in column B.

   (5) The closed-form equation for location is entered in C2 (see Figure 5.35). In *Excel*, the formula is:

   =$A$5+($A$7*B2)

   Copy the formula to the other cells in column C.

	A	B	C
1	Control Panel	Time	Car 1 Location
2		0	20
3	Car 1	1	20.9
4	*Start Location*	2	21.8
5	20	3	22.7
6	*Rate*	4	23.6
7	0.9	5	24.5
8		6	25.4
9	Car 2	7	26.3
10	*Start Location*	8	27.2
11	14	9	28.1
12	*Rate*	10	29
13	1.2	11	29.9
14		12	30.8
15	Time Increments	13	31.7
16	1	14	32.6
17		15	33.5

Figure 5.34.
Spreadsheet for Car 1 location.

	A	B	C
1	Control Panel	Time	Car 1 Location
2		0	=$A$5+($A$7*B2)
3	Car 1	=B2+$A$16	=$A$5+($A$7*B3)
4	*Start Location*	=B3+$A$16	=$A$5+($A$7*B4)
5	20	=B4+$A$16	=$A$5+($A$7*B5)
6	*Rate*	=B5+$A$16	=$A$5+($A$7*B6)
7	0.9	=B6+$A$16	=$A$5+($A$7*B7)
8		=B7+$A$16	=$A$5+($A$7*B8)
9	Car 2	=B8+$A$16	=$A$5+($A$7*B9)
10	*Start Location*	=B9+$A$16	=$A$5+($A$7*B10)
11	14	=B10+$A$16	=$A$5+($A$7*B11)
12	*Rate*	=B11+$A$16	=$A$5+($A$7*B12)
13	1.2	=B12+$A$16	=$A$5+($A$7*B13)
14		=B13+$A$16	=$A$5+($A$7*B14)
15	Time Increments	=B14+$A$16	=$A$5+($A$7*B15)
16	1	=B15+$A$16	=$A$5+($A$7*B16)
17		=B16+$A$16	=$A$5+($A$7*B17)

Figure 5.35.
Spreadsheet formulas for Car 1 movement.

a) Complete the spreadsheet to include the information for Car 2. The heading for Column D should be "Car 2 Location." Add the start location and the rate to the control panel for Car 2.

b) Explain how to use the spreadsheet to determine when Car 2 catches up with Car 1.

c) Suppose Car 1 increases its velocity to a rate of 1.05 miles per minute and Car 2 increases its velocity to 1.45 miles per minute. Predict when Car 2 catches us to Car 1. Make the necessary changes to the spreadsheet and determine when Car 2 catches Car 1.

d) Give an example of a recursive equation in the spreadsheet model.

e) Give an example of a closed-form equation in the spreadsheet model.

LESSON THREE

Up in Lights

KEY CONCEPTS

Rates of change

Variables
and constants

Linear functions

Recursive and
closed-form
representations

Elementary
programming

Matrices

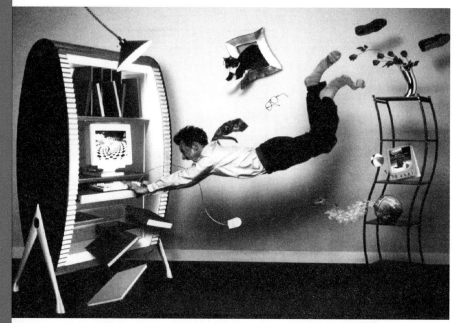

The Image Bank

PREPARATION READING

You're in Command

C omputers are not as smart as they seem. Yes, they can perform calculations at lightning speed, and they can store pages and pages of information on a pocket-size disk or peanut-size chip. Computers answer phones, monitor heart beats, adjust the speed of a car or a space shuttle, manage the flow of millions of phone conversations, and wash your dishes. The computer doesn't have a choice. It merely responds to the simple instructions of an operator and the complex instructions designed in the software by the computer programmer.

Computers do not create animated movies. People use computers and computer software to produce animation. An animated movie or virtual-reality simulation represent the combined efforts of the software user (the animator) and the software designer (the programmer) using available technology.

This lesson introduces you to the world of the computer programmer. In this lesson you will design simple programs to communicate with the graphing calculator. You begin with the basics: controlling the motion of points on the screen using particular statements and commands in a calculator program. How far you go and how much you amaze others will depend on your persistence and creativity.

The first activity introduces you to the basic language and structure of a calculator program. You create a program to animate a single dot or point horizontally across the screen. The second activity takes you to the next step: using matrices in a calculator program, you move a cluster of points horizontally across the screen. In the lessons that follow, you continue the process of building more complicated programs as you animate individual points and groups of points in a variety of directions.

CONSIDER:

1. Computer programs are used to perform repetitive procedures. A tax table, payment plan, or population projection involving repeated calculations can be produced quickly using a computer program. Many processes in life involve repetition. The trash collector repeats the same route every week. The calendar starts over on January 1 and follows the same sequence of dates each year (except leap year). The pistons in an automobile engine fire in the same order or sequence over and over again. Can you think of other processes that involve repetition? (Name at least two.)

ACTIVITY

4

CALCULATOR ANIMATION

Computer animation results from thousands of changes that occur simultaneously with each new frame. If you want to change a thousand points, then you have to know how to change or move one point. In this activity you learn how to make a single point move horizontally across the screen of a graphing calculator. Each numbered task summarizes an important step in the animation process.

PREPARE THE CALCULATOR TO DISPLAY GRAPHICS.

1. Restore the default settings to avoid unwanted complications.

   Make sure you are working with decimal numbers, not binary or hexadecimal.

   Make sure you are in function mode, not parametric or polar.

   Make sure you are in normal mode, not scientific or engineering.

   If your calculator displays grids, axes, and labels, turn these off.

   If your calculator displays statistical plots, turn them off.

   Clear the screen of calculations, graphs, tables, or other statements in preparation for drawing pictures. Your calculator may use the commands "ClrHome" or "Erase."

   Clear the screen memory of previous drawings. Your calculator may use the commands "ClrDraw" or "Erase."

2. Define the viewing window or the range.

   You want to make sure when you tell the calculator to display a particular point that the point will appear in the viewing window.

   The instructions in this lesson and throughout this unit assume that your minimum values for x and y are both 0. Your maximum x-value is 94. Your maximum y-value is 62.

CALCULATOR ANIMATION

You may change these numbers at any time. These numbers have been chosen to match a common number of pixels for a calculator viewscreen.

TURN ON A POINT AT A SPECIFIC LOCATION.

3. Draw a single point on the screen at location (47, 31).

   Your calculator may use the command "Pt-On (47, 31)" or "Pixon 47; 31" or something similar.

TURN OFF A POINT AT A SPECIFIC LOCATION.

4. Turn off the point that you just turned on at location (47, 31).

   Your calculator may use the command "Pt-Off (47, 31)" or "Pixoff 47; 31" or something similar.

5. Practice turning on and off several points on the screen.

   You may want to draw a line or a picture on your screen by turning on several points at the same time.

WRITE A PROGRAM TO TURN ON SEVERAL POINTS, ONE AT A TIME, TO DRAW A HORIZONTAL LINE.

6. Start a new program on your calculator. Give it the name *Line*.

   Use the first few lines of your program to prepare the calculator for drawing. You may want to create a mini-program that summarizes steps 1 and 2. Copy the mini-program into this and other animation programs you create.

   Make sure the command to clear the screen appears early in your program.

7. Include a program line to turn on the point at (15, 20). Exit from the edit mode, and run or "execute" your short program to make sure it does not have errors.

8. Modify your program. Select the menu item to "Edit" whenever you want to modify your program. Most calculators use the word EDIT or an abbreviation such as EDT.

ACTIVITY

4

CALCULATOR ANIMATION

Add four more lines to turn on four more points. Run your program.

9. Edit your program. Add five more lines to turn off the five points you turned on. Run your program. Watch carefully because the points will flash on and then off quickly.

The next step involves creating a **loop** in your program. A loop allows you to repeat the same set of instructions or commands several times. In an animation program you may want to turn on and off several points to simulate movement. You want to repeat the command to turn on a point followed by a command to turn off the same point.

10. Create a new program or modify your old program by deleting the commands to turn on and off the five separate points. If you create a new program you need to copy the commands from steps 1 and 2. Remember to include a program line to clear the screen.

11. Begin the loop. Most calculators have more than one command that may be used to begin a loop structure. This unit will use the "For" statement to begin a loop. You will need to consult your calculator manual to determine the statement that signals the end of the loop.

The For statement must include a starting number, ending number, and step amount. A variable is used as a counter with the For statement.

Your calculator may use the structure: For $(H, 1, 95, 2)$. The first time the program encounters the For statement, the value of H is 1. The value of H remains 1 until the statement signaling the end of the loop is reached. Then 2 is added to the value of H and the steps within the loop are repeated. When the value of H exceeds 95, the program exits the loop and moves on to the lines which follow the loop.

Your calculator may use a structure similar to BASIC programming language. "For $H = 1$ to 95; Step 2" is identical to "For $(H, 1, 95, 2)$."

ACTIVITY

CALCULATOR ANIMATION

4

12. Draw a series of points to produce a horizontal line. Draw the line at row 20 on your calculator screen.

You need to increment your counter by 1 or you will have spaces between the points on the line.

Change "For (H, 1, 95, 2)" to "For (H, 1, 95, 1)" or change step 2 to step 1 or make a similar change to your program.

Follow the For statement with a Draw command that turns on the point at (H, 20). Each time your program goes through the loop it will use a new value for H and turn on a different point.

Your calculator may use "Pt-On (H, 20)" or "Pixon H; 20."

13. Follow the Draw command with the statement that ends the loop.

Most calculator and computer programming languages use the word "End" to complete a loop. Add the End statement to your program.

14. Execute or "run" your program and watch it draw a horizontal line across the screen of the calculator. It should draw the line quickly.

15. To slow down the program, and the line drawn by the program, you can add a timing loop. A timing loop has a beginning statement and an end statement but no lines between the beginning and the end. The number of times the program increments the timing loop determines the speed at which the program runs.

Examples of timing loops are:

"For (H, 1, 100, 1)" followed by "End" or "For $H = 1$ to 100; Step 1" followed by "End."

Insert the two statements of the timing loop immediately following the command to turn on a point. (You now have two End statements in your program: one to end the timing loop and one to end the main program loop.)

ACTIVITY

CALCULATOR ANIMATION

4

16. Run your program with the timing loop added. The line is drawn more slowly. Try some variations for your *Line* program.

    Vary the speed at which the line is drawn.

    Draw two parallel lines at the same time.

    Draw a dotted line.

**WRITE A PROGRAM TO MAKE A SINGLE
POINT FLASH ON AND OFF SEVERAL TIMES.**

17. Start a new program with the name *Blinker*. Include the following statements in the *Blinker* program.

    Copy the program lines that prepare the calculator for graphics.

    Include a statement to clear the screen.

    Begin a loop with the For statement. Use the variable *H*. Start with 1. End with 50. Step by 1.

    Turn on the point at (30, 20).

    Begin a timing loop with a For statement.

    End the timing loop with an End statement.

    Turn off the point at (30, 20)

    Begin a timing loop with a For statement.

    End the timing loop with and End statement.

    End the main program loop using the word End. (Notice the program has two End statements.)

18. Try some other variations on the *Blinker* program.

    Change the rate at which the point flashes.

    Make more than one point blink at a time.

    Vary how long the point stays on compared to off.

ACTIVITY

CALCULATOR ANIMATION

4

**WRITE A PROGRAM TO MAKE A SINGLE POINT
APPEAR TO TRAVEL ACROSS THE SCREEN.**

19. Start a new program with the name *Dottie* or modify (Edit) your *Blinker* program. Begin the program with the lines which prepare the calculator for graphics.

20. In this program combine features of the *Blinker* and *Line* program together.

    Include a statement to clear the screen.

    Begin a loop with the For statement. Use the variable *H*. Start with 0. End with 94. Step by 1.

    Turn on the point at (*H*, 20).

    Begin a timing loop with a For statement.

    End the timing loop with an End statement.

    Turn off the point at (*H*, 20)

    End the main program loop.

21. Modify your *Dottie* program.

    Change the rate at which the point appears to move.

    Animate two points that travel across the screen at the same time.

    Change the animation so the moving point begins in the center of the screen.

22. Here are some challenges for the more advanced programmer:

    Animate a dot so it moves from right to left.

    Animate a dot so it moves in a vertical direction.

    Animate two moving dots, one vertical and one horizontal.

    Animate a figure or a letter.

INDIVIDUAL WORK 4

Out of the Loop

1. Suppose it takes 18 seconds for your calculator to animate a point from the left side of your calculator screen to the right side of the screen using the program *Hline*. The lines of a typical program *Hline* are shown, with line numbers for reference.

Program: *Hline*

   (1) : Prepare the calculator to display graphics.

   (2) : 0→Xmin:94→Xmax:0→Ymin:62→Ymax

   (3) : For(P,0,94,1)

   (4) : Pt-On(P,20,2)

   (5) : For(Q,1,50,1)

   (6) : End

   (7) : Pt-Off(P,20,2)

   (8) : End

   a) What change could you make to the program to make the point move faster?

   b) Describe another change you could make to the program to move the point faster.

   c) What is the velocity of the point in pixels per second?

2. Consider the following calculator program *Iw* (Numbers next to the program lines are for reference purposes):

Program: *Iw*

   (1) : Prepare the calculator to display graphics.

   (2) : 0→Xmin:94→Xmax:0→Ymin:62→Ymax

   (3) : For(P,0,45,1)

   (4) : 2P+3→H

   (5) : Pt-On(H,20,2)

   (6) : End

a) The loop begins with $P = 0$. What is the first point that is turned on?

b) Which point is being turned on when $P = 10$?

c) Describe what this program is designed to accomplish.

d) Suppose you change the number 20 in line (5) to 30. How does the animation change?

e) What changes could you make to this program to slow the point down?

3. Suppose you write a different computer program including the statement $H + 2 \rightarrow H$ that adds 2 to the value of H, the horizontal location of the moving point, each time the program passes through the loop statements.

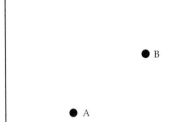

a) If 0 is the beginning value for H, then what is the value of H after the fifth time through the loop?

b) If 13 is the beginning value for H, what is the value of H after the eighth pass through the loop?

c) A recursive equation for $H + 2 \rightarrow H$ is $H_{new} = H_{old} + 2$. Write the closed-form equation for when H begins at 13. You will need to include a variable T that counts the number of times the program repeats the loop.

Figure 5.36.
Viewscreen for Item 4(a). [0,94] x [0,62]

d) Adjust the statement $H + 2 \rightarrow H$ so the point moves twice as far in the same amount of time (same number of passes through a loop).

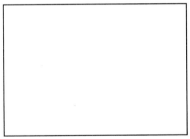

4. a) Observe the viewscreen in **Figure 5.36** and estimate the coordinates for points A and B.

b) Copy the viewscreen in **Figure 5.37** onto your paper. Draw the result of the program line: Pt-On (16, 40). Label the point C.

Figure 5.37.
Viewscreen for Item 4(b). [0,100] x [0,50]

c) Point D at (50, 20) is shown in **Figure 5.38**. Approximate appropriate values for Xmax and Ymax.

5. The program statement $M + 3 \rightarrow M$ adds 3 to M each time the statement is encountered in a loop process. $M + 3 \rightarrow M$ is equivalent to the recursive equation $M_{new} = M_{old} + 3$.

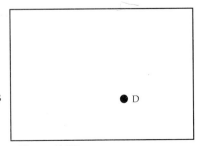

a) Write a recursive program statement that subtracts 2 each time.

Figure 5.38.
Viewscreen for Item 4(c). [0,?] x [0,?]

b) Write a recursive program statement to match: $M_t = M_{t-1} + 25$.

c) Write a recursive equation that adds 7.5 with each time.

d) Write a recursive equation to match the program statement:
$P + 200 \rightarrow P$.

6. You have animated a point on a calculator screen. The calculator screen has a horizontal dimension and a vertical dimension. You have been moving points in a horizontal line using programs similar to program *Hline*. (Numbers next to the program lines are for reference purposes.)

Program: *Hline*

(1) : Prepare the calculator to display graphics.

(2) : $0 \rightarrow$ Xmin:94$\rightarrow$Xmax:0$\rightarrow$Ymin:62$\rightarrow$Ymax

(3) : For(P,0,94,1)

(4) : Pt-On(P,20,2)

(5) : For(Q,1,50,1)

(6) : End

(7) : Pt-Off(P,20,2)

(8) : End

a) How can you change the program *Hline* to make the point move in a vertical direction?

b) What statements can you add to the program *Hline* to animate two points moving side by side in a horizontal direction?

c) What statements can you add to the program *Hline* to animate two points, one moving in a horizontal direction and the other in a vertical direction?

A loop in a calculator program models a situation that is repeated several times. Loops are useful for situations requiring **counters** and **accumulators**. Statements that use counters add 1 (or whatever constant you choose) recursively to a variable each time the loop is repeated. For example, in the statement $K + 1 \rightarrow K$, K is a counter that is incremented by 1 in a loop. Statements that use accumulators add some quantity recursively to a variable each time the loop is repeated. In the statement $L + (amount) \rightarrow L$, L is an accumulator in a loop. The amount added each time does not have to be the same, although it often is. Note that each time you repeat the process is called an **iteration** of the process.

There are several situations in life that can be modeled with a loop process. Each year when your birthday arrives you complete another iteration of a loop. You add 1 to your year counter. Each time a ferris wheel passes the point where you got on the ride, you complete another iteration of a loop. You add 1 more to the counter keeping track of the number of revolutions.

If you are involved in track activities at your school, you may run races around an oval track. Each lap around the track is an iteration. The counter keeps track of total laps. An accumulator is used to keep track of the total distance you run. If the distance around your track is 400 meters, then total distance is modeled by the accumulator: $D + 400 \rightarrow D$. If you place $20.00 in a piggy bank each week, your accumulated savings is modeled by the statement: $S + 20 \rightarrow S$.

In a calculator program you use the statement "For $(S, 1, T, 1)$" combined with "End" to create a loop that is repeated until S equals T. The variable S serves as a counter if 1 is added to S with each iteration of the loop. Other statements can be added to your loop to serve as counters.

7. Suppose Yutaka runs around a 400-meter track. The following statements appear within a calculator program describing his run.

   : $0 \rightarrow P$:$0 \rightarrow D$

   : For(S,1,200,1)

   : $P+1 \rightarrow P$ (counter)

   : $D+400 \rightarrow D$ (accumulator)

   : End

a) Complete a table of sample values as shown in **Figure 5.39**.

b) What is the total distance after 12 laps?

c) How would the table change if the statement For $(S, 1, 200, 1)$ were changed to For $(S, 1, 200, 2)$?

d) How would the table change if $D + 400 \rightarrow D$ became $D + 350 \rightarrow D$?

Loop variable S	Counter P (laps)	Total distance D (m)
	0	0
1	1	
2		
4		

Figure 5.39.
Table for Item 5(a).

e) Construct an arrow diagram to represent the action of the accumulator in one iteration of the following loop:

: For(T,1,50,1)

: D+100→D

: End

8. Thus far in this unit, motion has been along a horizontal line. Now consider vertical motion. Altitude and depth measure vertical distance and are related to vertical motion.

a) Jaime flies a hang glider. He is at an altitude of 1500 feet after 30 seconds and 1100 feet after 2 minutes. What is his rate of descent?

b) Carrie is scuba diving. She is at –20 feet after 12 minutes of dive time. If she rises at a rate of 6 feet per minute, when does she reach the surface?

c) The water level in a reservoir is 38 feet. Flood stage is 45 feet. If the water is rising at a constant rate of 0.75 feet per hour then how long will it take to reach flood stage?

9. Rafael hikes. He begins his day hike at 0800 (8:00 a.m.), starting at an elevation of 4200 feet. He reaches the 9500-foot peak at 1400 (2:00 p.m.). After a 1-hour rest he hikes down the mountain and returns to his starting point at 1730 (5:30 p.m.).

a) Calculate Rafael's rate of ascent in feet per hour.

b) Calculate his rate of descent in feet per hour.

c) Assuming he ascends at a constant rate, write an equation to represent Rafael's altitude h after t hours of climbing.

d) What was his altitude at 1130?

10. Rate of change is important to the modeling of contexts other than motion.

a) A bathtub with a capacity of 16 gallons is filling at a rate of 2 gallons per minute. You start timing when the tub contains 3 gallons. Write an equation describing this situation. Be sure to indicate what your variables are.

b) Natalie has saved $16 to buy a personal CD player that costs $129. She has an afternoon job at which she earns enough money after 3 weeks. If Natalie earns money at a constant rate during that time, what is her rate of earning (with appropriate units)?

c) If you know that Natalie works 2 hours per day, 5 days per week, express your answer to part (b) as a daily rate and as an hourly rate.

d) In 1980, there were no moose in New York's Adirondack State Park. By 1993 there were between 26 and 30 moose. Assume that all the moose arrived by migration at a constant rate. Determine reasonable estimates for the migration rate and justify your estimates.

e) Use your answer to part (d) to predict the number of moose in 2010 assuming conditions do not change.

11. a) If you light the points (10, 10), (10, 11), (10, 12), (10, 13), (10, 14), (10, 15), (11, 10), (12, 10), (13, 10), and (14, 10) all at the same time, what letter would you see?

b) Suppose the letter is animated and moves 2 points to the right. Identify the locations that are now lit.

12. The lights in **Figure 5.40** are displayed on a marquee sign.

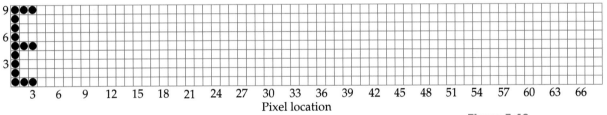

Figure 5.40.
Letter E marquee.

a) How do you make the letter move to the right?

b) Describe frame-by-frame changes you would make so the letter appears to shrink in size without moving to the right.

13. Imagine that you shade in your initial so that its left edge starts in column 25 in a marquee grid (see blank marquee grid in **Figure 5.41**).

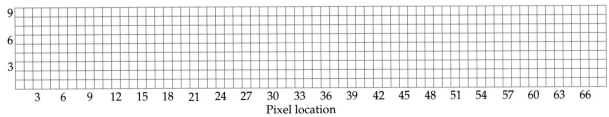

Figure 5.41.
Blank marquee grid.

a) Suppose the computer is set so that every 0.1 second your initial shifts 1 column to the right. Where is the left edge of your initial after 0.2 second? After 0.4 second? After 1.5 seconds?

b) How far does your initial move every second? What is the rate of this motion?

c) Look for a relationship between the location of the left edge of your initial and the amount of time it takes to get there. Fill in a table like the one in **Figure 5.42**.

Left edge	Timer
25	0.0
26	0.1

Figure 5.42.
Table for Item 12(c).

d) Explain how to find the location after 2.5 seconds

e) Based on your table and any other observations you have made, describe how to predict exactly where the left edge of your initial will be if someone tells you how long the timer has run.

f) Suppose that the sign continues to change its display every 0.1 second, but that you now want your initial to move twice as fast. How could you get that to happen? What exactly does it mean when you say your initial moves twice as fast?

g) With this new, faster initial, explain how to predict the location of the left edge of your initial when you know how long it has been moving. Follow your explanation with a symbolic equation.

h) Use your explanation or equation to complete the tables in **Figure 5.43**.

Old location	New location	Time (sec.)	Location (column)
25	27	0.0	25
27	29	0.1	27

Figure 5.43.
Tables for Item 12(h).

i) Write a recursive equation to represent the relationship between old location and new location. Use 0.1 sec. as the recursive period.

j) Which answers in parts (a–h) would be the same even if you decided to track a different column of your initial (and not the left-hand side)?

14. Suppose the relationship between the left edge of your initial and time can be described as follows: multiply the number of seconds that have passed by 2 and add that to 5 to find the left edge of your initial.

   a) Where does your initial start when the display is turned on?

   b) Where is your initial after 2 seconds have gone by?

   c) How far does your initial move in 1 second?

   d) Write a symbolic equation to represent the location of the left-hand side of the letter.

   e) How long will it take for your initial to reach column 20? What equation did you solve?

   f) How are velocity, location, and time related in general? Try to write a rule that uses V for velocity, L for location and T for time.

15. In Unit 2, *Secret Codes and the Power of Algebra*, you used matrices to shift the position numbers of a message. Suppose you made a matrix [A] that contained the x-coordinates of a letter in row 1 and the y-coordinates in row 2. If you could turn on, then off, every point represented by matrix [A], how would you make an entire letter move 2 pixels in a horizontal direction with each change in a screen?

16. A particular intersection has obstructed visibility from one direction. The Department of Transportation (DOT) has installed a yellow light that begins flashing whenever an approaching car passes over a sensor cable buried in the road. The speed limit on the road is 35 miles per hour, and the sensor is 500 feet from the intersection. Should the DOT install a sensor to turn the flashing light off, or should they just use a timer? (Questions to ask yourself are, "What is the purpose of the light? Why is the sensor 500 feet away? How fast should DOT assume drivers really travel?")

E-MOTION

5

You animated a single point in a horizontal direction in Activity 4. In this lesson you move a cluster of points in a horizontal direction. Just as letters appear to move across the screen of a marquee sign, so you will make a letter move across the screen of your graphing calculator. This is the next step on your way to developing more complicated animations.

Matrices can be used to make your calculator marquee more efficient. The matrices used in this animation are 2 x n with row 1 containing all the x-coordinates and row 2 containing all the y-coordinates. The matrix in **Figure 5.44** tells the calculator to position the letter E in the lower-left corner of the calculator screen.

Figure 5.44.
$$\begin{bmatrix} 1 & 1 & 1 & 1 & 1 & 2 & 2 & 2 & 3 & 3 & 3 \\ 1 & 2 & 3 & 4 & 5 & 1 & 3 & 5 & 1 & 3 & 5 \end{bmatrix}$$

1. Define a matrix in your calculator to store the points as listed above. Label the matrix [F].

2. Begin a new program called *Emotion*.

3. Remember to use the first few lines of your new program to restore the default settings, clear the screen, and set the minimum and maximum values for the window.

4. To display the letter E on the screen, you want to turn on all the points of the letter at about the same time. You need to create a For/End loop within the program that will quickly will turn on all eleven points by the time the loop is complete. To do so, refer to Activity 4, and add lines similar to the following:

   Begin a loop with a For statement. Use the variable N. Begin with 1. Step by 1. End with 11 (because there are 11 points in matrix [F]).

E-MOTION

The next line of the program turns on a different point with each pass through the loop. The coordinates for each point are selected from matrix [F].

Use the calculator command that selects a value from a designated position in a matrix. One calculator uses the combination [F] (2, 3) to identify the number in the second row, third column of matrix [F]. You need to use N to designate the column because you move to the coordinates in the next column every time you pass through the loop. One calculator uses the combination:

Pt-On ([F] (1,N), [F] (2,N))

Use the End statement to end the loop.

5. Discuss with your group how to make the letter E move from left to right across the screen. Organize a frame-by-frame map for the first few frames. Discuss what changes need to happen from frame to frame.

6. Add the necessary steps to the program *Emotion* to make the letter E appear to move from left to right across the screen.

7. Make the letter E move from right to left across the screen.

8. Make a short message or the word "HI" travel across the screen.

INDIVIDUAL WORK 5

Moving with Ease

1. a) In a matrix, record the coordinates of the points that are currently "on" in the marquee in **Figure 5.45**.

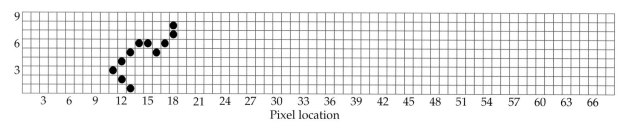

Pixel location

Figure 5.45.
Marquee for
Item 1(a).

b) What letter is represented by the following matrix?

$$
\begin{bmatrix}
33 & 33 & 34 & 34 & 35 & 35 & 36 & 37 & 38 & 39 & 39 & 40 & 40 & 41 & 41 \\
08 & 07 & 06 & 05 & 04 & 03 & 02 & 01 & 02 & 03 & 04 & 05 & 06 & 07 & 08
\end{bmatrix}
$$

c) In matrix form, write the coordinates to represent the letter *M*. The bottom left corner of the letter *M* should start at (19, 1).

2. Consider each page of a flip-book marquee as a frame in an animation. The flip book is a simple and old-fashioned way to produce the illusion of motion. The artist carefully draws each page or frame of the flip book so that it is slightly different from the previous page. When several pages are viewed in rapid succession (or fanned), the objects in the drawings appear to move. You can construct a flip book to simulate the movement of words across a marquee screen.

   a) Suppose each letter in your flip-book marquee shifts by 2 pixels or columns from one frame to the next. If you flip the pages at a rate of 7 frames per second, how fast do the letters move in pixels per second?

   b) Describe two ways to make the message appear to move faster.

   c) How would you make the motion of the letters on the marquee appear smoother?

   d) The marquee displays discrete frames. Is it possible to make a continuous marquee sign? Explain your answer.

3. All points in a marquee move the same distance from one frame to the next. Select the point at location (14, 6) on a marquee. The point moves at a rate of 3 pixels per frame and 4 frames per second.

   a) Write a closed-form equation to represent the horizontal position *h* of the point after *f* frames.

b) Write a recursive equation using 1 frame as the recursive period.

c) Write a closed-form equation to represent the horizontal location h of the point after t seconds.

d) Write a recursive equation using 1 sec. as the recursive period.

4. Zeke has a savings account at a local bank. He deposits the same amount of money each week. The equation $T = 15w + 924$ represents his accumulated total T after w weeks.

a) Find the accumulated total after 8 weeks.

b) When does the total reach $2000? What equation do you solve?

c) Write a recursive equation for his total.

d) Explain how the situation is different if the accumulated total is represented by $T = 20w + 924$.

The following sample calculator program is for use with Items 5–7.

Program: *Moveees*

(1) : Prepare the calculator to display graphics.

(2) : Set the minimum and maximum values for your window.

(3) : [F]→[H]

(4) : For(P,1,80,1)

(5) : For(N,1,11,1)

(6) : Pt-On([H](1,N),[H](2,N))

(7) : End

(8) : For(M,1,100,1)

(9) : End

(10) : For(N,1,11,1)

(11) : Pt-Off([H](1,N),[H](2,N))

(12) : End

(13) : [H]+[G]→[H]

(14) : End

$$[F] = \begin{bmatrix} 1 & 1 & 1 & 1 & 1 & 2 & 2 & 2 & 3 & 3 & 3 \\ 1 & 2 & 3 & 4 & 5 & 1 & 3 & 5 & 1 & 3 & 5 \end{bmatrix} \quad [G] = \begin{bmatrix} 1 & 1 & 1 & 1 & 1 & 1 & 1 & 1 & 1 & 1 & 1 \\ 0 & 0 & 0 & 0 & 0 & 0 & 0 & 0 & 0 & 0 & 0 \end{bmatrix}$$

5. Refer to the program *Moveees*.

   a) What does this program do?

   b) What changes do you make to the program if you want the letter to move 2 pixels with each frame instead of 1?

   c) What changes do you make to the program if you want the letter to travel farther across the screen?

   d) What changes do you make to the program if you want a different letter displayed?

   e) Describe the purpose of lines (8) and (9).

6. Refer to the program *Moveees*.

   a) Change line (4) to For (*P*, 1, 80, 4). Describe how the animation changes.

   b) Change line (5) to For (*N*, 1, 8, 1). Describe what is different.

   c) Change line (4) to For (*P*, 40, 80, 1). Describe how the animation changes.

   d) Switch rows 1 and 2 in matrix [G]. Describe what is different.

7. Refer to the program *Moveees*.

   a) Describe changes you would make to the program or matrices to cause the animation of the letter *E* to begin in the lower-right corner of the screen.

   b) Describe changes you would make to the program or matrices to cause the letter *E* to move from right to left.

Refer to program *EEoption* for your work on Items 8 and 9.

Program: *EEoption*

(1) : Prepare the calculator to display graphics.

(2) : Set the minimum and maximum values for your window.

(3) : For(P,0,80,1)

(4) : For(N,1,11,1)

(5) : [F](1,N)→H

(6) : [F](2,N)→V

(7) : Pt-On(H+P,V)

(8) : End

(9) : For(M,1,100,1)

(10) : End

(11) : For(N,1,11,1)

(12) : [F](1,N)→H

(13) : [F](2,N)→V

(14) : Pt-Off(H+P,V)

(15) : End

(16) : End

$$[F] = \begin{bmatrix} 1 & 1 & 1 & 1 & 1 & 2 & 2 & 2 & 3 & 3 & 3 \\ 1 & 2 & 3 & 4 & 5 & 1 & 3 & 5 & 1 & 3 & 5 \end{bmatrix}$$

8. Refer to the program *EEoption*.

a) What does this program do?

b) A loop begins at line (4). Which line ends that loop?

c) Change line (4) to For (*P*, 1, 80, 2). Describe what is different about the animation.

d) Why do you no longer need the line [H] + [G]→[H], which is in *Moveees*?

9. Refer to the program *EEoption*

   a) Describe what happens when you change line (7) to Pt-On (H + 2P, V) and line (14) to Pt-Off (H + 2P, V).

   b) Describe what happens when you change line (7) to Pt-On (H, V + P) and line (14) to Pt-Off (H, V + P).

   c) Describe what happens when you change line (7) to Pt-On (H + 2P + 3, V) and line (14) to Pt-Off (H + 2P + 3, V).

   d) Describe what happens if you eliminate lines (11–15) from the program.

10. a) What letter is represented by the matrix [A]?

$$[A] = \begin{bmatrix} 6 & 6 & 6 & 6 & 7 & 7 & 7 & 8 & 8 & 8 & 8 \\ 1 & 3 & 4 & 5 & 1 & 3 & 5 & 1 & 2 & 3 & 5 \end{bmatrix}$$

   b) Write the result of [A] + 2*[B]. Explain how this changes the letter that appears on the screen.

$$[B] = \begin{bmatrix} 0 & 0 & 0 & 0 & 0 & 0 & 0 & 0 & 0 & 0 & 0 \\ 1 & 1 & 1 & 1 & 1 & 1 & 1 & 1 & 1 & 1 & 1 \end{bmatrix}$$

   c) Write the result of 2*[A] + [B]. Explain how this changes the letter that appears on the screen.

   d) Suppose the 0's in the first row of [B] are all changed to 1's. What is the effect of adding [B] to [A]?

11. The point (X, Y) represents the location of a point on a letter when the motion begins. P is the counter that determines the frame.

   a) The point moves 1 pixel to the right with each frame. Write an equation to represent the horizontal location H of the point after P frames.

   b) Suppose the point moves 2 pixels to the right with each frame. Write an equation to represent the horizontal location H of the point after P frames.

   c) Suppose the point begins at (4, 5) and moves 2 pixels to the right with each frame. When (at which frame) does it reach horizontal location 88? What equation do you solve?

   d) When does the point reach horizontal location 89?

12. A balloon moves vertically at the same time a car moves horizontally. The tables in **Figure 5.46** identify the locations of each object at particular times.

Car

Time (min.)	Horizontal (mi.)	Vertical (mi.)
0	14	0
1	14.8	0
2	15.6	0
4	17.2	0

Balloon

Time (min.)	Horizontal (mi.)	Vertical (mi.)
0	0	6
1	0	56
2	0	106
4	0	206

Figure 5.46.
Tables for car and balloon.

a) At what time does the balloon reach a height of 125 feet? What equation do you solve?

b) At what time does the car reach a horizontal location of 21.7 miles? What equation do you solve?

c) Write a recursive equation to represent the motion of the car or the balloon using 1 min. as the recursive period.

d) Describe the graph of vertical position versus time for the balloon. Explain how to graph vertical position versus time on the graphing calculator.

13. Thus far, almost all motion has been horizontal at a constant rate.

a) Describe how to make a word stretch.

b) Describe how to make a message accelerate as it moves across the screen.

LESSON FOUR
Escalating Motion

KEY CONCEPTS

Rates of change

Variables and constants

Linear functions

Elementary programming

Recursive and closed-form representations

Parametric equations

Graphs of functions

Matrix operations

Systems of linear equations

The Image Bank

PREPARATION READING

Pulled in Two Directions

*I*magine a world in which everything moves in one direction in a straight line. No right or left. No up or down. No stairs or escalators or hills. The one and only street is straight. Cars don't need steering wheels. There is no turning back. These are characteristics of a one-dimensional world.

You do not live in a one-dimensional world. Animation and life involve movement in more than one direction. You have created a model for animating a single point and a group of points in a horizontal direction. The model is too simple. You need a model that allows for a full range of motion.

In Lesson 3, you identified equations and wrote calculator programs to represent horizontal or vertical movement. In this lesson you first modify the model by animating a single point so it moves in a diagonal direction. You have to be able to move one point in any direction before you can animate a bunch of points in any direction.

You must modify the equation for motion so that it results in diagonal motion. Time is the explanatory variable in a motion equation. Location is the response variable. When motion is horizontal, the response variable is x or horizontal location. You represent horizontal motion with an equation like $x = 3 + 2t$ or $x_t = x_{t-1} + 2$.

When motion is vertical, the letter y is commonly used for vertical location. When motion is diagonal, there is a change in both the x- and y-coordinates for location. How do you represent a change in horizontal *and* vertical location when time is also a variable? Can you make a table? Can you write a symbolic equation so technology can be used?

Equations that separate motion into horizontal and vertical components while including time as a third variable are called **parametric equations**. In this lesson you use parametric equations to represent location with two variables that depend on a third variable, time. The graphing calculator simplifies your work with parametric equations.

CONSIDER:

1. Ramps and stairs are both used to get from one height to another. Which does a better job representing the diagonal motion of a point? Explain your answer.

2. An Etch-a-Sketch® is a drawing toy. By turning one knob you can draw horizontal lines. Another knob is used for vertical lines. Why is it a challenge to draw a diagonal line?

3. How do you make a dot move diagonally on a calculator or computer screen?

FOLLOW THE PATH

6

A graph can represent the path of an object moving diagonally. You can write an equation, using *x* and *y*, to represent the path by applying what you learned in the Unit 4, *Prediction*. However, the graph in **Figure 5.47** does not tell you where the object begins, where it ends, or where it is after 3 seconds. Neither the equation nor the graph include the variable time. This is an example of a **path graph** (also known as a state-space graph). The graph shows where the object has been, but not when or even in what direction.

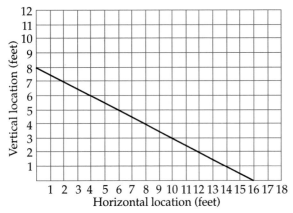

Figure 5.47.
Path graph.

In Lesson 2, you learned that an object that moves horizontally can be represented by the equation $x = a + bt$. The variable *y* is not needed when only horizontal position changes with respect to time. Similarly, an object that moves vertically can be represented by the equation $y = c + dt$. The variable *x* is not needed when only the vertical position changes. Diagonal movement requires a combination of *x*, *y*, and *t*. One way to represent all three variables is with a **time-lapse graph**. A time-lapse graph adds sample times to the graph of the path (see **Figure 5.48**).

Another way to combine location and time is with parametric equations. The purpose of this activity is to introduce you to parametric equations to represent motion. You will use parametric equations and the graphing calculator to produce time-lapse graphs and three-column tables (see **Figure 5.49**). Include parametric equations with your mathematical toolbox.

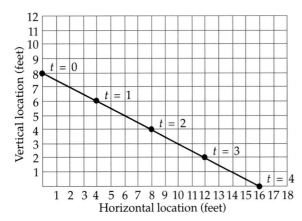

Figure 5.48.
Time-lapse graph.

ACTIVITY

FOLLOW THE PATH

6

1. Use the time-lapse graph in Figure 5.48 to complete a table like that in Figure 5.49.

Time (sec.)	Horizontal location (ft.)	Vertical location (ft.)
0		
1		
2		
4		

Figure 5.49.
Table for time-lapse graph.

2. Write an equation to represent the path of the object, y in terms of x. Use the closed form $y = mx + b$. Explain how you determine the answer.

3. Describe how you find the location of the object at $t = 2.5$ sec. Include in your explanation reasons that you cannot use the equation from Item 2.

4. The equation for the path of the object does not include time. Write closed-form equations to represent position in terms of time.

5. a) Create two new graphs on paper. Graph horizontal location versus time on the first graph and vertical location versus time on the second graph. These two graphs use time as the explanatory variable. They are called **time-series graphs**. They provide information about the motion but do not show the motion directly. (If you use the graphing calculator instead of paper you may need to adjust your variables to match the calculator. The t in your motion equation becomes X in the calculator. The x in your motion equation becomes Y in the calculator. Be careful.)

   b) Explain how to use the two graphs to find the location of the object at $t = 2.5$ seconds.

FOLLOW THE PATH

Pairs of equations such as $x = a + bt$ and $y = c + dt$ are examples of parametric equations. One equation, $x = a + bt$, expresses the horizontal location in terms of time. The second equation, $y = c + dt$, expresses the vertical location in terms of time. When combined, the two equations represent time and location on a grid.

Parametric equations may be used in contexts involving one explanatory variable and two response variables. In Unit 4, *Prediction*, you examined several situations that could be expressed as parametric equations. For example, femur length and ulna length could be response variables for the explanatory variable, height. Manatee deaths and powerboat registrations are response variables to the explanatory variable, year. The surface area of a cube and the volume of a cube are response variables to the explanatory variable, side length.

6. The motion of an object is represented in the table in **Figure 5.50.** (Time is measured in seconds, locations are in feet.)

Time (sec.)	Horizontal location (ft.)	Vertical location (ft.)
0	1	6
1	3	7
2	5	8
3	7	9
4	9	10
5	11	11
6	13	12
7	15	13

Figure 5.50.
Table for Item 6.

ACTIVITY

FOLLOW THE PATH

6

a) Which columns from the table do you use to plot the actual path of the motion? Why?

b) On a piece of graph paper, plot the path of the motion. Because this is a path graph, you need to graph vertical location versus horizontal location. Label your axes carefully.

c) Write a closed-form equation to represent the path of the object. Use the form $y = mx + b$.

d) Create the time-series graphs to represent horizontal and vertical location.

e) Explain how to use the time-series graphs to determine the time and the exact location of the object on the grid when the horizontal location reaches 21.

f) What equation would you solve to determine the time when the object reaches a horizontal location of 21? Solve the equation.

INDIVIDUAL WORK 6

The Second Story

1. Use the graphing calculator and your parametric equations from Activity 6, Item 4 to graph the path of the object. Change the calculator mode to parametric. Enter the horizontal equation as x and the vertical equation as y.

   Use the TRACE or TABLE feature and compare with the table in Activity 6, Item 1. Sketch your calculator viewscreen on paper.

2. a) Create a three-column table (like **Figure 5.51**) from the following word descriptions. Describe the resulting motion in your own words.

   Horizontal location = 3 × time + 2

   Vertical location = 6 × time + 3

Time (sec.)	Horizontal location (ft.)	Vertical location (ft.)
0		
2		
3		
5		

Figure 5.51.
Table for Item 2(a).

   b) Write the corresponding recursive equations using 1 sec. as the recursive period.

   c) How would doubling the horizontal velocity affect the table?

   d) Write closed-form equations for an object with the same horizontal and vertical velocities but starting at (15, 28).

3. An object on an animated escalator starts at a point 20 feet above the floor and 12 feet to the right of a wall. It descends at a rate of 1 foot per frame while moving to the right at a rate of 2 feet per frame. The reference point or origin for this reference system is against the wall on the bottom floor.

   a) Plot the motion of the object with a time-lapse graph. (A time-lapse graph adds sample times to the graph of vertical location versus horizontal location.) Prepare a table, if necessary, before constructing your graph on paper.

b) Write closed-form equations that describe the horizontal and vertical components of the escalator motion. Remember, the animation shows the escalator is descending.

c) What equation should you solve if you want to find the frame when the point on the escalator reaches the ground level? Solve the equation.

d) What is the horizontal location when the escalator reaches the ground level? Explain or show how you determined your answer.

e) Suppose you graphed the path graph (vertical location versus horizontal location) and both time-series graphs (vertical location versus frame and horizontal location versus frame). The point (8, 12) is a point on one of the three graphs. Which one? What does the point (8, 12) tell you about the "where" or the "when" regarding the motion of a point on the escalator?

f) Suppose the motion of the escalator were reversed. Write parametric equations (closed form) for a point that travels from the bottom of the escalator ($t = 0$) to the top.

4. Suppose the vertical location of a point on another escalator is represented by the closed-form equation $y = 30 - 3t$. Measurements are in feet and seconds.

a) What is the slope of the line $y = 30 - 3t$? What does the slope tell you about the motion of the escalator?

b) What is the y-intercept for the line $y = 30 - 3t$? What does the y-intercept tell you about the motion of the point?

c) Explain how to determine the time needed to go from the top of the escalator on one floor to the bottom of the escalator one floor below.

5. An object or point moves horizontally at a rate of 2 meters per second and vertically at 1.5 meters per second. It is located at (8, 5.5) when $t = 0$ seconds.

a) Write closed-form equations for both the horizontal and vertical components of the motion.

b) Write recursive equations using 1 sec. as the recursive period.

c) Where is the object at $t = 9$ seconds?

d) Does the object pass through the point (32, 18)? Explain how you determine your answer.

6. The control panel for parametric equations must include the starting location and rate for both horizontal and vertical components (**Figure 5.52**).

Component	Starting location (m)	Rate (m/sec.)
Horizontal	$x = 4$	3
Vertical	$y = 12$	0.75

Figure 5.52.
Parametric control panel.

a) Where is the object at $t = 6.5$ seconds?

b) When does the object reach the horizontal position of 20 meters? Show how you determine the answer.

7. What goes up must come down! The conveyor belt from a warehouse to its loading platform starts 50 feet to the right of and 25 feet above the loading floor. It drops 3 feet every second and travels to the left twice as fast as it falls.

a) Write closed-form equations to represent the horizontal and vertical positions of a box on the conveyor in terms of how long it has been on the belt.

b) Complete a three-column table like the one in **Figure 5.53** to represent some of the locations of the box on its way down.

Time	Horizontal	Vertical
0		
1		
2		
4		
6		

Figure 5.53.
Three-column table for Item 7(b).

c) Draw and label two time-series graphs (horizontal location versus time and vertical location versus time) and one path graph (vertical location versus horizontal location) of the motion.

d) Describe how to use these graphs to determine the location of the box after it has been on the belt for 6 seconds.

e) Describe how to use the graphs to determine the time and location at which the box reaches the loading platform, located one foot above the floor.

8. The horizontal velocity (ground speed) of an airplane is 440 feet per second (300 miles per hour). The plane is descending at a rate of 15 feet per second from an altitude of 12,000 feet.

a) Describe this motion using equations, tables, and graphs. Explain how all these representations are related, and indicate how each can be useful.

b) Find an equation for the path graph. How is this equation related to the component equations?

9. The following recursive equations describe the motion of a point:

$$x_t = x_{t-1} + 3 \qquad\qquad x_0 = 5$$

$$y_t = y_{t-1} - 2 \qquad\qquad y_0 = 15$$

Distance is measured in miles. Time is measured in hours.

a) Create and complete a table like that in **Figure 5.54** to represent sample times and locations.

Time (hr.)	Horizontal location (mi.)	Vertical location (mi.)
0		
1		
2		
3		
4		

Figure 5.54.
Blank table for Item 9(a).

b) Write closed-form equations for horizontal location and vertical location.

c) Write an equation to represent the path of the object. That is, write an equation for y in terms of x.

10. The following closed-form equations describe the motion of a point:

$$x = 10 - 2t$$

$$y = 4 + 6t$$

a) Write recursive equations for x and y.

b) Write an equation to represent the path of the object using 1 unit as the recursive period.

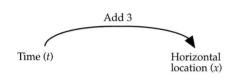

Figure 5.55.
Arrow diagram for time and horizontal location.

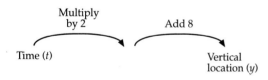

Figure 5.56.
Inverse of Figure 5.55.

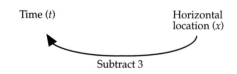

Figure 5.57.
Arrow diagram for time and vertical location.

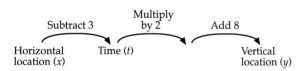

Figure 5.58.
Combined arrow diagram for substitution process.

Symbolic methods may be used to convert parametric equations to a single closed-form equation in y and x. Use the equations $x = 3 + t$ and $y = 8 + 2t$ as an example. The arrow diagrams in **Figures 5.55–58** illustrate the steps to take.

The arrow diagram for $x = 3 + t$ begins with t and ends with x.

The inverse of $x = 3 + t$ begins with x and ends with t (Figure 5.56).

The arrow diagram for $y = 2t + 8$ begins with t and ends with y (Figure 5.57).

The equation for the path combines the two components. The arrow diagram in Figure 5.58 illustrates how the inverse of the horizontal component followed by the vertical component begins with x and ends with y.

Solve the first equation, $x = 3 + t$, for t by reversing the process and creating the inverse. The opposite of adding 3 is subtracting 3 from both sides of the equation, creating $t = x - 3$ as the result.

The process for the second equation begins with t and ends with y. Begin with $x - 3$ in place of t, follow the steps in the process, and set the result equal to y:

Multiply $(x - 3)$ by 2 to get $2(x - 3)$. Add 8 to get $2(x - 3) + 8$. Use the distributive property and remove parentheses. $2x - 6 + 8$ simplifies to $2x + 2$. The final result: $y = 2x + 2$.

Here is one more example to illustrate the process. Convert the parametric equations $x = 4 + 2t$ and $y = 3t - 10$ to a closed-form equation representing y with respect to x.

Change $x = 4 + 2t$ by reversing the steps and operations.

Subtract 4 $\qquad\qquad\qquad x - 4 = 2t$

Divide by 2 $\qquad\qquad\quad x/2 - 4/2 = 2t/2$

Simplify the results $\qquad 0.5x - 2 = t$

Substitute $0.5x - 2$ for t in the vertical component. Use the distributive property and simplify:

$$y = 3(0.5x - 2) - 10$$

$$y = 1.5x - 6 - 10$$

$$y = 1.5x - 16$$

11. Use symbolic methods to convert the parametric equations to closed-form equations representing y in terms of x. Draw an arrow diagram to illustrate the steps.

a) $x = 6 + t$

$\quad y = 5t - 1$

b) $x = 2t + 5$

$\quad y = 4t + 7$

The arrow diagram is done for you as **Figure 5.59**.

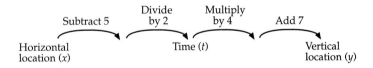

Figure 5.59.
Arrow diagram for Item 11(b).

c) $x = 0.5t - 8$

$\quad y = 3t - 16$

Throughout *Mathematics: Modeling Our World*, you are encouraged to practice useful skills and procedures and to discover your own methods for solving problems. Mathematicians develop several methods for solving or changing equations, and then they choose the appropriate method for a particular problem. Finding the fastest, easiest, or most reliable method is an important task. In this case, step-by-step procedures are not necessarily the fastest or easiest way to eliminate the third variable, t.

On page 452, you were shown how to change the parametric equations $x = 3 + t$ and $y = 8 + 2t$ to the form $y = mx + b$. Here is another approach:

Notice the first equation has a t (or $1t$) and the second equation has $2t$. You want to eliminate the t by substitution. Multiply the first equation by 2 so that $x = 3 + t$ becomes $2x = 6 + 2t$. Now subtract 6 from both sides of the equation to obtain $2x - 6 = 2t$. This means, $2x - 6$ is equivalent to $2t$. Replace $2t$ in the second equation with $2x - 6$. The second equation, $y = 8 + 2t$, becomes $y = 8 + (2x - 6)$. Simplify $y = 8 + 2x - 6$ by subtracting 6 from 8 to get $y = 2x + 2$.

12. Try this new approach or invent your own approach to convert the following pairs of parametric equations to closed-form equations representing y in terms of x.

   a) $x = 6 + t$ $y = 5t - 1$

   b) $x = 2t + 5$ $y = 4t + 7$

   c) $x = 0.5t - 8$ $y = 3t - 16$

13. a) Apply the method you practiced in Item 11 to convert the parametric equations $x = a + bt$ and $y = c + dt$ to a closed-form equation using y in terms of x.

   Begin by solving $x = a + bt$ for t by reversing the steps (subtract a, divide by b).

   b) Suppose your pair of parametric equations are $x = 5$ and $y = 3 + 2t$. What difficulty might you experience when you write an equation to represent the path? What is the equation for the path?

14. The following equations represent the motion of an airplane flying at a level altitude across a region mapped with a grid. Time is measured in minutes and location is measured in miles.

   $$x = 180 - 3t$$

   $$y = 75 + 2t$$

   x represents the horizontal position on the grid or the location in the east-west direction from your reference point.

   y represents the "vertical" position on the grid or the location in the north-south direction from your reference point.

   For example, the point (100, –50) is east and south of the reference point.

a) What is the horizontal rate of change (the east-west direction)?

b) What is the "vertical" rate of change (the north-south direction)?

c) What is the slope of the line representing the path?

d) When is the horizontal or east-west location 40? What equation do you solve?

e) When is the vertical or north-south location 150? What equation do you solve?

f) Where is the airplane at $t = 15$ minutes? Include references to north, south, east, or west in your answer.

g) Is the point (168, 83) a point on the graph of the path? Explain your answer.

h) Explain the meaning of the point (10, 150) in reference to the equation $x = 180 - 3t$.

i) Explain the meaning of the point (120, 115) on the graph of the path.

15. The two time-series graphs (horizontal location versus time and vertical location versus time) in **Figure 5.60** represent the motion of an object.

   a) Explain how to find the location of the object at $t = 3$ seconds.

   b) Where does the object begin (at $t = 0$ seconds)?

   c) Describe the path of the object or prepare a graph showing the path of the object.

16. Create a spreadsheet to model the motion represented by the equations:

   $$x = 9 + 4t$$

   $$y = 20 - 2.5t$$

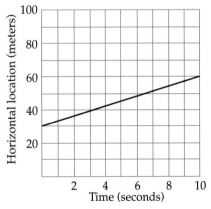

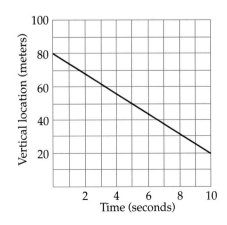

Figure 5.60.
Time-series graphs for Item 15.

ACTIVITY

7

PARAMETRIC PLAY

All parametric equations of the form $x = a + bt$ and $y = c + dt$ produce graphs with similar features. The letters a, b, c, and d are the constants or control numbers. A change in the motion or starting location of an object causes one or more of the control numbers to change. Conversely, when you change a control number, the motion changes. Graphs reflect those changes.

The purpose of this activity is to analyze and generalize the relationship among the control numbers, the situation, and the graph. Then you will be able to describe the location and motion of any point that follows a linear path at a constant velocity. You performed a similar analysis of the control numbers in the equation $y = mx + b$ in Unit 4, *Prediction* (Lesson 1, Activity 2, page 297).

1. The following parametric equations pinpoint location (x, y) at time (t) for a moving object. Graph the parametric equations using a graphing calculator. Experiment with different windows to obtain the best view of the graph.

    $x = 3 + 1.5t$

    $y = 2 + 4t$

2. The variables are x, y, and t. The numbers 3, 1.5, 2, and 4 are the control numbers: $a = 3$, $b = 1.5$, $c = 2$, and $d = 4$. Experiment with different parametric equations by changing the control numbers in a systematic way. How does changing the numbers affect the location and motion of a point? What stays the same and what changes? Write a report on your findings.

    Hint: You may recall from your work in Unit 4, *Prediction*, that if you change all the numbers at once, it is difficult to see how the motion is affected by the individual numbers. Try a systematic approach for gathering information. For example, start by concentrating on the role of the number in the 3 position. How does changing the 3 affect the motion of the point? What effect do negative control numbers have on the location and motion of the point? Develop the rest of your investigation plan yourself.

ACTIVITY

PARAMETRIC PLAY

7

In your report, include a description of your investigation plan. Generalize your results. Suppose $x = a + bt$ and $y = c + dt$ represent the motion of an object. In your conclusions, summarize the role each number (a, b, c, and d) plays in the location and movement of the point. In terms of the motion of the point, what stays the same and what changes as each number in the equations is altered? Include some specific examples of equations that you tried, what happened, and why you think it happened.

Apply your general rule and observations to the following items.

3. Refer to a path graph of $x = 3 + 1.5t$, $y = 2 + 4t$.

   a) Describe how the motion of the object would change if you replaced 3 with 8.

   b) Describe how the motion of the object would change if you replaced 1.5 with 0.5.

   c) Which control number(s) would you change to make the path of the object slope downward instead of upward?

4. An object follows a diagonal path described by the parametric equations:

   $x = 8 + 3t$

   $y = 20 - 2t$

   a) How would you change the equations to make the object to start higher?

   b) How would you change the equations if you want the object to start farther to the left?

   c) How would you change the equations to make the object move faster to the right?

   d) How would you change the equations to make the object go up or down faster?

INDIVIDUAL WORK 7

Constant Changes

1. The parametric equations $x = 5t + 3$ and $y = 6t + 1$ represent the location of a moving object. Measurements for distance and location are in meters and measurement of time is in seconds.

   a) What are the constants in these equations? Explain the meaning of each of the constants.

   b) What happens to the motion of the object if you change 5 to 2?

   c) What happens to the motion of the object if you change 1 to 8?

   d) Identify the explanatory variable(s) and the response variable(s).

2. Use the control panel information in **Figure 5.61** to answer the following items.

Component	Starting location (pixel)	Rate (pixels per sec.)
Horizontal	10	2.5
Vertical	40	-6

Figure 5.61.
Control panel information for Item 2.

   a) Write closed-form equations to represent horizontal and vertical location.

   b) Write recursive equations to represent horizontal and vertical location. Identify the recursive period.

   c) Write an equation to represent the path of the object.

   d) What is the slope of the path, and what does it mean?

3. Consider the following parametric equations. Time is measured in hours. Distance and location are measured in kilometers.

$$x = -3t + 4$$
$$y = -1.5t - 6$$

   a) Complete a control panel like that in **Figure 5.62**.

Component	Starting location	Rate
Horizontal		
Vertical		

Figure 5.62.
Blank control panel for Item 3(a).

b) (3, –5) is a point on the graph of $x = -3t + 4$. Explain the meaning of the point (3, –5).

c) Explain how the motion of the object and the path of the object are affected if you change –3 to +3.

4. Arrow diagrams (**Figure 5.63**) describe the horizontal and vertical motion of a train traveling across an area mapped with a grid. Time is measured in minutes. Distance and location are measured in miles. In this situation, horizontal location refers to locations to the east or west of a reference point and is represented by the variable x. Vertical location refers to locations to the north or south of a reference point and is represented by the variable y.

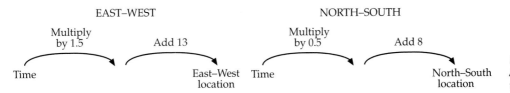

Figure 5.63. Arrow diagrams for traveling train.

a) What is the grid location at $t = 2$ minutes?

b) Complete a control panel.

c) Write closed-form and recursive equations to represent the control panel information. Use x for horizontal or east-west position and y for vertical or north-south position. Indicate the recursive period.

d) Is grid location (22, 11) located near the railroad tracks? Explain your answer.

5. A helicopter was detected at the locations indicated in **Figure 5.64** (mapped with a grid, like the train in Item 4).

Time (min.)	Horizontal position (east-west) (km)	Vertical position (north-south) (km)
1	5	8
3	10	15
6	15	24
10	24	35
15	32	50
20	43	66

Figure 5.64. A helicopter's location.

a) Determine the equation of the best-fit line for the graph of horizontal (east-west) position versus time.

b) Determine the equation of the best-fit line for the graph of vertical (north-south) position versus time.

c) Determine the equation of the best-fit line for the path of the helicopter.

d) Describe how to project the location of the helicopter at $t = 12$ minutes.

e) Parametric equations allow you to represent the horizontal (x) location and vertical (y) location at a particular time (t). What equation might you add to communicate the altitude of the helicopter above the ground?

6. Arrow diagrams identify the relationship between the variables x, y, and t (**Figure 5.65**).

Figure 5.65.
Arrow diagrams for Item 6.

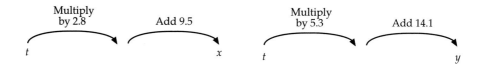

a) Solve for t when $x = 38.4$.

b) Solve for t when $y = 92.6$.

c) Solve for y when $x = 45.8$.

7. The parametric equations $x = 3 - 2t$ and $y = 4 + t$ represent a moving object that does not pass through the point (5, 8).

   a) Change one (and only one) of the control numbers so the object will now pass through the point (5, 8). Explain how you know your answer is correct.

   b) Are other answers possible? Why or why not?

   c) In the context of animation, why is it important to make a moving object pass through a particular point?

8. Recall your work in Unit 4, *Prediction*. Explain how the relationship between powerboat registrations, manatee deaths, and year can be modeled using parametric equations. Describe the advantages to using parametric equations.

9. In Activity 7, you changed the constants one at a time in the equations $x = a + bt$ and $y = c + dt$. You discovered the effect that each constant has on the motion and location of the object. In each case, the object follows a linear path.

   Suppose you replace t with t^2 in one or both of the equations. Describe how the path and time-series graphs are affected by changing t to t^2. Investigate by changing one, then the other, then both equations to $x = a + bt^2$ and $y = c + dt^2$. Systematically change the constants a, b, c, and d, one at a time. Write the observations you make and the conclusions of your investigation.

10. Equations of the form $x_t = x_{t-1} + b$. and $y_t = y_{t-1} + d$ are recursive representations of parametric equations. As part of the control panel you must also indicate the starting point: $x_0 = a$ and $y_0 = c$. Parametric equations of the form $x = a + bt$ and $y = c + dt$ are closed-form representations. Show how to convert the recursive equations to closed-form equations in parametric settings.

The purpose of this activity is to create a calculator or computer program that will animate a point in a diagonal direction.

1. Discuss with your group how to make a point move diagonally. Plan your steps, then create a new calculator program called *Diagonal*. Expect to show the results of your program to the class.

   Write the program lines that work for your group.

Construct calculator programs for each of the following challenges:

2. If you used recursive equations in your program *Diagonal*, create a modified version of *Diagonal* using closed-form equations. If you used closed-form equations, create a modified version using recursive equations.

3. Adapt the program *Misspt* (miss point) to the language of your graphing calculator or computer. The program displays an animated point missing a target point. Modify the program instructions so the animated point passes directly through the target point. Change no more than two of the control numbers.

DIAGONAL MOVES

Program: *Misspt*

(1) Prepare the calculator to display graphics.

(2) $0 \rightarrow$ Xmin:94$\rightarrow$Xmax:0$\rightarrow$Ymin:62$\rightarrow$Ymax

(3) Turn on the point at (40, 30).

(4) Begin the main loop. Use P as the variable. Start at 1. Step by 1. End at 50.

(5) $X=2P+10$ or $2P+10 \rightarrow X$

(6) $Y=-2P+70$ or $-2P+70 \rightarrow Y$

(7) Turn on the point at (X, Y).

(8) Insert a timing loop to delay the animation.

(9) Turn off the point at (X, Y).

(10) End the main loop.

4. Modify your diagonal point animation so the point moves faster by skipping over points.

5. Modify your program *Diagonal* so the point leaves a trail. When the animation is complete you will see the path left by the point.

6. Animate a point so it moves diagonally and bounces off a wall.

7. Animate a point that reverses its direction and retraces its path. (You may want to have it leave a trail and then reverse the direction and erase the trail.)

8. Animate two points in a diagonal direction so that one point chases the other point.

9. Create your own interesting and challenging variation.

INDIVIDUAL WORK 8

Corner to Corner

1. The intent of this program is to animate two points whose paths cross so that the points collide in the center of the screen.

Program: *Diadots*

(1): Prepare the calculator to display graphics.

(2): 0→Xmin:94→Xmax:0→Ymin:62→Ymax

(3): 0→H:0→V

(4): For(P,1,60,1)

(5): H→H+1

(6): V→V+1

(7): Pt-On(H,V,2)

(8): Pt-On(94+H,V,2)

(9): For(T,1,50,1):End

(10): Pt-Off(H,V,2)

(11): Pt-Off(94+H,V,2)

(12): End

a) In the program *Diadots*, identify the lines that contain mistakes.

b) Fix the mistakes. Rewrite the lines correctly.

2. Refer to the example program *Points* when you answer the following items.

   Program: *Points*

   (1): Prepare the calculator to display graphics.

   (2): 0→Xmin:94→Xmax:0→Ymin:62→Ymax

   (3): 0→X:0→Y

   (4): For(P,0,50,1)

   (5): 2P+3→X

   (6): P+2→Y

   (7): Pt-On(X,Y,2)

   (8): For(T,1,200,1):End

   (9): Pt-Off(X,Y,2)

   (10): End

   a) Describe the animation you will see on the calculator screen when you execute the program *Points*.

   b) Change line (4) to For $(P, 10, 50, 2)$. Explain how the animation changes.

   c) Change line (5) to $P + 3 \rightarrow X$. Explain how the animation changes.

   d) Change line (10) to For $(T, 1, 50, 1)$. Explain how the animation changes.

   e) What changes would you make to the program so that the point travels down and to the right instead of up and to the right?

3. a) Write a program called *Sixer* that will plot the following points on the calculator screen:

   (5, 3), (11, 5), (17, 10), (23, 13), (30, 16), (35, 20).

   b) Use regression analysis to find the best-fit line for the points listed in part (a).

   c) Edit your program *Sixer*. Write a program that will plot the six points and then show an animated point following the path of the best-fit line.

4. Write a program named *Twopts* that will animate two points at the same time. Have one of the points move along a horizontal line while the other moves along a diagonal line.

5. Write a program named *Rvsept* to animate a point that begins in the upper-right corner of the screen and travels toward the lower-left corner.

6. Animate a point to follow a curved path rather than a straight-line path.

7. Write a program to draw a square.

8. Electronic equipment detected activity throughout an area mapped by a grid. The data representing detections are recorded in the table in **Figure 5.66**.

Time (min.)	2	2.5	3	4	6
Detection locations	(3, 8) (1,2) (4, 1)	(5, 10)	(5, 9) (7, 11) (8, 6)	(11, 14) (9, 18)	(20, 20) (13, 21) (6, 12) (15, 13)

Figure 5.66.
Table for Item 8.

a) Assume the object travels in a straight line. Which combination of points represents a moving object? Why?

b) Project the location of the object at $t = 10$ minutes.

9. Suppose you graph the parametric equations $x = 25 - 2t$ and $y = 18 + 0.5t$.

a) Which point is not on the graph: (21, 19), (17, 20) or (5, 23)?

b) The point (9, 21) is not on the graph. Describe changes you could make to one or both component equations so (9, 21) is a point on the graph.

t	x	y
2	6	12
5	15	24

Figure 5.67.
Data for Item 11.

10. When is the object modeled by the equations $x = 50 + 1.25t$ and $y = 38 + 0.75t$ closest to the point (100, 75)? How do you know?

11. Use the sample data (**Figure 5.67**) for the location of a particular object and write parametric equations to fit the data.

12. The motion of an animated point is described by the arrow diagrams in **Figure 5.68**. Time is measured in seconds. Distance and position are measured in pixels.

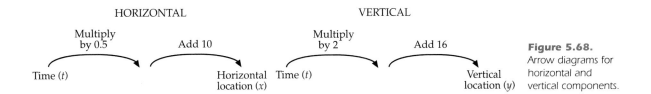

HORIZONTAL   VERTICAL

Multiply by 0.5   Add 10   Multiply by 2   Add 16

Time (t)   Horizontal   Time (t)   Vertical
                  location (x)   location (y)

Figure 5.68.
Arrow diagrams for horizontal and vertical components.

a) Write closed-form and recursive equations to represent the motion. Identify the recursive period.

b) Where is the point at $t = 4.5$ seconds?

c) When does the point reach the location (18, 48)?

d) The point (8, 14) is a solution for the equation $x = 0.5t + 10$. Explain the meaning of the point.

e) What is the slope of the equation $x = 0.5t + 10$? What does it mean?

f) Write an equation to represent the path of the point.

g) The constants or control numbers are 0.5, 10, 2, and 16. Explain how the motion of the point changes if you replace 0.5 with 4.

13. An animated point moves 2 pixels per frame in the horizontal direction and 3 pixels per frame in the vertical direction. The frames change at a rate of 10 frames per second. After 2 seconds the point is at (50, 75).

a) Where is the point at $t = 0$ seconds?

b) When does the point reach location (132, 198)? Explain your answer.

c) When does the point reach location (119, 178)? Explain your answer.

ACTIVITY

9

ESCA-LETTER

The purpose of this activity is to create a calculator or computer program to animate a cluster of points in a diagonal direction.

1. With your group, discuss how to move a cluster of points diagonally. The cluster of points should represent a letter of the alphabet. Plan your steps before you create a new program on the calculator. Expect to show the results of your program to the class. Write the steps for the calculator program that worked for you.

Complete one or more of the following challenges. Write the program lines that accomplish the task described.

2. Modify your program so the letter is tilted as it moves diagonally.

3. Modify your program to make a word move diagonally.

4. Modify your program so the letter or word reverses direction after a certain point.

5. Modify your program so the letter or word bounces off a wall.

6. Create your own interesting and challenging variation.

INDIVIDUAL WORK 9

Up and Over

1. a) In a matrix, write the coordinates of the points to display to create the letter *R* on a calculator screen (**Figure 5.69**).

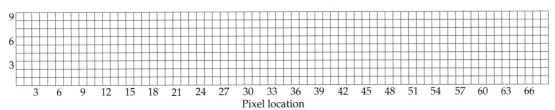

Pixel location

Figure 5.69.
Grid for calculator screen.

   b) In a matrix, write the coordinates for points to display an arrow.

   c) In a matrix, write the coordinates for points to display a smiling face.

2. Refer to the program *EEoption* to answer the following items.

   Program: *EEoption*

   (1): Prepare the calculator to display graphics.

   (2): 0→Xmin:94→Xmax:0→Ymin:62→Ymax

   (3): For(P,0,80,1)

   (4): For(N,1,11,1)

   (5): [F](1,N)→H

   (6): [F](2,N)→V

   (7): Pt-On(H+P,V)

   (8): End

   (9): For(M,1,100,1)

   (10): End

   (11):For(N,1,11,1)

   (12): [F](1,N)→H

   (13): [F](2,N)→V

   (14): Pt-Off(H+P,V)

   (15): End

   (16): End

$$[F] = \begin{bmatrix} 1\ 1\ 1\ 1\ 1\ 2\ 2\ 2\ 3\ 3\ 3 \\ 1\ 2\ 3\ 4\ 5\ 1\ 3\ 5\ 1\ 3\ 5 \end{bmatrix}$$

a) Change line (7) to Pt-On $(H + P, V + P)$ and line (14) to Pt-Off $(H + P, V + P)$. Describe how the motion changes.

b) Change line (7) to Pt-On $(H, V + P)$ and line (14) to Pt-Off $(H, V + P)$. Describe how the motion changes.

c) What lines would you add to the program or change in the program to make the letter travel down and to the left?

3. Refer to the program *Diadots* when you answer the following items.

Program: *Diadots*

(1): Prepare the calculator to display graphics.

(2): 0→Xmin:94→Xmax:0→Ymin:62→Ymax

(3): 0→H:0→V

(4): For(P,0,60,1)

(5): H+1→H

(6): V+1→V

(7): Pt-On(H,V,2)

(8): Pt-On(94–H,V,2)

(9): For(T,1,50,1):End

(10): Pt-Off(H,V,2)

(11): Pt-Off(94–H,V,2)

(12): End

a) Where do the two points collide?

b) Where would the two points collide if you changed line (6) to $V + 2 \to V$?

c) Change the program so the points cross paths without colliding.

4. Suppose you want to animate a point that starts at (12, 25) and ends at (60, 97) in 10 steps or frames.

a) Where is the point after 1 step? Explain how your determine the answer.

b) Where is the point after 3 steps? Explain your answer.

c) Generalize how to determine the displacement for each step when the motion is from (a, b) to (c, d) in t steps.

d) Interpret your answer to part (c) in terms of the equations for the motion and their time-series graphs.

LESSON FIVE

Intersections and Interceptions

KEY CONCEPTS

Rates of change

Linear functions

Elementary programming

Parametric equations

Graphs of functions

Matrices

Systems of linear equations

The Image Bank

PREPARATION READING

Yield the Right-of-Way

Animating a car is easy compared to animating a running person. The parts of the car move together in the same direction. The arms and legs of the human figure move in different directions at the same time.

Animation involves the movement of several points and objects in different directions at the same time. You began with the simplest model and animated one point or one object in a horizontal direction only. Then you animated one point or one object in a diagonal direction by using parametric equations. Now you apply what you learned in previous lessons to animate two points with different motions. If you can animate or track two points with different motions, then you can animate as many points as you like.

Mathematical models involving motion have a variety of applications. There are numerous situations in life where people track moving objects. Air-traffic controllers track airplanes to make sure planes don't cross paths at the same time. Missile-defense systems track objects with the intention of intercepting them. A ball player wants to get to a particular location on a playing field at the same time a baseball, football, or soccer ball arrives. In each application, you look at different ways to predict the path and motion of an object, just like you do in animation.

What can you do to change the path or motion of a point? The control numbers a, b, c, and d for the parametric equations $x = a + bt$ and $y = c + dt$ determine the starting location and velocity of an object traveling in a line. Change the control numbers, and you change the starting location or velocity.

How do you determine if two moving objects are going to collide? Investigating the movements of two points leads to systems of equations. In this lesson you solve systems of parametric equations as you determine whether or not two objects collide.

Apply what you have learned about parametric equations and calculator programming. Discover important mathematics as you take another step in the modeling process and move closer to creating your own animation.

CONSIDER:

1. Explain how two points can cross paths without colliding.

2. Suppose the motion of one object is described by the equations $x = 3 + 4t$ and $y = -5 + 10t$. The motion of a second object is described by the equations $x = 7 - 3t$ and $y = -4 + 4t$. How do you determine if the two objects collide?

3. The horizontal velocity of an object is the rate of change of horizontal location with respect to time. Vertical velocity is the rate of change of vertical location with respect to time. How do you determine the actual velocity of the object if all you know are the horizontal and vertical rates? Is the actual velocity more than, less than, or equal to, the sum of the horizontal and vertical rates?

ACTIVITY

COLLISION COURSE

10

Two animals in a cartoon appear to be on a collision course. By applying parametric equations, you can investigate this situation to determine whether their paths will cross and whether they will collide. Use the partial data in **Figure 5.70** in your investigation. Prepare a report to summarize your discoveries.

Animal A		
Time (min.)	Horizontal (km)	Vertical (km)
0	0	–4
2	4	2
4	8	8
6	12	14

Animal B		
Time (min.)	Horizontal (km)	Vertical (km)
1	0	2
2	2	3
3	4	4
4	6	5

Figure 5.70.
Cartoon animal data.

Your group investigation report must include the following:

- One time-lapse graph displaying the path for Animal A and the path for Animal B.

- Two sets of parametric equations, one set for each animal. (This information will allow the computer to perform a simulation of the action in real-time.)

- A written argument that documents your team's procedure and defends your answers to the items below. This will be valuable as a reference for future investigations.

1. Is each animal moving at a constant rate? How do you know?

2. Is each animal traveling in a straight line? How do you know?

3. Do the animals' paths cross? If so, determine where they cross, and show how you determine the point of intersection.

ACTIVITY

10

COLLISION COURSE

4. Do the animals collide? If so, determine when they collide. If not, explain how to determine when the animals are closest to one another.

5. At $t = 1$ minute, a third animal is sent from a cartoon building (located at (10, 3) to intercept one of the first two animals. This third animal moves with a constant velocity in a straight line.

   a) Identify the control numbers required for the third animal to intercept the first animal at exactly $t = 3$ minutes.

   b) Identify the control numbers required for the third animal to intercept the second animal at exactly $t = 3$ minutes.

   c) Will the third animal require a greater velocity to intercept the first animal or the second animal? Give reasons to support your answer.

6. Write a calculator animation program called *Rabitrdr* to model the motion of the first two animals. Run the program and demonstrate the collision or near-collision of the two animals.

7. Develop a spreadsheet model that keeps track of the location of each animal and the horizontal and vertical distance between the moving animals. Reserve the first column of your spreadsheet for control panel information for the two moving animals.

   a) The table in Figure 5.70 gives the location of the animals at 1- and 2-minute intervals. Design the spreadsheet so the time interval between data points can be adjusted in the control panel.

   b) Explain how to use the spreadsheet to determine if the animals collide.

   c) How does the choice of time interval affect your investigation of the distance between the two animals?

INDIVIDUAL WORK 10

Crossing the Line

1. a) Use the following parametric equations to complete a table like that in **Figure 5.71**:

   $x = 5 + 8t$

   $y = 27 - 6t$

   b) Graph time-series and path graphs of this situation, then write a verbal description of the motion.

Time (sec.)	x (ft.)	y (ft.)

   Figure 5.71. Table for Item 1(a).

2. Write a set of parametric equations to match the data in the table in **Figure 5.72**. Include both recursive and closed-form versions. Use the form x_{new} and y_{new} to start the recursive equations.

Time	x	y
0	15	10
1	17	13
2	19	16
3	21	19
5	25	25

   Figure 5.72. Table for Item 2.

3. The movement of a point is described by the parametric equations:

   $x = 9 + 2t$

   $y = 30 - 3t$

   a) What is the location of the point at $t = 8$?

   b) Explain how to determine whether the object passes through the location (15, 20).

4. An object follows a diagonal path described by the parametric equations:

   $x = 8 + 3t$

   $y = 20 - 2t$

a) How would you change the equations to make the object start higher?

b) How would you change the equations if you want the object to start farther to the left?

c) How would you change the equations to make the object move faster to the right?

d) How would you change the equations to make the object go up or down faster?

5. The locations of two objects are described by the parametric equations:

Object 1: $x = 1 + 9t$

$y = 10 + 3t$

Object 2: $x = 10 + 9t$

$y = 10 + 5t$

a) Which object starts farther to the left? Explain how you made your decision.

b) Which object rises or falls the fastest? Explain how you made your decision.

c) Do the objects' paths cross? Verify your answer graphically.

You can use symbolic methods to determine the point of intersection of two lines. Suppose you want to find where the paths of two objects will cross. Begin by writing equations to represent the paths for the objects. (Use symbolic methods developed in Lesson 3 or use linear regression with points on the path of each object.)

EXAMPLE

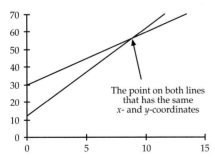

Suppose the path of the first object is $y = 3x + 30$, and the path of the second object is $y = 5x + 12$. At the point of intersection, the x- and y-coordinates are the same for both objects (**Figure 5.73**). The equation $y = 3x + 30$ means y is identical to $3x + 30$ for all points on the first line and the equation $y = 5x + 12$ means y is identical to $5x + 12$ for all points on the second line.

The point on both lines that has the same x- and y-coordinates

Figure 5.73.
Graph showing the intersection of $y = 5x + 12$ and $y = 3x + 30$.

At the point of intersection where the y-coordinates are the same, the expressions $3x + 30$ and $5x + 12$ must also be the same. That is,

$$3x + 30 = 5x + 12.$$

The solution to the equation $3x + 30 = 5x + 12$ is $x = 9$.

You have determined the x-coordinate where the two lines cross. Now find the y-coordinate.

Each equation describes a relationship between the x- and y-coordinate for each point on the line. For the first equation, $y = 3x + 30$, the value of y when $x = 9$ is found by substituting 9 for x: $y = 3(9) + 30$ or $y = 57$.

Check your answer by determining the y-coordinate when $x = 9$ for the second equation, $y = 5x + 12$. Replace x with 9: $y = 5(9) + 12$ or $y = 57$.

The point (9, 57) is a point on both lines because it is a solution to the equation for each line.

6. Practice using symbolic methods to find the points of intersection for the following sets of linear equations.

   a) $y = 5x + 32$ and $y = 9x + 16$.

   b) $y = 18 - 3x$ and $y = 2x - 22$

   c) $y = -2x + 42$ and $y = 7x - 48$

   d) Solve the equation $12x + 43 = 5x + 15$ (you don't need to find a y-value this time).

   e) Solve: $-6 - 2.5m = 6m + 34.5$

7. The motion of Object 1 is modeled by the parametric equations: $x = t + 6$ and $y = 4t + 10$. The motion of Object 2 is modeled by the parametric equations: $x = 3t + 2$ and $y = 6t + 6$. Time is measured in seconds, and distance or location is measured in feet.

   a) Write equations to represent the paths for the two objects.

   b) Describe how to find the location where the paths cross, then find it.

   c) When does each object reach the point where the paths cross?

   d) How do you determine if the two objects collide?

8. Prepare a spreadsheet program to confirm your answers for Item 7.

9. Write a system of parametric equations for two points that collide at location (8, 15) after 4 seconds.

10. The motion of two objects is being tracked and the data are recorded in the tables in **Figure 5.74**. Do the two objects collide? If yes, when? How do you know? If no, how do you know? Explain how you determine your answer.

Object 1

t	0	2	5	6	8
x	−1	1	4	5	6
y	−7	−1	9	11	18

Object 2

t	0	1	3	4	10
x	−8	−4	−1	0	12
y	7	8	10	12	17

Figure 5.74.
Data for Item 10.

11. How can two objects pass through the same location at the same time and not collide? Use the tables in **Figure 5.75** to answer the following items.

Animal 1

Frame	1	2	5	9
Location	(4, 3)	(6, 9)	(12, 27)	(20, 51)

Animal 2

Frame	2	3	8	10
Location	(1, 19)	(5, 21)	(25, 31)	(33, 35)

Figure 5.75.
Data for Item 11.

a) Where do the paths cross?

b) When do both animals arrive at the place where their paths cross?

c) Why don't they collide?

12. Objects 1 and 2 follow nonlinear paths. Do they collide? Where? When?

Object 1: $x = t + 3$ and $y = t^2$

Object 2: $x = 2t + 1$ and $y = 7 - t^2$

The purpose of this activity is to create a calculator or computer program that will animate two points in different directions.

Choose one or more of the tasks from Items 1–5, or create your own. Discuss with your partner or group how you will track two different points. Then plan and program your animation. Expect to show the results of your program to the class.

1. Animate a chase in which one point is trying to catch the other point. The points must follow different paths that eventually cross. This is similar to a predator chasing a prey. Will the predator catch the prey?

2. Two points move diagonally. One leaves before the other. The second point is sent out from a different location at a faster rate to intercept the first point.

3. Two points start at different locations and race to a common destination.

4. Bring the letters *H* and *I* together from two different places following diagonal paths to form the greeting, "HI."

5. Make one point follow a linear path while the other point follows a curved path.

INDIVIDUAL WORK 11

Gotcha

1. Refer to the sample program *Intercep* when you answer the items below.

   Program: *Intercep*

   (1): Prepare the calculator to display graphics.

   (2): 0→Xmin:94→Xmax:0→Ymin:62→Ymax

   (3): For(T,0,50,1)

   (4): 2T+5→M

   (5): T+30→N

   (6): T+10→R

   (7): 3T+4→S

   (8): Pt-On(M,N,2)

   (9): Pt-On(R,S,2)

   (10): For(Z,1,200,1):End

   (11):Pt-Off(M,N,2)

   (12): Pt-Off(R,S,2)

   (13): End

   a) Describe what program *Intercep* is designed to do.

   b) How would the animation change if you eliminated lines (11) and (12)?

   c) Change line (7) to $T + 50 \rightarrow R$. How does the animation change?

   d) Suppose you want to increase the rate of the first point only. What changes do you make to the program?

   e) Line (7) is an example of a closed-form equation in a calculator program. Describe the changes you would need to make in the program if you wanted to make line (7) a recursive equation.

2. One point begins at location (23, 17) and moves horizontally 1 meter per second and vertically 4 meters per second. A second point begins at (58, 29).

First Point

Component	Starting location	Rate
Horizontal	$x=23$ meters	1m/sec.
Vertical	$y=17$ meters	4m/sec.

Second Point

Component	Starting location	Rate
Horizontal	$x=58$ meters	?
Vertical	$y=29$ meters	?

Figure 5.76.
Motion control
panels for Item 2.

a) Determine the horizontal and vertical rate of the second point so that it intercepts (collides with) the second point in exactly 10 seconds. Summarize using motion control panels (as in **Figure 5.76**).

b) Where do the two points collide?

c) Produce tables and a time-lapse graph, on paper, to verify your results for parts (a) and (b).

3. Determine whether or not the moving objects represented by the following parametric equations collide. Explain how you determined your answer.

Object 1: $x = t + 5$ and $y = 2t + 6$.

Object 2: $x = 3t - 5$ and $y = 4t - 4$.

4. Model Item 3 using a spreadsheet program. Explain how your spreadsheet program may be used to determine whether or not the two objects collide.

5. Object 1 begins movement at (2, 8) and moves in a horizontal direction at 1 meter per second and in a vertical direction at 3 meters per second. Object 2 begins movement at (–4, 0) and moves in a horizontal direction at 2 meters per second and in a vertical direction at 4 meters per second.

a) Do the paths cross? If so, where?

b) Do the objects collide? If so, when? If not, when are they closest to one another?

c) If the objects do not collide, change one of the control numbers so that they do collide.

6. Recall in Unit 2, *Secret Codes and the Power of Algebra*, that shift ciphers form a family of functions.

a) What do the parametric equations $x = 2 + t$ and $y = 3 + 2t$ have in common with the parametric equations $x = 3 + t$ and $y = 4 + 2t$?

b) What do the parametric equations $x = 2 + t$ and $y = 3 + 2t$ have in common with the parametric equations $x = 2 + 2t$ and $y = 3 + 4t$?

c) Why may the parametric equations $x = 2 + t$ and $y = 3 + 2t$ be considered members of the same family as $x = 2 - 4t$ and $y = 3 + 3t$?

LESSON SIX

Fireworks

KEY CONCEPTS

Rates of change

Linear functions

Elementary programming

Parametric equations

Graphs of functions

Matrices

The Image Bank

PREPARATION READING

The Grand Illusion

Cartoons, marquee signs, morphs, virtual reality, and video games all have something in common: they create the illusion of motion. Frame by frame, hundreds and thousands of points change, producing images that "move." Behind each move of each point is mathematics.

You first learned how to make a single point move in a horizontal direction. Then you used parametric equations to produce diagonal motion. Matrices were introduced to simplify moving a cluster of points. The task became more complex when you moved two points in different directions at the same time. Each lesson has brought you one step closer to simulating real motion.

In the first lesson, you discussed how to produce the next frame of an animated ball bouncing or of fireworks exploding. In this lesson, you carry out those instructions. Combine your own creativity with the mathematical tools of closed-form equations, recursive equations, parametric equations, programming processes, and matrices to produce a simple animation.

Your group's objective is to program your calculators to animate a star-shaped firework. Compare your answers and programs with those of other members of your group. Help other group members find errors if a program does not work as it should.

In order to animate a complete star-burst fireworks display, you need to break the problem down into smaller parts and solve them. The approach outlined below includes animating the launch of the booster flare, building a star shape, designing its collapse (star to point is easier than point to star), and reversing the collapse to form the exploding star.

1. First, the firework must be boosted up high into the sky.

   a) Several times in this unit you have moved a point from one location to another. This same process can be used for multiple points to animate the star firework. If you know the starting and ending locations, what other information is needed in animating motion?

   b) Create and run the program *Flare* to animate a point rising from (0, 16) to (46, 30).

Collapsing a star back to its center point is very similar to the booster-flare process. Eventually you will reverse the steps that lead to a collapsing star and create an exploding-star sequence. The advantage of beginning with the collapse from a star to a single point is that it is easier to check the initial locations.

2. a) First, you need a star. On a piece of graph paper, carefully place points in a grid to form a star shape centered at (46, 30). Decide, as a group, on the number of points that will make up your star. Write the coordinates for each of the points in your star.

   b) Write equations that move all of your assigned points to the center at (46, 30) in 10 steps. (You need one set of equations for each point.)

YOU'RE THE STAR

Compare your equations with those of other group members. Notice that they differ only in the displacement values and starting locations. This suggests that time could be saved by using matrix notation. For example, matrix [A] could store the coordinates of all starting locations, matrix [B] could store velocity information for all points, and matrix [C] could give "current locations" of all points. Other schemes are possible, too. In matrix notation, then, the recursive parametric equations are shown in **Figure 5.77**.

$$\begin{bmatrix} \text{new } X \\ \text{new } Y \end{bmatrix} = \begin{bmatrix} \text{displacement } X \\ \text{displacement } Y \end{bmatrix} + \begin{bmatrix} \text{old } X \\ \text{old } Y \end{bmatrix}$$

Figure 5.77.
Matrix representation for adding displacements.

3. Use your calculator to write a program called *Star* that collapses your group's star back to its center at (46,30) in ten steps. Consider using matrices to store the *x*- and *y*-coordinates of the points you are animating and their respective control numbers. Remember to include a timing loop to keep the points on briefly.

4. When your collapsing star program is working properly, modify your program *Star* so that it begins at (46, 30) as the central point and expands in 10 steps to be a star.

5. Edit your *Star* program so that its first line is the name of your booster-flare program, *Flare*, or copy all of the lines for *Flare* at the beginning of your *Star* program so that the entire booster flare and starburst are produced by running the new program.

EXTENSIONS

6. Since gravity works to slow the rise of the flare over time, how could you change the lines in your booster flare program to create a more realistic booster animation? Would this require constant or varying velocities?

7. How could you change your program so that it would "morph" from any matrix [A] to any other matrix [D]?

INDIVIDUAL WORK 12

Calculator Animation Project

The field of computer animation is fascinating. Computer programming offers a variety of challenges and opportunities for you to create computer software that make tasks easier for people or games that are fun to play. This unit may be a beginning for you.

Create your own simple and unique animation on the graphing calculator.

Include the following components in your calculator animation:

- A matrix containing the initial coordinates for the object.

- Parametric equations, either recursive or closed-form.

- A timing loop to slow down the animation so it can be observed.

- A diagonal path for the object that changes sometime during the animation or a point-by-point change in shape (like the firework program).

- Your program must be unique and show your creativity.

On a separate piece of paper, submit the following with your calculator program:

- The lines of the program with a note next to each line explaining the purpose of the line. For example:

  $: H + P \rightarrow H$ This line increases the H-coordinate by the amount P.

  $:$ For $(P, 1, 50, 2)$ This line begins a loop. The initial value of P is 1. The statement adds 2 to P with each iteration of the loop until P reaches or exceeds 50.

- The initial location of the object mapped on grid paper to represent a calculator screen. Fill in the pixels.

- A summary of what you learned while doing this project. Include difficulties and challenges you had to overcome.

- A description of changes you would like to make to your animation to make it more complicated or "real" if you had the time to learn more math. This is, in essence, a proposal for your next animation project if there were a unit called *Advanced Animation*.

ANIMATION

SPECIAL EFFECTS

**UNIT
SUMMARY**

Wrapping Up Unit Five

1. The arrow diagrams in **Figure 5.78** describe the horizontal and vertical motion of an object. Distance and location are measured in meters. Time is measured in seconds.

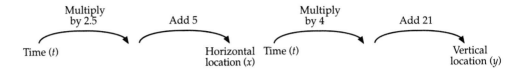

Figure 5.78.
Arrow diagrams for Item 1.

a) Complete a motion control panel (**Figure 5.79**) to identify the control numbers affecting the movement of the object. Include units.

Component	Starting location	Rate
Horizontal		
Vertical		

Figure 5.79.
Control panel for Item 1(a).

b) Write recursive and closed-form parametric equations.

c) Identify the location of the object at $t = 19$.

d) Write an equation to represent the path of the object.

e) When does the object pass through the point (20, 45)? What two equations could you solve to answer this question?

f) What would it mean if the two equations you solved in part (e) gave you different answers?

g) The point (8, 53) is a point on the graph of vertical location versus time. What is the meaning of the point?

h) Is the equation for the path a linear function? How do you know?

i) Change 5 to 12, and explain how the animation will be different.

j) Change 4 to 2, and explain what happens.

k) Explain how to change the direction of the slant of the path. How do you make the object go down and to the right instead of up and to the right?

l) Prepare a three-column table and a time-lapse graph to represent the motion of this object for up to 10 seconds.

m) Would this object collide with an object whose motion is represented by the parametric equations $x = 5t + 1$ and $y = 8t + 16$? How do you know?

Object 1					
Time (min.)	0	4	6	10	15
Location (km)	(1, 5)	(25, 13)	(38, 18)	(60, 25)	(90, 35)

Object 2					
Time (min.)	2	3	8	11	20
Location (km)	(4, 0)	(12, 4)	(52, 24)	(75, 35)	(148, 72)

Figure 5.80.
Sample locations for two moving objects.

2. You have been tracking the movement of two objects across an area mapped by a grid. Sample data are recorded in the tables in **Figure 5.80**. The x-coordinate represents horizontal location or east-west distance from the reference point. The y-coordinate represents vertical location or north-south distance from the reference point.

a) Are both objects traveling in straight lines? How do you know?

b) Write equations for both objects to represent horizontal location with respect to time, vertical location with respect to time, and vertical location with respect to horizontal location.

c) Do the objects collide? When? How do you know?

d) Which object passes closest to (45, 30)? When?

3. Practice skills related to closed-form parametric equations, recursive parametric equations, equations representing path, displacement, and velocity:

a) Change $x = 3 + t$ and $y = 2t + 1$ to the form $y = mx + b$. Show your method.

b) Change $x = 5t - 15$ and $y = 3t + 12$ to recursive form.

c) Change $x_t = x_{t-1} + 4$, $x_0 = 9$ and $y_t = y_{t-1} - 16$, $y_0 = 23$ to closed-form parametric equations.

d) Calculate the vertical and horizontal rates in pixels per second for a point that changes 5 vertical pixels per frame and 2 horizontal pixels per frame and is animated at a rate of 8 frames per second.

e) Calculate the vertical and horizontal displacement after 12 seconds for the points described in part (d).

f) Write an equation to represent the horizontal position of a point after f frames if the point moves from (160, 84) to (50, 132) in 15 frames.

4. Alter the fireworks program developed by your group in Activity 12 to do one of the following:

a) Change the speed of the animation.

b) Add at least three more points to the star.

c) Change the location where the firework explodes.

Write the program lines that you changed.

5. Suppose $x_t = x_{t-1} + 4$ represents the column location for a point in a marquee. The point begins motion at column 22 when $t = 0$. Time is measured in seconds.

a) Describe the motion in words.

b) Where is the point after 18 seconds? How do you know?

c) Write the closed-form equation.

d) When does the point reach column 76? What equation do you solve?

6. An object starts at the location (5, 3) and moves vertically at twice the rate of its horizontal movement. Decide on the component velocities you want to use and the display rate (frames per second), then write both recursive and closed-form parametric equations to represent this motion.

7. You are given three-column tables in parts (a–c) (**Figures 5.81, 5.82, 5.83**). Write a set of parametric equations for each situation. At least one of these sets should be recursive and at least one set should be in closed form. Assume that listed times are for consecutive frames and are measured in seconds.

a)

Time	Horizontal	Vertical
0	3	2
1	5	5
2	7	8
3	9	11

Figure 5.81

b)

Time	Horizontal	Vertical
0	0	10
1	2	7
2	4	4
3	6	1

Figure 5.82

c)

Time	Horizontal	Vertical
0	8	4
0.5	7	10
1	6	16
1.5	5	22

Figure 5.83

8. Look more closely at the three-column table in Figure 5.81, and do the following:

   a) Prepare a time-lapse graph.

   b) Write a description of the motion represented by the graph.

   c) Explain how to find the location of the object at a time not given in the data table, then find the location at $t = 20$.

   d) Explain how to find the amount of time needed for the object to reach a particular location for which the horizontal coordinate is not in the data table. Use your method to find the value of t when $x = 27$.

   e) Determine whether the object will pass through the point (19, 40). Explain how you arrive at your answer.

9. Given a set of parametric equations, complete each table (**Figures 5.84, 5.85, 5.86**). Assume all times are in seconds.

a) $x = 3 + 2t$

   $y = 1 + 4t$

   (1 frame/sec.)

Time	Horizontal	Vertical
0		
1		
2		
3		

Figure 5.84.
Blank table for Item 8(a).

b) $x = 10 - 2t$

   $y = 5 + 3t$

   (2 frame/sec.)

Time	Horizontal	Vertical
0		
0.5		
1		
4		

Figure 5.85.
Blank table for Item 8(b).

c) $x = 6t$

   $y = 1 - 2.5t$

   (1 frame/sec.)

Time	Horizontal	Vertical
0		
1		
2		
3		

Figure 5.86.
Blank table for Item 8(c).

10. Use the table from Figure 5.84 to complete the following tasks:

a) Prepare a graph comparing vertical location (y) to time (t). Plot the four points from the data table. This is a time-series graph.

b) Prepare a second graph to compare horizontal location (x) with time (t). Plot the four points from the data table. This is also a time-series graph.

c) Prepare a third graph to compare horizontal and vertical location. Plot the four points from the data table. This is a graph of the path.

Since points on the graph between adjacent data points "make sense" for this situation, you may connect the points in each of the graphs in parts (a–c).

d) Refer to your graph from part (a). What change(s) would you need to make in the control numbers of the original parametric equations if you wanted the graph to be steeper?

e) If the time-series graph for part (a) were steeper, what changes would you see in the graph for part (c)? In the time-series graph for part (b)?

f) What changes could you make to the control numbers of the original parametric equations so the time-series graphs for parts (a) and (b) would change and the graph for part (c) would remain unchanged?

g) Write an equation in the form $y = mx + b$ for the path graph in part (c).

11. Prepare a three-column table and a time-lapse graph for each of the following sets of parametric equations. Your table must include at least four different times (t) of your own choosing.

a) $x = t^2$ and $y = 2t + 1$

b) $x = 4$ and $y = 3$

c) $x = 2t$ and $y = -2$

d) Write a verbal description of the path of motion for each of the three tables you just completed.

e) Write a set of parametric equations that would create a vertical path.

12. Use the table in **Figure 5.87** to answer the items that follow. Assume that time is in seconds and that the display rate is one frame per second.

a) Prepare a time-series graph of vertical location (y). Plot the representative five points from the data table. Note that the numbers from the x-column will not be used.

t	x	y
0	3	2
1	6	4
2	9	6
3	12	8
4	15	10

Figure 5.87.
Table for Item 11.

b) Prepare a second time-series graph next to your first graph to compare horizontal location (x) with time (t). Plot the representative five points from the data table. You do not need to use the numbers from the y-column of the data table.

c) Prepare a graph of the path. Plot the five points from the data table.

Since motion is continuous (without jumps), points on the graph between adjacent data points make sense for this situation, and you may connect the points in each of the graphs in parts (a–c).

d) Write both recursive and closed-form equations for this motion.

e) Refer to your graph from part (a). What change(s) would you need to make in the original table if you wanted the points to have a steeper linear pattern?

f) If the points for part (a) were changed to follow a steeper path, what changes would you see in the graph for parts (b) and (c)?

g) What changes could you make to the original table so that the graphs for parts (a) and (b) would change and the graph for part (c) would remain unchanged?

13. People who live in places with mountains and snow know about skiing. They are also familiar with ski lifts, which are used to carry skiers from the bottom of a hill to the top. A person who is carried to the top of the hill by the ski lift then skis to the bottom of the hill and takes the lift back up to the top once again.

An animator wishes to show the movement of the ski lift taking skiers up the hill. The ski lift rises 10 feet every second. Its horizontal movement is 20 feet every second. Assume the ski lift begins at (0, 0).

a) Write a set of parametric equations to describe the movement of the ski lift.

b) Prepare a three-column table (like that in **Figure 5.88**) with sample data to be used in graphing.

c) Construct a time-series graph of the horizontal position of the ski lift. Remember to label your axes.

d) Construct a time-series graph of the vertical position of the ski lift.

Time	Horizontal	Vertical

Figure 5.88. Three-column table.

e) Construct a time-lapse graph of this motion, with horizontal position on the horizontal axis and vertical position on the vertical axis.

f) Describe how the graphs in parts (c–e) would change if you changed the vertical climb rate to 15 ft./sec., leaving its horizontal rate unchanged.

g) Describe how the graphs in parts (c–e) would change if you changed the horizontal rate to 30 ft./sec., leaving its vertical rate 10 ft./sec.

h) If the path graphed in part (e) were steeper, describe how the two graphs from parts (c) and (d) would change.

i) How would you change the time-lapse graph to show that the ski lift was traveling faster in the vertical direction? Horizontal direction? Both vertical and horizontal at the same time?

j) The points in the horizontal-position time-series graph slope upward as you move from left to right. What would it mean if the points sloped downward? What would it mean if the points were level or horizontal?

k) How would the graphs from parts (c–e) change if the ski lift were descending at a rate of 10 ft./sec. with the horizontal rate still 20 ft./sec.?

Mathematical Summary

A nimation is the result of small changes in location for thousands of points over several frames. The computer animator uses mathematics to keep track of the location of each point and communicate the type of change.

A reference system is necessary to locate points. When a point moves only in a horizontal direction, a single number can be used to identify the horizontal distance along a number line. Each location x is paired with a time t.

As a point moves, the location changes by an amount called displacement. The amount of displacement depends on the velocity of the point and the amount of time.

$$\text{displacement} = \text{velocity} \times \text{time}$$

Units of time may be measured in real time (seconds, minutes, hours) or animation time (frames). The rate at which a point appears to move depends on the displacement from one frame to the next and on the number of frames displayed per second.

The changing location of a point along a line may be represented with two types of equations.

Closed-form equations relate one quantity to another quantity. In the context of motion, closed-form equations may be used to identify the current location in terms of time or frame number. The current location is the starting location increased by the displacement that has occurred since the point started moving.

$$\text{current location} = \text{starting location} + \text{displacement}$$

$$\text{current location} = \text{starting location} + \text{velocity} \times \text{time}$$

$x = a + bt$ where a is the starting location at $t = 0$ and b is the velocity.

Recursive equations relate one quantity to a previous or next value of the same quantity. In the context of motion, recursive equations identify the current location with respect to the previous location based on a corresponding change of one unit of time. The displacement is calculated from one unit of time to the next.

$$\text{current location} = \text{previous location} + \text{displacement}$$

$$\text{current location} = \text{previous location} + \text{velocity} \times 1$$

$x_{\text{current}} = x_{\text{previous}} + \text{displacement}; \; x_{\text{initial}} = \text{location at time } 0$

$x_t = x_{t-1} + b; \; x_0 = a$

A starting location must be specified when you use recursive equations.

Diagonal movement, or movement in two dimensions, requires a reference system using two variables such as (x, y) to identify the location of a point. An equation of the form $y = mx + b$ may be used to identify the path of an object, but the equation of the path does not tell when the point passes through a particular location.

Parametric equations are used to identify the location of an object at a particular time. Location is determined by the combination of a horizontal component and a vertical component. Both closed-form and recursive equations may be used to find the horizontal and vertical coordinates of a point at a given time (remember, time may be measured in seconds or frames).

Closed form:

$$x = a + bt$$

$$y = c + dt$$

Recursive form:

$$x_t = x_{t-1} + b, x_0 = a$$

$$y_t = y_{t-1} + d, y_0 = c$$

The letters a, b, c, and d are called constants or control numbers. The letters a and c identify the starting location for the object (at $t = 0$). The letter b represents the horizontal velocity, while the letter d represents the vertical velocity. Control numbers may be listed in a control panel (see **Figure 5.89**).

Component	Starting location (units)	Rate (units per time)
Horizontal	$x = a$	b
Vertical	$y = c$	d

Figure 5.89.
Sample parametric control panel.

A negative horizontal velocity means that the point is moving from right to left. A negative vertical velocity means that the point is moving downward. When b and d are both positive, the point moves upward to the right. When b and d are both negative, the point moves downward to the left.

You can alter the path of the point or object, the "where," and the location at a particular time, the "when," by changing one or more of the control numbers.

Symbolic methods have been introduced as a way to convert from one form of an equation to another and from one type of representation to another. Parametric equations with variables x, y, and t can be converted to a single equation with x and y. The first equation is solved for t by reversing the steps and applying the inverse of the operations used. The new expression for t, now written in terms of x, is substituted for t in the second equation. The arrow diagrams in **Figures 5.90, 5.91, and 5.92** and generalized equations illustrate this process.

Begin with time as the explanatory variable and horizontal location as the response variable (Figure 5.90).

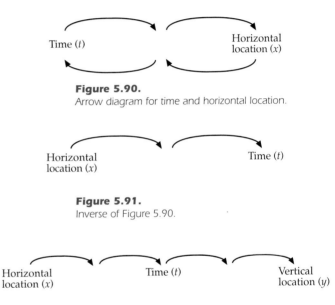

Figure 5.90.
Arrow diagram for time and horizontal location.

Apply the inverse operations and show horizontal location in the position of explanatory variable and time in the position of response variable (Figure 5.91).

Figure 5.91.
Inverse of Figure 5.90.

While time is the response variable in this inverse process, it becomes the explanatory variable in the next step, the substitution step (Figure 5.92).

In the final, combined equation, horizontal location is the explanatory variable and vertical location is the response variable.

Figure 5.92.
Arrow diagrams to show substitution process for time.

The process illustrated in the arrow diagrams in Figures 5.90–5.92 may be generalized using symbolic equations and the substitution method. Solve the first equation of the form $x = a + bt$ for t. Substitute the resulting expression for t in the place of t in the second equation.

Using a specific example, $x = 3 + 2t$ and $y = -1 + 4t$ may be converted to $y = 2x - 7$ by following the steps illustrated in the arrow diagrams:

$$x = 3 + 2t \qquad\qquad y = -1 + 4t$$

$$x - 3 = 2t \qquad\qquad y = -1 + 4(0.5x - 1.5)$$

$$t = \frac{1}{2}(x - 3) \qquad\qquad y = 2x - 7$$

$$t = 0.5x - 1.5$$

Symbolic methods may be used to find the point of intersection of the paths of two objects. If $y = ax + b$ and $y = cx + d$ represent the paths of two different objects, then the y-coordinate for both objects is identical where their paths cross. The x-coordinate of the point of intersection may be found using substitution and solving the equation:

$$a + bx = c + dx.$$

Usually the first step in solving an equation similar to $a + bx = c + dx$ is to collect the x terms on one side of the equation and combine these terms. The resulting equation to be solved is similar to equations you solved in Units 2, 3, and 4 (*Secret Codes and the Power of Algebra, Landsat,* and *Prediction*).

Using a specific example, solve to find the point of intersection for $y = 5x + 24$ and $y = 2x + 45$. Substitute for the y-values to obtain the equation: $5x + 24 = 2x + 45$.

To change $5x + 24 = 2x + 45$ to equivalent equations until you obtain $1x$ or x, subtract $2x$, subtract 24, divide by 3:

$$5x + 24 = 2x + 45.$$

Subtract $2x$ from both sides to get the x terms together:

$$5x - 2x + 24 = 45.$$

Combine $5x - 2x$ to get $3x$:

$$3x + 24 = 45.$$

Subtract 24 from both sides:

$$3x = 45 - 24,$$

$$3x = 21.$$

Divide both sides by 3:

$$1x = 21/3 \text{ or } 1x = 7 \text{ or } x = 7.$$

Animation involves the movement of many points at the same time. Several points are joined together to form a letter or object. The coordinates of several related points may be organized in a matrix. In this unit you used a 2 x n matrix to list the coordinates of all the points in your animated figure.

The graphing calculator is a useful tool for understanding how important mathematics is to animation. The programming language of the calculator is similar to the programming language used by designers of animation software. The Pt-On(x, y) allows you to light up the pixel at location (x, y), and the Pt-Off(x, y) allows you to turn off the same pixel. The command Pt-On([A] (1, 3), [A] (2, 3)) would light up the point with x-coordinate found in the first row, third column of matrix [A] and y-coordinate found in the second row, third column of matrix [A].

Loops are used for steps that are repeated several times in a program or process. For most loops in programs, it is necessary to identify a "counter" variable, together with its starting and ending values and the size of the steps by which to count.

Closed-form and recursive equations have a different format when included in a calculator program.

$$x = a + bt \text{ becomes } a+b*T \rightarrow X.$$

$$x_t = x_{t-1} + b \text{ becomes } X+b \rightarrow X.$$

These simple commands and equations are used to animate points and clusters of points on a calculator screen.

Movement in life is continuous. Movement simulated by animation or video is discrete. The difference between continuous and discrete is like the difference between riding your bicycle down a smooth ramp or riding it down stairs (well, not quite that bad!).

Mathematics is the key to computer animation. There is a lot that is happening and a lot that is changing in the field of computer animation, so get moving!

Glossary

ACCUMULATOR:
A variable that keeps track of total amounts. It is most often used in a loop process to store total amounts that increase (or decrease) with each pass through the loop.

CLOSED-FORM EQUATIONS:
Equations that allow you to find the value of one variable given the value of the other variable. Equations of the form $c = 2p + 4$, $y = 0.5x + 10$, and $x = -3 + 5t$ are examples of closed-form equations.

COUNTER:
A variable to which a constant (usually 1) is added; it keeps track of the total times a process is performed.

DISPLACEMENT:
Distance in a particular direction. Displacement of −2 meters and +2 meters represent the same distance, but in opposite directions.

FRAME:
A picture in a set of pictures that are displayed sequentially. Motion pictures on a television screen normally display 30 frames each second.

ITERATION:
Each instance that a process is repeated. A process repeated over and over is called an iterative process.

LOOP:
A programming structure that allows a process to be repeated several times. The most common loop used in this unit is the For/End loop.

PARAMETRIC EQUATIONS:
Two or more equations, each relating a different response variable to the same explanatory variable.

PATH GRAPH:
The graph of y versus x. In other contexts, the path graph is often referred to as the state-space graph.

RATE OF CHANGE:
Another way to think about the concept of slope, it is a ratio or fraction that compares the change in one quantity to the change in another. In animation, it is often used to show a change in location with respect to a change in time.

RECURSIVE EQUATIONS:
Indicate the relationship between the current value of a variable based on the previous value of the same variable using a constant increment in the explanatory variable. Recursive equations require designation of an initial condition.

REFERENCE POINT:
The zero point in a reference system.

REFERENCE SYSTEM:
A system for designating a location using a reference point and a consistent measuring scale and direction to indicate distance or displacement from the reference point.

STARTING LOCATION:
The location of an object when $t = 0$ in the chosen reference system for time.

TIME-LAPSE GRAPH:
A graph of y versus x that includes sample times displayed on the graph to show when a moving object passes through a particular location on the graph.

TIME-SERIES GRAPH:
Any graph or equation that uses time as the explanatory variable.

VELOCITY:
The ratio of the displacement of an object to the duration (time) of that displacement. It is the rate at which an object moves in a specific direction.

UNIT

6

Wildlife

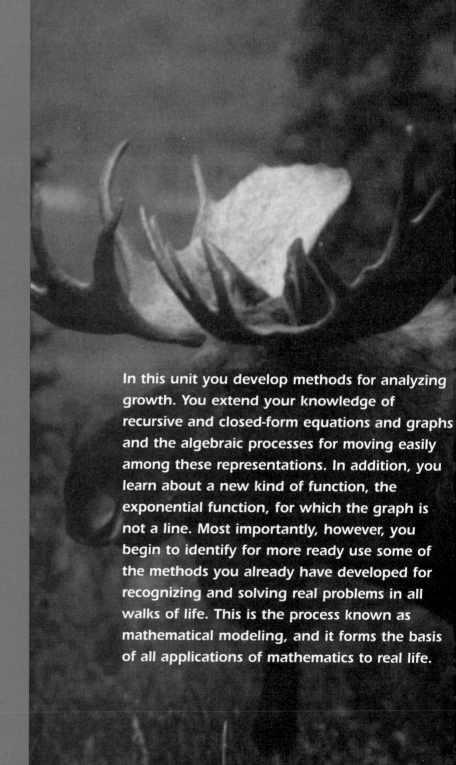

In this unit you develop methods for analyzing growth. You extend your knowledge of recursive and closed-form equations and graphs and the algebraic processes for moving easily among these representations. In addition, you learn about a new kind of function, the exponential function, for which the graph is not a line. Most importantly, however, you begin to identify for more ready use some of the methods you already have developed for recognizing and solving real problems in all walks of life. This is the process known as mathematical modeling, and it forms the basis of all applications of mathematics to real life.

WILDLIFE

Often situations that appear to be quite different are actually similar in some way. For example, wildlife management, investing money, and dealing with warm liquids on cool days do not seem to have very much in common.

Wildlife managers are charged with the responsibility of understanding the natural world and with helping environmental systems survive. One facet of that work is observing and recording population data for various populations within a system. They need to be able to answer the following questions: How do wildlife populations change through time? Is it possible to make reasonable predictions about how large a population will become in the future? Should humans intervene? If so, when and how?

Many individuals and corporations invest their money. Banks are one type of institution with which you are familiar. One of their main tasks is to accept money from depositors, invest it, and return some portion to the depositor. To encourage people to save with them, banks pay interest; the more interest they pay, the more likely they are to gain new customers. The people who invest money with banks need to answer the following questions: How do balances in simple savings accounts change through time? How long will it take for an initial deposit to become large enough for some predetermined purpose? How does the size of the initial deposit affect growth?

A warm drink or cup of soup is good on a cold day. But have you ever let something sit out to cool off a little, only to return to find it too cold to be good? How long is long enough?

Although they are very different in many respects, each of these situations is the same mathematically. In this unit you develop an understanding of the mathematics behind these situations so that you can apply it to the many phenomena that share this mathematical similarity.

LESSON ONE

What Was the Question?

KEY CONCEPTS

The mathematical modeling process

Linear functions

Function notation

Parametric graphs

Additive processes

Graphical, numerical, and symbolic representations

PREPARATION READING

Bullwinkle Returns

*A*dirondack State Park is a six-million-acre wilderness area in upstate New York. Prior to 1980, the last moose recorded in the park had been shot in 1861. After 1980, however, some moose were again seen there. In 1988, it was estimated that between 15 and 20 moose were in the park. In 1993, new estimates put the number at 25 to 30 moose. The New York State Environmental Conservation Department (ECD) conducted a survey at that time to determine what policies the public favored. A majority of the people surveyed favored a "gradual increase in the moose population as the animals migrate from nearby New England states and Canada and an expansion of their numbers through natural reproduction." Conservationists suggested moving 100 moose into the park over a three-year period. The ECD determined that such a plan would cost $1.3 million.

Put yourself in the position of commissioner of the ECD, and suppose that you must make a recommendation to the governor about this situation.

CONSIDER:

1. Pose a specific question concerning these data to which you would like an answer.

2. What additional information do you need in order to be able to answer your question?

ACTIVITY

1

THE ANATOMY OF A MODEL

Much of mathematics is the study of patterns. Patterns in numbers or calculations are clearly mathematical. But there are other patterns that are just as important. In fact, the way you *use* math is one of the most important features of mathematics. In this unit, you not only add to your collection of math tools, you also consider systematic approaches that can help you make progress in unfamiliar situations.

Think back to the first five units of this course. In Unit 1, you figured out how to make a fair decision when a lot of people were involved. In Unit 2, you sent secret messages that a friend could understand, but that outsiders could not easily read. In other units, you looked at the earth from a great distance, made predictions from fragmentary data, and animated cartoons.

What common thread runs through all the work you have done in these varied settings? You worked with words, numbers, tables, matrices, arrow diagrams, equations, graphs, geometric figures, and data. Those are some of the mathematical things you studied, the "tools of your trade." How were those tools used? What are some common features of the thinking—the problem solving—you did?

The process of beginning with a situation and gaining understanding about that situation is generally referred to as "modeling." If the understanding comes about through the use of mathematics, the process is known as **mathematical modeling**. You have been doing mathematical modeling since you began the year. Here is one general summary of the main steps in modeling.

THE MODELING PROCESS

Step 1.
Identify a situation: Notice something that you wish to understand, and pose a well-defined question indicating exactly what you wish to know.

ACTIVITY

THE ANATOMY OF A MODEL

1

Step 2.

Simplify the situation: List the key features (and relationships among those features) that you wish to include for consideration. These are the assumptions on which your model will rest. Also note features and relationships you choose to ignore for now.

Step 3.

Build the model: Interpret in mathematical terms the features and relationships you have chosen. (Define variables, write equations, draw shapes, gather data, measure objects, calculate probabilities, etc.)

Step 4.

Evaluate and revise the model. Go back to the original situation and see if results of mathematical work make sense. If so, use the model until new information becomes available or assumptions change. If not, reconsider the assumptions you made in step 2 and revise them to be more realistic.

One general principle that guides every modeler is hidden in the second step: keep it simple. In general, all models ignore something, and first-draft models usually ignore several things. Good models are simple enough to be understood by someone else, but are still accurate.

Except for the third step, these steps can be applied to many tasks: first decide what you're trying to do; narrow your focus to a few things that really matter; and then evaluate how well you did and make necessary adjustments. This advice is helpful not just for solving mathematical problems, but for writing essays, preparing a new recipe, or developing a new play in your favorite sport. Rarely does the first attempt at doing something unfamiliar result in the best you can do.

These steps are really just a set of broad guidelines. Each step can involve several smaller steps that depend on the particular situation you are investigating. For example, in the third step, after putting things into mathematical form, you may need to explore the mathematics a bit. That is, you may need to solve equations, transform graphs or figures, or move to other representations in your original model. And there is certainly no rule saying that the steps must be carried out in the exact order listed here. For example, you might wish to evaluate and revise well before you get to the end of the process. Even the problem you are solving can change as you learn more. The more you learn, the more questions you may have.

Look at these steps with your work on voting methods in mind. You wanted to understand decision-making. That desire led to the question, "How can a group reach a decision in a fair manner?" The most obvious feature of this situation is that it is composed of a number of individual people involved in the decision-making process and a number of individual selections from which the choice is to be made. You assumed that each individual decision-maker had a ranking of the choices, from best to worst, with no ties. Of course, "fair" had to be defined. You assumed that the opinion of each decision-maker should be included in an equal manner. You also assumed that the final decision should not end up being the worst choice for a majority of the group. However, when you considered the first model—everyone votes only for their first choice, highest vote total wins—you discovered that this last assumption was sometimes violated. This result was not desirable, so you needed to revise the model.

In fact, you went through this model-and-revise cycle several times. Each model was an improvement over its predecessor, but none was perfect. This did not mean the work was worthless. Remember, the goal is insight, understanding. Do you know more about what to watch for in elections? Can you avoid being manipulated by others? Absolutely.

1. Review your work in earlier units and identify as many of the steps of the modeling process as you can. Be sure you can state the goal of each investigation as a specific question whose answer you sought.

2. For the moose problem described in the preparation reading, pose a specific question, the answer to which would be useful in your role as commissioner.

INDIVIDUAL WORK 1

Round One

1. Describe a difficult decision you have made recently. Discuss how each step in the modeling process is reflected in your thinking in making that decision.

2. The media have reported stories of students carrying weapons to school, brandishing them at teachers and peers. School crime and violence affect rural, suburban, and urban communities, and public and private schools. Consider the following information.

   • The Department of Justice recently reported that nationally 100,000 guns are brought to school each day.

   • A National Crime Survey released in May 1991 found that about 50% of all violent crimes against youths aged 12–19 occur in school buildings, on school property, or on adjacent streets. Twelve percent of the violent crimes in school buildings involve an offender with a weapon.

   • The Centers for Disease Control and Prevention state that 1 in 5 school children reports carrying a weapon of some type. About 1 in 20 (5.3%) said they had brought a gun to school. Almost 1 in 3 students in 31 Illinois high schools said they had brought a weapon to school for self-protection at some time during their high-school career.

   • Sixty percent of all administrators, nationwide, maintain that guns have never been involved in incidents at their schools.

   [Nordland 1992] [Stephens 1994]

   a) What does all this mean for your school? Pose a specific question concerning these data to which you would like an answer.

   b) What additional information would you like to have in order to answer your question?

3. In June 1994, British tabloids fanned fears in England with headlines such as "Eaten Alive" and "Killer Bug Ate My Face." The stories told of a "deadly flesh-eating bacteria" that killed 11 people in England and Wales in 1994. These fears spread to the United States when it was reported that a handful of cases—and at least two deaths—had occurred in the U. S. that same year.

The disease is passed by human-to-human contact, and its favorite route of entry is through an open wound. The streptococcus bacterium is very common (it causes "strep" throat), but the publicized cases involve a rare variant of the germ, called strep-A. If caught early, the infection is easily treated with antibiotics. The bug is so virulent, however, that it cannot be taken lightly.

Strep-A may now afflict 10,000 to 50,000 Americans per year; perhaps 3000 die. However, this microbe has been around for years, and no one thinks it will suddenly cause an epidemic. Your chances of getting a life-threatening strep-A infection are roughly comparable to those of being hit by lightning.

a) Pose a specific question concerning these data to which you would like an answer.

b) What additional information would you like to have in order to answer your question?

4. a) In Unit 5, *Animation/Special Effects*, you studied recursive equations for describing an object's location. Describe the motion having the following recursive representation:

$$x_{next} = x_{current} + 4, \, x_{initial} = 7.$$

(Distance is measured in meters; 1 frame = 1 second; initial time = 0.)

b) Give at least two other representations of this motion, and write a sentence about each one indicating when it would be useful.

c) One way to use your calculator with recursive descriptions of processes is known as "home screen iteration." It may be used with any calculator that has a key for the answer to the previous calculation. For many calculators, the key is labeled ANS. To iterate means to repeat, so you might guess that the method involves repeating something on the home screen. In fact, that's correct. Here is how to use it.

First, type the initial value and press ENTER. (For the situation in 4(a), press 7 ENTER.)

Next, type the recursive equation, using ANS as your current value, and press ENTER. (For 4(a), press ANS+ 4 ENTER.)

If your calculator has a replay feature, repeatedly pressing the ENTER key lets you step through the process. Each ENTER is another step.

Figure 6.1 shows one student's screen after three steps.

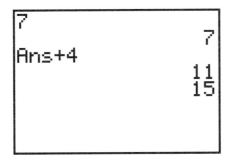

Figure 6.1.
Home screen iteration.

Use home screen iteration to find the location of the object in 4(a) after 20 steps. Check using some other method.

d) Every recursive description is based on the period of the recursion, which is the size of one step of the counter. In the description in 4(a), the period was 1 second. Consider a motion described recursively by the equation

$$x_{next} = x_{current} + 8, x_{initial} = 7.$$

(Distance is measured in meters; 1 frame = 2 seconds; initial time = 0.)

What is the period of this recursive description? How is the motion different from that described in 4(a)? Be specific.

5. a) At a book store, the first pencil you buy costs 25¢. After that, though, each additional pencil costs only 17¢. Write a recursive equation describing the total amount charged for buying pencils, ignoring tax.

   b) Find the corresponding closed-form equation for this situation, and use it to determine how much 503 pencils would cost.

   c) Make two graphs, one for the recursive description, showing next cost v. current cost, and the other for the closed-form description, showing cost v. pencils. Then explain where information about pencil pricing is shown in the graphs and equations.

6. a) Even if you did not answer Item 5, read through its parts. How is the phrase "ignoring tax" related to the modeling process?

   b) Where does the work done in answering the parts of Item 5 fit into the modeling process?

7. Fadi was just beginning to work on a problem when his brother spilled ink on the data table. **Figure 6.2** shows what was left of the information and the question he was trying to answer. Help him solve the problem.

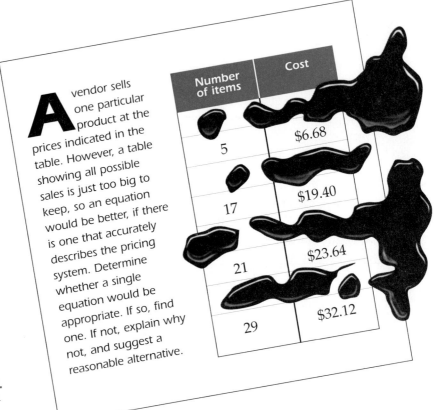

A vendor sells one particular product at the prices indicated in the table. However, a table showing all possible sales is just too big to keep, so an equation would be better, if there is one that accurately describes the pricing system. Determine whether a single equation would be appropriate. If so, find one. If not, explain why not, and suggest a reasonable alternative.

Number of items	Cost
5	$6.68
17	$19.40
21	$23.64
29	$32.12

Figure 6.2.
Pricing table.

In the situations you have examined thus far, you have dealt with the relationship between one quantity and another. For example, you have considered location and time (Item 4), and cost and number of pencils (Item 5). Tables do an excellent job of letting you know how the two quantities are paired. But what if you want to talk about just one specific case, one pair?

There are two standard notations with which you should be familiar. The first, the **ordered pair**, you have already seen and used quite a bit. For example, the pair (2, 42) means that 2 pencils cost 42¢ in the context of Item 5. Ordered pairs are always written in parentheses, and units are usually omitted from the pair. However, units need to be made clear to the reader, or the pair becomes meaningless. You have used this notation throughout your graphing work.

The second notation, called **function notation,** is new. Recall that the equations you have used since Unit 2, *Secret Codes and the Power of Algebra,* define functions. With function notation, you can label one of the quantities so that you may better remember what it represents. For example, with the pencil example, you might write $c(2) = 42$. This is read as, "*c* of 2 is 42," or "*c* at 2 is 42,"or more fully, "The value of *c* when *p* is 2 is 42." The fact that 2 is the number of pencils (*p*) has to be known in advance, and you have to know that *c* represents cost in cents. You can use any letter, word, or symbol you like. Thus, the number in parentheses is the value of the independent variable (number of pencils, here), and the other number is the value of the named quantity (cost).

Two notes are important here. First, the parentheses in function notation do *not* mean multiplication. Instead, they are telling you that you are looking at something called *c* when something else is 2. And that's the second idea. Think of the $c(2)$ as being a single quantity. Remember, it's actually just a different way of writing 42.

8. a) For motion, state in words the meaning of the ordered pair (0, 7).

   b) What does $x(5) = 27$ mean in the context of Item 4 and function notation?

   c) Write (0, 7) in function notation.

   d) What does (17, 19.40) mean in the context of Item 7?

   e) For Item 5, which of the following are correct?

   (1, 25) (4, 76) (0, 8) $c(3) = 59$ $c(6) = 93$ $c(14) = 246$

   f) For Item 7, write two correct pairs, first using ordered pairs and then using function notation.

9. Find an equation of the line containing the two points (0, 15) and (5, 25). Explain your method and interpret the equation.

10. a) The graph in **Figure 6.3** relates temperatures in two different systems. Use the graph to approximate the temperature in degrees Celsius that corresponds to a temperature of 98.6°F. Explain in words how you used the graph.

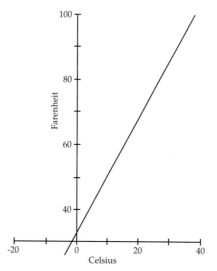

Figure 6.3.
Fahrenheit temperatures equivalent to given Celsius temperatures.

b) Approximate the slope of the line in the graph in Figure 6.3. (Be sure to read the axes carefully.) What are the units for the slope? What does the slope tell you about the two temperature scales?

c) Reading from the graph, list two ordered pairs describing equivalent Celsius and Fahrenheit readings. Then write the same information in function notation.

d) Write an equation describing the relationship between Fahrenheit and Celsius temperatures.

11. a) The parametric equations $y = 24 + 0.47t$ and $x = t$ describe the motion of an object in the plane. Describe the motion in words. Be sure to include interpretations for all numbers in the equation and indicate appropriate units of measure for each quantity.

b) Use the parametric mode on your calculator to graph these equations. Sketch the graph, and indicate what it tells you about the motion. Be sure to mention all important information.

c) Find an equation for the path of motion. (If possible, use both graphical and algebraic methods to find the equation and compare the results from the two methods.) Then use the normal graphing mode on your calculator to graph your equation and check that the graph is the same as the one you obtained in part (b).

d) Compare the original parametric equations for this motion to your answer to part (c). Explain why, based on only the original equations, the results in part (c) are not surprising.

12. a) The equation $h = 7 + 4t$ describes the motion of an object along a horizontal line. Graph the equation in the usual way (not parametrically) on your calculator by using y for h and x for t. Describe your graph and explain what its main features tell you about the actual motion (select reasonable units of measure as needed).

   b) Now change your calculator to parametric mode. The graph in part (a) had t on the horizontal axis and h on the vertical axis. Therefore, in parametric equations, try $x = t$ and $y = 7 + 4t$. Sketch the corresponding graph and compare it to the graph from part (a). (Adjust the t-values the calculator uses in its display so that the same values are used as were used for x-values in part (a).)

13. a) In the equation $y = 24 + 0.47x$, x represents the number of small candies in a package, and y represents the weight of the package (in grams). Explain what each other number in the equation tells you about the pack of candy. Be sure to include appropriate units of measure.

   b) Graph this equation, and compare it to the graph you obtained in Item 11(b).

   c) Your answers to Items 12(b) and 13(b) might suggest that the parametric mode can be used to graph any equation that you can graph normally (not just those applying to motion). If that is true, explain how to do it, and demonstrate by graphing the equation $y = 0.5x^2$, first in the usual mode and then parametrically. If it is not true, explain by showing an example of an equation that can be graphed in the usual way but not parametrically.

14. a) Since the recursive form of the motion in Item 4 involves only the addition of a constant to the current location, it is referred to as an **additive process**. A process described by a recursive equation of the form "next = current + constant" is an additive process. If an object's motion is described by an additive process, what does that tell you about the actual physical motion itself?

   b) Processes other than motion can also be additive. Read the situations described in Items 5, 7, 11, 12, and 13(a). Which of these are additive processes? Explain.

c) For motion, being additive has a specific meaning. Explain what additivity tells you about each of the situations you identified in part (b). What does additivity tell you about the closed-form graphs?

d) Suppose in Item 10 that you wanted to make a table of corresponding temperature values, counting 1°C for each row in your table. Would that generate a table of Fahrenheit temperatures that could be described by an additive process? If so, write its recursive equation. If not, explain how you know.

Note that any recursive description necessarily generates only a discrete set of values, so graphs showing only what comes from the recursive description always consist of sets of isolated points. The recursive process involves stepping from current to next with nothing in between. However, for both motion and temperatures, the quantities location and temperature can take on values between those produced by the recursive equation.

One advantage of closed-form equations is that they may be used to graph *all* points within a particular relationship, not just those that fall "on the steps." Of course, this means that closed-form equations can sometimes give you unrealistic values. For example, in the pencil-pricing situation, the closed-form equation produces a cost for 2.3 pencils, but it is extremely doubtful that anyone would really use it!

Remember, the most important calculator you have is the one between your ears. When modeling, always check to be sure that what you are doing makes sense, both mathematically and contextually.

LESSON TWO

A First Model

KEY CONCEPTS

The mathematical modeling process

Recursive and closed-form descriptions

Inequalities

Simulation

Translations

Scale changes

PREPARATION READING

Moose Facts

*I*n Activity 1, you formulated one or more questions whose answers would help you, as the commissioner, decide whether to bring in more moose. No doubt some of the questions you asked were relatively simple and could be answered with just a little research. Others will become the focus of your model-building experience. Some questions may not have simple answers and may require very complex mathematical models or other approaches. For example, the question of what is desirable is one that mathematics cannot answer directly. Mathematical models may help describe quantities of interest, but mathematics cannot decide whether a particular number is good or bad.

CONSIDER:

The moose scenario is based on a short article from *The New York Times* (April 7, 1993). The article did not answer the following questions.

1. The ECD is a state agency trying to make decisions about spending $1.3 million in state funds. Whom might they have surveyed to get opinions about their decision?

2. What questions might they have asked on the survey?

3. Why might conservationists favor moving 100 moose into the park?

4. What other groups of people might have an interest in the number of moose in the park?

Questions such as those above are important, but they generally cannot be answered by a simple mathematical model. Therefore, turn your attention to important questions that *can* be dealt with mathematically.

The first step in the modeling process is to identify a situation of interest. One question that you or your classmates may have posed to help make the commissioner's decision is "How many moose will be in the park if 100 (or 0) additional moose are brought in?"

The second part of the modeling process is to make simplifying assumptions that assist you in describing the problem mathematically.

In general, you normally have two choices for each feature you identify. The first, and usually simplest, option is to make an assumption about the feature. You may assume that it does not contribute much to the situation and ignore it. (That *is* an assumption.) You may also assume some particular behavior or value and build your model on that assumption.

The other option is to avoid making any assumptions about the particular feature and instead make it a variable. Making the feature a variable is usually a bit more complex than making a single assumption about it, but that process gives a better final description of the effect of the feature. Generally, though, it is wise to limit such variables to one per modeling cycle.

> For each feature you identify, you generally must either:
> - make an assumption, or
> - treat it as a variable to observe.

As noted, the question of understanding the effect of bringing in 100 moose (or not) has been identified. Here are some facts about moose, some of which are repeated from Lesson 1.

- Moose are the largest living members of the deer (Cervid) family.

- At birth, a moose calf weighs 24 to 35 lb.

- Moose can survive in snow up to 3 1/2 ft. deep.

- An Alaskan bull moose may be 7 ft. tall and weigh up to 1800 lb.

- Young bulls are more likely to travel far from their birthplace and are sometimes driven away by stronger bulls.

- The moose's only predator in Adirondack State Park is the black bear.

- Many deer carry a parasite benign to deer but usually fatal to moose. The deer pass parasite larvae to foliage that is eaten by snails. Moose may consume the snails while eating plants.

- Well-developed moose antlers may weigh 55 to 75 lb.

- The average life span of a moose in the wild is 15 to 23 years.

- A moose can gallop up to 35 mi./hr. for short distances.

- Moose do not tolerate prolonged temperatures above 75°F.

- An adult moose consumes 35 to 60 lb. of plant materials daily.

- Moose were first seen in Adirondack State Park after 1980.

- In 1988, it was estimated that 15 to 20 moose were in the park.

- In 1993, new estimates put the number at 25 to 30 moose in the park.

- The conservationist group that advocated importing 100 moose projected that such action would result in a total population of about 1300 moose in Adirondack State Park in 20 years (i.e., in 2013).

ACTIVITY

2

MIGRATING MOOSE

1. Identify the most relevant factors listed in the preparation reading. Because there are so many factors involved in this problem, it may be necessary to make additional simplifying assumptions later in the modeling process. Remember—the simpler, the better.

2. Reread the question posed above the list of facts. Note that time has not been mentioned. It is likely that the effects of the decision to bring in additional moose (or not) will differ over time. So, even though it may not have been discussed directly before, time is one of the features of the situation. Decide whether it should be dealt with as an assumption or be treated as a variable, and defend your decision.

3. Carry out the second and third steps of the modeling process. As clearly as possible, state all the assumptions that you believe may be helpful. Then, using only your assumptions, construct mathematical representations of the growth of the moose population. Use as many representations as you think help explain what is going on, but be sure different representations are based on the same assumptions. Consider tables, graphs, equations, arrow diagrams, etc.

4. Use your model(s) to describe the situation if 100 additional moose were to be added, as suggested by the conservationists. Be sure to show explicitly how your model is used.

5. Use any unused information from the original facts list to carry out the last step of the modeling process, evaluating your model. Describe both your method and your results.

INDIVIDUAL WORK 2

Analyzing Migration

Y ou first encountered additive models for change in Unit 5, *Animation/Special Effects*. The initial model for the growth of the moose population is also additive. The following items provide a close look at some of the properties of additive processes in general, examine other situations in which additive models might be appropriate, and explore variations of additive models of the Adirondack State Park moose population.

1. JaMelle examined the given information about moose and made the following assumptions.

   - Population changes only by migration of new male moose into the park.

   - The same number of moose arrive each year.

   - The migration rate is the slowest that is also consistent with the 1988 and 1993 observations.

   - There are no deaths.

   - No additional moose are brought into Adirondack State Park.

   a) What population values did JaMelle assume for 1988 and 1993? Explain. Write your answer both in ordered pair and function notation.

   b) What migration rate did she use? Be sure to include proper units.

   c) Write a single sentence describing mathematically the exact process by which the moose population increases. Is this an additive process?

   d) Rewrite your answer to part (c) as a recursive equation, stating the next population in terms of the current population. Include information about starting the process, too.

   e) Though you really don't need your calculator here, use home screen iteration to determine the population for the year 2013.

   f) Use your answers to part (a) to determine a closed-form equation for this migration model. Without drawing its graph, what kind of graph would you expect for this equation?

g) Look at the equations you wrote for parts (d) and (f). In part (d), your variables were next population and current population. In part (f), the variables were year and current population. Altogether, you used three variables. Make a table with a separate column for each quantity that appeared in your equations. Complete rows representing the populations predicted by JaMelle's model for the years 1988 through 1998.

h) Graph each pair of quantities in your table. For each pair, explain how you decided which variable should be on the horizontal axis and which belongs on the vertical axis.

i) For each graph, use methods you learned in Unit 4, *Prediction*, to find an equation describing the pattern. Are any of the equations surprising?

j) The model you developed using JaMelle's assumptions is an additive model. Explain which assumption(s) forced that to be true.

2. Mark decided to model the moose population using essentially the same assumptions, with the following change:

- The migration rate is the fastest that is also consistent with the 1988 and 1993 observations.

Build this model, recursively and in closed form, using graphs and tables as appropriate. Use home screen iteration to predict the population for 2013.

Inequality notation is a mathematical statement of a directional relationship between two numbers. For example, saying that the number of moose in Adirondack State Park in 1988 is between 15 and 20 can be written using the inequality $15 \le p(1988) \le 20$, provided that everyone understands that p represents the population of moose.

This inequality is a contraction of two separate sentences: $15 \le p(1988)$ and $p(1988) \le 20$. In such a contraction, the inequalities must have the same sense; that is, they must point in the same direction. Thus you can say that $20 \ge p(1988) \ge 15$, but you would not say something like $15 \le 20 \ge 18$. As you might suspect, sentences involving $<$ or $>$, either by themselves or with $\le$ or $\ge$, follow the same rules.

3. a) Write an inequality describing the observed moose population in 1993.

b) Write an inequality describing possible rates of change (in moose per year) that are consistent with the 1988 and 1993 observations and satisfy JaMelle's other assumptions (see Item 1).

c) Adopt JaMelle's basic assumptions, and use the results of part (b) to write an inequality about the year in which the first moose returned to the park after all the years with no moose. Explain your method.

d) Keep the same first four assumptions, but change the last one to allow 100 moose to be added at the end of 1993. Write an inequality about the predicted moose population in 2013.

e) Explain what the statement $15 \le p < 20$ means if p represents the moose population at a particular instant. What does it mean if p represents the length (in centimeters) of a pencil? Other than the units of measure, what is the fundamental difference between these two meanings?

4. Glynnis baby-sits and wants to begin a regular savings plan. One bank offers young depositors an account on which there are no service charges and no interest. Thus the money is available if needed, but it is not invested. Glynnis opens such an account with an initial deposit of $50. Each month after that, she adds exactly $25 to the account.

a) List the assumptions that seem to have been made and then build both recursive and closed-form mathematical models (equations) representing this situation.

b) Is this an additive model? Explain why or why not, based only on the initial assumptions you listed in part (a).

c) Make a table with four columns labeled: time in bank, current balance, next balance, and change in balance (see **Figure 6.4**). Use the table to record the results of simulating one year's worth of activity in Glynnis's account, reflecting your initial assumptions. Then explain how your answer to part (b) is visible in the table. (Note: Do your accounting after each deposit. The first row has been completed as an example.)

Time in bank (mo.)	Current balance ($)	Next balance ($)	Change in balance ($)
0	50	75	+25
1	75		

Figure 6.4.
Glynnis's bank account information.

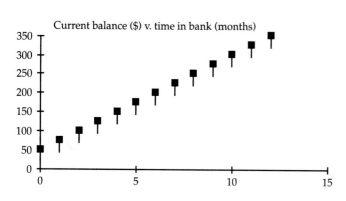

Figure 6.5.
Signposts on the bank
balance graph.

d) **Figure 6.5** shows a graph of current balance versus month. Glynnis has added little vertical segments to the points of the graph, creating what appear to be "signposts." Describe the information that these signposts give you about the savings situation. Be specific, relating your response to your work in part (a) as closely as possible. What would it look like if the signposts were drawn along the *x*-axis instead of hanging from the actual points?

e) Glynnis described her current balance/next balance record-keeping system to Chris. He thought about it for a bit and then suggested that she could use prior balance and current balance as her recursive variables. Do you agree? If so, write the corresponding recursive equation. If not, explain carefully.

Age (days)	Weight (lb.)	Age (weeks)	Weight (lb.)
0	6000	0	6000
1	6200	1	7400
2	6400	2	8800
3	6600	3	10,200
4	6800	4	11,600
5	7000	5	13,000

Figure 6.6.
Records of whale weights
using different time scales.

5. It is true that blue whales are born weighing approximately 3 tons, and they grow at a rate of about 200 lb. per day for the first few months of life. **Figure 6.6** shows excerpts from growth records kept by two different (hypothetical) research institutes for a single whale.

a) Copy these tables and add columns for the change in weight. Then write recursive equations describing these two sets of data, and explain whether they represent additive models.

b) Graph the weight-versus-time data using the units stated in the tables. For each graph, find the corresponding closed-form equation representation. Explain how each could be more valuable than the other.

c) Explain how the slopes of the graphs in part (b) are related to your recursive equations in part (a). Show the addition of weight from point to point on your graphs using the signpost scheme used in Item 4.

d) Extend the graph of the first data set so that it includes several weeks of time. Then calculate the slope of that line using points that are one week apart, and show the related changes on the graph.

e) Use unit-conversion methods to relate the slopes of the two data sets.

6. A candle is 1 foot high and burns at a constant rate. It takes exactly 1 1/2 hours for the candle to be fully consumed. Graph the height of the candle versus time, and find a closed-form equation for the graph. Then write at least two different recursive equations for the height of the candle. (Hint: Consider different time periods for the recursive forms.)

Items 1–6 illustrate the defining characteristic of additive models, which is based on the recursive description of such models.

> Additive models are characterized by the fact that in recursive calculations the change in values from step to step is constant.

7. For each of the following closed-form equations, construct a short four-column table showing the explanatory variable, current and next values of the response variable, and the difference between next and current values. (See Item 4.) Then identify the model as additive or not. Keep as many decimals in your calculations as needed to identify the form of the model.

a) area $= 0.5(\text{length})^2$

b) distance $= 55 \times \text{hours}$

c) $y = 3x + 4$

d) population $= 500/\text{tagged}$

8. Write the recursive equations for three different additive processes. Each of your equations should have two variable quantities (something$_{next}$ and something$_{current}$). Make a recursive graph of each equation: graph something$_{next}$ versus something$_{current}$. Describe how the

geometry of the graphs is related to the algebra of your equations and then formulate what you think is a general rule for the graphs of recursive equations for additive processes.

9. Identify each process as additive or not. Explain your answers carefully.

a) $p_{next} = p_{current} + 28$

b) $p_{next} = p_{current} - 28$

c) $p_{next} = 2 \times p_{current} + 28$

d) $p_{next} = 1.3 \times p_{current}$

e) $p_{current} = p_{prior} + 1.7$

f) $p_{current} = p_{prior} - 1.7$

10. **Figure 6.7** shows several graphs. For each one, approximate its equation if you can. (If you can't, tell why you can't.) Then indicate whether it could be the graph of a recursive equation for an additive process and whether it could (also or instead) be the graph of the closed-form equation for an additive process. Justify your claims.

Figure 6.7.
Graphs of five models.

a)
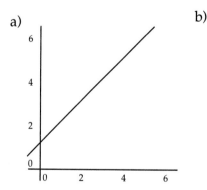

b)
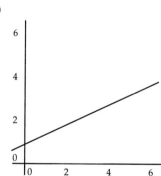

c)
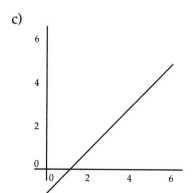

d)
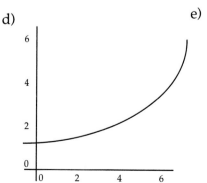

e)
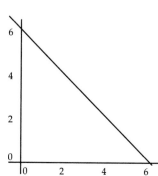

TECHNO-MODELING

Remember that an additive model is one in which a constant is added to one variable as the other variable increases by 1. The variable that increases steadily by 1 is the independent variable; it is graphed on the horizontal axis in a closed-form graph and serves as the counter in the recursive form.

You have seen that the closed-form graph of every additive model is always a straight line and that the slope of that line is exactly equal to the amount of the addition in the recursive form. In symbols, if the recursive form is $p_{next} = p_{current} + m$, then m will also be the slope of the closed-form representation of p versus the counter. Thus the closed-form equation will be something like $p = mx + b$, where x is the value of the counter and b is related to the initial values of the variables.

Notice that the units for m differ in the recursive and closed forms. For the recursive model, m has the same units as the dependent variable because it is added to that variable. It represents the *amount* of change in that variable in one count. For the closed-form model, m has the units of the dependent variable divided by the units of the independent variable. It is the slope or the rate of change of one variable with respect to the other.

Here is one closed-form representation of the moose population over time, using the migration-only assumption:

$$p_{future} = (\text{migration rate} \times \text{time}) + p_{initial}.$$

Time represents the number of years since moose first returned and is the independent variable for this model. Since time starts when moose first return, $p_{initial}$ is 0 in this closed-form expression. As was the case in Unit 5, *Animation/Special Effects* this equation still has the general form:

future amount = initial amount + change in amount.

Equation representations are great for many purposes. And each of the main forms, recursive and closed-form, is the best form in some situations.

ACTIVITY

3

TECHNO-MODELING

CONSIDER:

1. For what purposes might the recursive form be preferred?

2. When might you prefer the closed form?

One thing that models allow you to do is to play out a scenario. You did that in class to predict the moose populations for the year 2013. You probably did so again in approximating the time at which moose began returning to the park. You may have used a closed-form equation or built a table of values (by hand or using a calculator or spreadsheet), or counted on your fingers. You have a lot of options. However, whatever the method you use, when you enact a situation, you are doing a **simulation.** Simulation is an important tool for modeling. It can help build a model as well as help check one.

In this activity you learn to use simulation in what-if situations. In addition, you examine how modelers can compare models that use different (but related) variables.

1. Consider the pencil-pricing model that you first examined in Lesson 1. Recall that the first pencil cost 25¢, but each additional pencil cost only 17¢. This is an additive model, described recursively as $c_{next} = c_{current} + 17$, $c_{initial} = 25$. Note here that the explanatory variable is the number of pencils and that the initial value corresponds to 1 pencil, not to 0, as is frequently the case. For now, though, pretend that you know neither this equation nor the corresponding closed-form equation. Here are some ways you might simulate this situation to gain some understanding if you had never seen it before.

   a) Physical. Get several pencils and a lot of change. If these are not readily available, make substitutes. For example, strips of paper could be pencils and smaller squares of paper could be coins. Then act out the assumptions—the cost rules—for various numbers of pencils. Be systematic.

TECHNO-MODELING

ACTIVITY

3

Start with one, then two, and so on. Record the relationship between pencils and price in an organized manner, and then look for patterns. For this situation (pencils and price), what are the strengths and weaknesses of this method? For what kinds of situations might this kind of hands-on simulation be more useful?

b) Calculator. Home screen iteration is a good way to simulate simple models. For the pencil situation, enter the initial price (25). Note that with this method you need to keep track of certain things in your head. Units is one of them, so remember cents. What else must you track mentally? Now enter the formula ANS + 17 on the home screen. Repeated pressing of the ENTER key produces a sequence of numbers. What do these numbers represent? Record your results, and compare them to those you obtained in part (a).

c) Computer. One of the best tools for simulating recursive models is a computer spreadsheet. Open your spreadsheet. Set up a table with column headings "Pencils" and "Total Cost" in cells A1 and B1, respectively. In the next cell of the Pencils column (A2), enter the number 1. In the adjacent cell (B2), under the heading Total Cost, enter 25 (or 0.25, if you prefer to work in dollars instead of cents). This sets the initial conditions. In the second cell of the Pencils column (A3), enter the formula to add 1 to the previous number; for many spreadsheet programs a formula such as =A2+1 or +A2+1 does the job. In the adjacent Total Costs cell (B3), enter the formula to add 17 to the previous number, such as =B2+17. Then extend these last two formulas down a few lines. Compare the results of your spreadsheet to those from the previous two methods. Comment on strengths and weaknesses of this method.

ACTIVITY

TECHNO-MODELING

3

d) Relying on only a simulation model (or two), determine the cost of 25 pencils if the store changed its prices so that the first pencil still cost 25¢, but each additional pencil cost only 13¢. That is, find $c(25)$.

e) Relying again on only a simulation model (or two), determine $c(78)$ if the store changes its prices so that the first pencil costs 27¢ and each additional pencil costs 17¢.

f) Still relying on only a simulation model (or two), determine the cost of each additional pencil if $c(1) = 23$ (cents) and the store owner wants $c(50)$ to be $6.60.

g) The store owner now wants to price pens in a similar fashion. Relying only on a simulation model (or two), determine the cost of the first pen and the (lower) cost of each additional pen if the store owner wants $c(50)$ to be $19.75.

h) Comment on the usefulness of the various types of simulations for investigating variations of the original setting.

As you have seen, calculators and computers can help make predictions far into the future when working from a recursive model. You also know that closed-form equations permit such predictions even more easily, requiring just a few calculations to complete. However, one drawback to using closed-form equations is the need to agree on the actual labels for your explanatory variable. For example, in order to use a closed-form equation for the moose population, you need to know how the equation represents a particular year, say 1988. Is it recorded as year 0, year 88, year 1988, or something else?

Year	Current population
0	18
1	20
2	22
3	24
4	26

Figure 6.8.
Tara's model.

Year	Current population
88	18
89	20
90	22
91	24
92	26

Figure 6.9.
Karim's model.

2. **Figures 6.8 and 6.9** show the results of parts of simulations of Tara's and Karim's migration-only models for the moose population.

a) Add two columns to each table to show, for each year, the next population and the change in population. Use that information to decide whether the models are additive, and explain your decision.

ACTIVITY

TECHNO-MODELING

3

b) Write recursive and closed-form equations to describe the two models.

c) Graph the closed-form equations from part (b) and use the signpost idea to show clearly on your graph how to see your answer to part (a).

d) Tara decided to answer part (c) using the parametric graphing mode of her calculator. This is the setup she used: $x_1 = t$, $y_1 = 18 + 2(t - 0)$. She set the graphing window so that t took on values between 0 and 25 (so $0 \le t \le 25$) in steps of 1, and the values of x and y were described by $0 \le x_1 \le 30$ and $0 \le y_1 \le 70$. Try it, discuss whether it does the job, and explain your reasoning.

e) Karim liked Tara's parametric idea, but he wanted to make it work for his model. He reasoned as follows. "Since my time values are just 88 more than each of Tara's time values and her time is represented by both t and x_1, I should be able to add 88 to t or x_1 to get time values for my model. And the population values are identical." This is the setup he used: $x_2 = x_1 + 88$, $y_2 = y_1$. He set the graphing window so that, again, $0 \le t \le 25$ and $0 \le y \le 70$. However, since the values of x were his time values, he chose $80 \le x_2 \le 110$. Again, try it and discuss whether or not it seems correct. Explain your reasoning.

f) Adjust the graphing window so that both Tara's and Karim's graphs can be seen at the same time. Describe the geometry of the graphs. How are they alike, and how do they differ? Be as specific as possible. Trace, moving back and forth between the graphs, to help describe their relationship.

g) Go back to the parametric equations used by Tara and Karim. Since some of the variables in those equations were defined to be the same (for example, $x_1 = t$), they may be substituted for each other. Rewrite the equations so that y_2 is expressed in terms of x_2 and constants only. Compare your result with your work in part (b).

INDIVIDUAL WORK 3

Additive Models

1. Have you ever wondered how the seven cities of ancient Troy came to be built on top of one another? Read the following excerpt from the book, *Rubbish! The Archeology of Garbage*, to find out.

A human being's first inclination is always to dump. From prehistory through the present day, dumping has been the means of disposal favored everywhere, including within cities. Archaeological excavations of hard packed dirt and clay floors—the most common type of ancient living surface—usually recover an amplitude of small finds, suggesting that many bits of garbage that fell on the floor were trampled into the dirt or were brushed into corners and along the edge of walls by the traffic patterns of the occupants. . . . The archaeologist C. W. Blegen, who dug into Bronze Age Troy during the 1950s, found that the floors of its buildings had periodically become so littered with animal bones and small artifacts that "even the least squeamish household felt that something had to be done." This was normally accomplished, Blegen discovered, "not by sweeping out the offensive accumulation, but by bringing in a good supply of fresh clean clay and spreading it out thickly to cover the noxious deposit. In many a house, as demonstrated by the clearly marked stratification, this process was repeated time after time until the level of the floor rose so high that it was necessary to raise the roof and rebuild the doorway." Eventually, of course, buildings had to be demolished altogether, the old mud-brick walls knocked in to serve as the foundations of new mud-brick buildings. Over time the ancient cities of the Middle East rose high above the surrounding plains on massive mounds, called tells, which contained the ascending remains of centuries, even millennia, of prior occupation. In 1973 Charles Gunnerson, a civil engineer with the U.S. Department of Commerce . . . calculated that the elevation . . . [of] Troy was about 4.7 feet per century."

[Rathje and Murphy 1992]

Build one or more simulation models of this situation. Be sure to state your assumptions clearly. Create a table that could be used to put Troy's elevation change in terms the average modern person could understand easily. Would the change be noticeable?

2. a) Recall that in Item 4 of Individual Work 2, Glynnis opened her account with $50, made monthly deposits of $25, and received no interest. Suppose that she made her initial deposit at the end of May. Since May is the fifth month of the year, she would like her table of balances to indicate the proper month, at least for the first year. Thus, instead of the initial balance being for month 0, she would like it to be for month 5. Make a table of values indicating this change in labeling.

   b) Graph the balance v. month number for both the old method of keeping records and the new method. Describe how the two graphs are related.

   c) Find closed-form equations for the two graphs in part (b).

   d) Use parametric equations to graph the two methods on the same axes. Note that month numbers in the new system are exactly 5 more than in the old system. Write your parametric equations and sketch their graphs to confirm that they agree with part (b).

   e) Combine your parametric equations using substitution to write a closed-form equation for the balance in terms of the new month numbers. How does this compare to your answer to part (c)?

3. Imran is examining equations whose graphs are lines. Recall that you learned about translations first in Unit 2, *Secret Codes and the Power of Algebra,* and then when translating images in Unit 3, *Landsat.* Now, Imran wants to be able to do the same thing to graphs. He knows the graph of the equation $y = 3x + 5$ is a line.

   a) Write a pair of parametric equations that have the same graph as this equation.

   b) Tell Imran how he could write a new set of parametric equations that would have the effect of moving his line 5 units to the right. Graph and trace to verify that corresponding points really have been translated exactly 5 units to the right.

   c) Explain how he could use parametric equations to translate his original equation to the left by 3.2 units. Again, graph and trace as a check.

d) Write a general rule describing how to write parametric equations of a line to translate it exactly k units to the right.

e) Use substitution to convert your answers to parts (b) and (c) to closed-form equations of lines. Graph your new equations to verify that they have the same graphs as the parametric equations from which they came.

f) Without doing any work, write the closed-form equation of Imran's line if he translates it to the right by 6 units.

g) Write a general rule describing how to rewrite the closed-form equation of a line to translate it exactly k units to the right.

h) Graph the equation $y = 0.5x^2$. Does your answer to part (g) apply to this nonlinear graph?

4. Write closed-form equations for each of the indicated translations of the given equations.

a) $y = 3x - 4$. Translate right 1.5 units.

b) $y = 1.5x + 7$. Translate right 4 units.

c) $3y + 2x = 10$. Translate left 1 unit.

5. For each of the pairs of graphs in **Figures 6.10–6.14**, explain how their equations must be related.

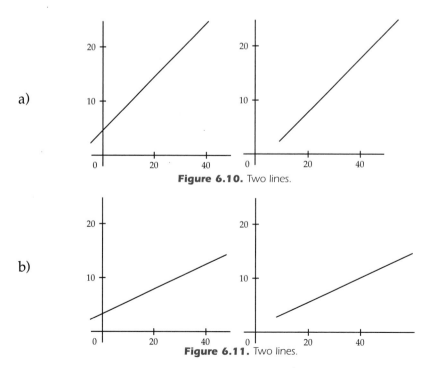

a)

Figure 6.10. Two lines.

b)

Figure 6.11. Two lines.

c)

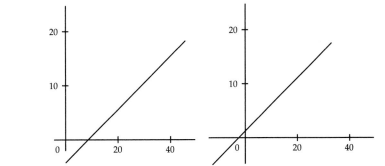

Figure 6.12. Two lines.

d)

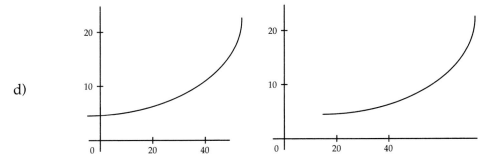

Figure 6.13. Two curves.

e)

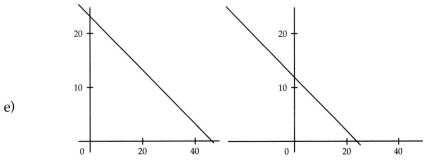

Figure 6.14. Two lines.

6. In Item 5 of Individual Work 2 (Lesson 2), two research institutes kept different kinds of growth information on a baby blue whale. Recall that one group measured time in days; the other used weeks. Thinking of the two scales, then, one group would use 7 tic marks on the axis for a week and the other group would use 1 tic mark for a week. From a graphical point of view, the daily model is wider than the weekly model; it shows more detail.

a) Write closed-form equations for the weight-versus-time models used by the two institutes.

b) Graph these two equations on the same set of axes. (Note that the horizontal axis can be labeled "time," but the units are different for the two lines. One uses weeks and the other uses days.)

c) In describing the geometry of the two lines you might first say that one line is taller, or steeper than the other. However, since the geometric difference is caused by changes in the time (horizontal) scale, there should also be a description that draws attention to that fact. Describe the effect in "horizontal" terms. (Hint: Compare several pairs of points with common y-values.) This kind of transformation is known as a **scale change transformation**.

d) Write parametric equations with x as time and y as weight for the weekly model. Graph your equations in the window $[0, 100] \times [-2000, 25{,}000]$ to verify agreement with the closed-form equation's graph.

e) Assuming that x_1 and y_1 form your parametric definition in part (d), enter the new equations $x_2 = 7x_1$ and $y_2 = y_1$. Explain how these equations are related to the daily model. Then, graph your equation (with the weekly graph) in the same window you used in part (d) to verify that it agrees with your earlier graph of the daily model. Explain the horizontal geometry of this graph relative to the graph in part (d).

f) Use substitution to write the parametric equations in part (e) in closed form by eliminating the parameter t. Compare this result to your equation in part (a).

g) Suppose that a third institute had kept records monthly (1 month = 4 weeks). Explain how their graph would compare to those you sketched above. Is it wider or narrower? How much? Write parametric equations relating their graph to the weekly model and then write the appropriate closed-form equation.

h) Generalize the relationship between rescaling the horizontal variable and the effect of the scale change transformation on a graph. Be as specific as possible.

7. Write closed-form equations for each of the indicated scale change transformations of the related graphs. (You may wish to check by graphing both the given equation and the transformed equation in the same window.)

a) $y = 3x - 4$. Make 1.5 times wider.

b) $y = 1.5x + 7$. Make 4 times narrower.

c) $3y + 2x = 10$. Make 2 times wider.

8. The additive moose population models discussed thus far have been unable to match all the population information that was given at the start of this lesson. That means that at least one assumption needs to be changed. Here are the assumptions that were listed in Item 1 of Individual Work 2, which led to the additive models you have examined thus far.

• Population changes only by migration of new male moose into the park.

• The same number of moose arrive each year.

• There are no deaths.

• No additional moose are brought into the park.

For each assumption, comment on whether changing it might be a reasonable adjustment to compensate for the particular inaccuracies observed with these models. If more than one change might be helpful, discuss which you would change first. If you have ideas about how to change the assumptions (not just which ones), discuss those, too.

LESSON THREE

Population Explosion

KEY CONCEPTS

The mathematical modeling process

Sensitivity

Additive processes

Relative rate of growth

Exponential functions

PREPARATION READING

Moose Are Sensitive, Too

*I*n Lesson 2, you examined several simple (probably additive) models of moose population growth. You began working on the last phase of the modeling process:

- Evaluate and revise the model. Go back to the original situation and see if results of your mathematical work make sense. If so, use the model until new information becomes available or assumptions change. If not, reconsider the assumptions you made in step 2 and revise them to be more realistic.

"Evaluate" has more than one meaning here. One is obvious: Check the numerical predictions that your model produces against the known numbers. However, evaluation also involves something known as **sensitivity analysis**. In the event that your first evaluation is not satisfactory, sensitivity analysis is important as you consider possible changes in your assumptions. It seeks to quantify how the predictions your model makes would change as you change particular assumptions. If small changes in assumptions make large differences in the final results, the model is said to be sensitive to that change and you will need to be sure those assumptions are as accurate as possible. On the other hand, if changing a particular assumption has very little effect, you can wait until the rest of the model is in good shape before dealing with it, or you can choose to ignore it altogether. Remember, if two models seem to do equally well numerically, the simpler model is the better one!

Key assumptions included in the additive models of the moose population are listed in Item 8 of Individual Work 3 and are repeated here. You were asked to rethink each of these assumptions, looking in particular at how they affected the numerical predictions of the additive models.

ADDITIVITY ASSUMPTIONS:

- Population changes only by migration of new male moose into the park.

- The same number of moose arrive each year.

- There are no deaths.

- No additional moose are brought into the park.

Other assumptions (or facts) were provided to you as a basis for making the assumptions listed here. Some of the facts you may have considered in making your assumptions are as follows.

FACTS SUPPORTING ADDITIVITY ASSUMPTIONS:

- Young bulls are sometimes driven away by stronger bulls and are more likely to travel far from their birthplace.

- The only predator of the moose in Adirondack State Park is the black bear.

- The average life span of a moose in the wild is 15 to 23 years.

None of these facts, however, provide you with any information that will allow you to complete the evaluation stage of the modeling process; you can't check your model against any of these statements. Other facts were more quantitative. Some are listed here.

QUANTITATIVE FACTS:

- After 1980, moose were first seen in Adirondack State Park.

- In 1988, it was estimated that 15 to 20 moose were in the park.

- In 1993, new estimates put the number at 25 to 30 moose in the park.

- The conservationist group that advocated importing 100 moose projected that such action would result in a total population of about 1300 moose in the park in 20 years (2013).

Now it is time to evaluate the additive models. Here are some questions you might need to answer through the course of this lesson.

- Were the errors in the models' predicted populations small enough that small changes in the facts or in the assumptions could correct the problem?

- Which assumptions show the most sensitivity? That is, what changes in assumptions would probably result in the largest changes in the predictions of the models?

- What facts (other than the numerical observations and predictions themselves) support changes you might make in the additivity assumptions?

ACTIVITY

ROUND TWO

4

1. Explicitly carry out the first kind of evaluation on one or more additive models of moose population growth in Adirondack State Park. Be sure to state exactly which models you are evaluating and exactly what facts you are using to check your models. Describe the results of your evaluation carefully.

2. List the four Additivity Assumptions (see the preparation reading) in order of decreasing sensitivity. The one most likely to be able to change any deficiencies you found during your evaluation should be listed first, then the next most important, and so on.

3. Here is a way to carry out a quantitative evaluation of the sensitivity of additive models. By their very nature, additive models are necessarily described by two fundamental equations: a recursion rule and an initial value. The two control numbers, then, are the amount added and the initial value. Without changing the properties embodied in the four additivity assumptions, experiment with these two control numbers to see how sensitive the model is to them. In particular, how much effect does a specific change in one of these values have in the model's predicted population for the year 2013? (Note: In order to be able to identify what change caused what effect, be sure to experiment with only one kind of change at a time. For example, compare the results of several different recursion rules, all using the same initial value. Then change the initial value and try the same recursion rules again.)

4. Based on your work in this activity, summarize your confidence in the assertion that growth of the moose population in Adirondack State Park governed by an additive process can lead to populations reasonably consistent with the three reported values in the original fact sheet. Explain carefully.

INDIVIDUAL WORK 4

Changing the Rules

1. In Unit 4, *Prediction*, you describe the number of manatee deaths in Florida over a period of time by the linear model $y = 0.125x - 41.4$, where y is manatee deaths and x is the number of powerboats registered (in thousands). Each of the coefficients (control numbers) in the equation have been rounded. The domain of the data is roughly $425 \le x \le 725$. This item examines the sensitivity of predictions to the two control numbers in this equation.

   a) Use the given equation to predict the number of manatee deaths for $x = 425$, $x = 575$, and $x = 725$. Round your answers to the nearest integer.

   b) Test the sensitivity of these predictions to changes in the vertical intercept, –41.4, by replacing –41.4 in turn by –33 and –49 and repeating the same three predictions. (These numbers represent about a 10% increase or decrease, respectively.)

   c) Now test the sensitivity of the original predictions to changes in the slope. Use –41.4 as the vertical intercept, but replace the slope with 0.1 and 0.15 (representing a 20% decrease or increase).

   d) Summarize your observations from parts (b) and (c). In particular, which seems to have the largest effect on predictions, changes in the slope or in the intercept?

2. a) Karen has a scatter plot of data in the window [0, 100] x [0, 400] and wishes to use a linear equation to provide a rough description of the data. She does not need an exact least squares model but wants to have a reasonably good estimate of the slope of the equation (see Item 1). Here is her plan. First, she will draw a reasonable line on the scatter plot. Then she will select two points on the line (whether or not they are data points). Finally, she will subtract the y-values and the x-values and then divide to get the slope. Is that plan mathematically valid?

   b) Since Karen is working from a graph, the values she reads for the coordinates of the points may be slightly incorrect. Assume that she can be sure of reading each value only to the nearest integer. Thus, each value may have as much as 0.5 units of error. Suppose the values she uses are (40, 170) and (50, 200). Write an equation containing these two points.

c) Write an equation describing the steepest line consistent with the possible errors in reading these coordinates.

d) Repeat part (c) for the line of least slope based on these two readings.

e) Check that the two points (10, 80) and (90, 320) lie on the same line as the points (40, 170) and (50, 200). Repeat the same kinds of computations as those in parts (c) and (d), this time using the two readings (10, 80) and (90, 320).

f) Graph the equations from parts (b), (c), and (d) on the same axes using Karen's window. Then graph the equations from parts (b) and (e) together using the same window. Comment on the sensitivity of linear equations to the locations of the points used to compute them. (That is, how does how far apart you select your two points affect the resulting equation if there is uncertainty in the coordinates you use?)

3. Review your work with additive models in Lesson 2 and your evaluation of them in Activity 4. You have already identified the assumptions that you believe are most likely to help you improve on the additive models. For the assumption that you selected as being the first to modify, write as many different new assumptions to replace it as you can. Incorporate the following conditions.

- Each new assumption should correct the problems with the additive models, taking into account the other information contained in the fact sheets.

- Each new assumption should be as precisely worded as possible; avoid the possibility that your statements can be misinterpreted.

- Each new assumption should be as quantifiable as possible. Remember, your goal is to produce something that can be translated into operational mathematics.

There are two fundamentally different
approaches that you might take in revising the
original additive models in order to correct for
the difficulties you have identified thus far. One approach is to
do the mathematics first and then interpret the results in the context of the problem as an evaluation. That is, make changes in the
equations for the model (without thinking about what those
changes mean in the context) until the numbers seem to work
out; then see if you can make sense out of the equations you have
created. A second approach is to go back to the context first,
modify the assumptions you made initially about it, and see how
those changes affect the mathematics.

The second approach generally does a better job of keeping the
analysis meaningful, but it also requires that you have a pretty
good knowledge of the contextual issues. The first approach is
appropriate in situations in which little or no contextual information is available. In that case, by matching the model's data to
reality, the structure of the model itself might be able to provide
clues about the actual, previously unknown, nature of the context, and real advances in learning can take place. Thus, both
methods are important.

Realistically, the line between the two approaches is not sharp;
contextual assumptions can motivate mathematical ideas, which
can then be explored without revisiting the context for a while. In
this activity you will examine contextual issues first. You may
produce a mathematical model based on your decisions in that
investigation. On the other hand, you may want to build a model
by adjusting the original equations for the additive model, saving
the ideas from your contextual considerations as a check.

Based on your work up to this point, it should be clear that
migration at a constant rate will not satisfy all the numerical
requirements you have. Thus, there seem to be two choices.
Either something other than migration is involved, or the migration rate changes.

NEW MODELS

1. Identify a reasonable process other than migration.

2. If no new process is involved but migration rates increase, what in the environment could cause that migration rate to change?

3. Without additional information, which of the two possibilities in Items 1 and 2 seems to be the more reasonable basis for new assumptions for the 20 years following 1993?

For the purposes of this activity, the assumption that all changes in the moose population are due only to migration will be replaced with the assumption that baby moose are born in the park.

Other decisions need to be made, too. In working with the new assumptions, should you keep the assumption of migration, or should births be studied without including migration? What about deaths? You know two principles of modeling on which to base these decisions. Fortunately, they both indicate the same approaches.

The first principle is that of simplicity. At first, every model should be as simple as reasonably possible. Complexity can be added as understanding develops.

The second principle is that of studying one thing at a time. By looking at only one factor at work, you can be sure that you know what causes the effects you observe.

Based on these two ideals, then, migration and deaths will be ignored for the present model. However, keep these decisions in mind; you might want to return later and add migration or deaths back into the mix.

So what do all these changes mean about the mathematical model? Remember, the goal is to model the growth of the size of the moose population in Adirondack State Park.

ACTIVITY

5

NEW MODELS

4. What data would be useful in predicting the amount of growth next year in the moose herd?

5. Of the data on the fact sheet in Lesson 2, which bit of information would you consider to be most important if you were to try to predict the moose population for the year 1994 using the birth assumption? Why?

6. What additional assumptions not included on the fact sheet would you need to make in order to predict the population for 1994?

7. Make whatever reasonable assumptions you need, based on your answer to Item 6, to predict the population at the end of 1994 if the additional 100 moose are not brought into the park. Be sure to state your assumptions carefully and explain why you think they are reasonable. Also show how you used those assumptions to make your prediction.

8. Recall that the conservationist group's prediction for the year 2013 was based on the additional assumption that 100 more moose would be brought to the park. Rework Item 7 to reflect that assumption. (Pretend that the additional moose arrived on December 31, 1993.) Again, list and defend any additional assumptions you need in order to make your prediction for 1994.

9. Summarize the logic of your models in Items 7 and 8. List the important assumptions you made and how they were used. Explain how you could extend your predictions several more years.

10. If you have not already done so, use your solutions to Items 7 and 8 and your summary in Item 9 to write a recursive model for moose population growth in Adirondack State Park. Don't worry about specific numbers yet, but do write the equations in a form that reflects the assumptions and common sense. Explain the connection between the form of your equation and the assumptions that have been made.

INDIVIDUAL WORK 5

Next

1. Write a closed-form linear equation consistent with each of the following statements.

   a) $x(3) = 5$, $x(9) = 7$

   b) $p(4) = 13$, $p(8) = 26$

2. Find the equation of the line that contains the two given points.

   a) $(3, 7)$, $(9, 21)$

   b) $(4, 28)$, $(8, 18)$

The overall appearance of a graph is sometimes sufficient to provide useful information. Examine the following graphs to determine what they do and do not tell you.

3. a) Decide whether each of the graphs in **Figures 6.15–6.20** could be the recursive graph of an additive process. Explain by writing an appropriate recursive equation.

   b) Decide whether each of the graphs in Figures 6.15–6.20 could be the closed-form graph of an additive process. Explain carefully, using the definition of additive.

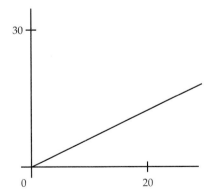

Figure 6.15.

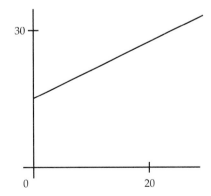

Figure 6.16.

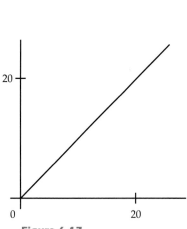

Figure 6.17.

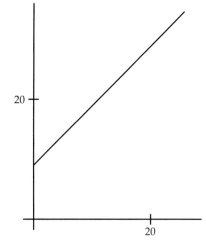

Figure 6.18.

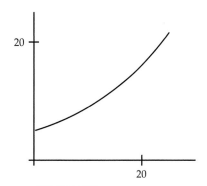

Figure 6.19.

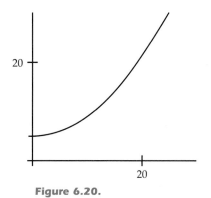

Figure 6.20.

Century	Population
0	120
1	124.7
2	129.4
3	134.1
4	138.8
5	143.5
6	148.2

Figure 6.21.
Population data.

Day	Population
0	1
1	5
2	13
3	25
4	41
5	61
6	85

Figure 6.22.
Population data.

Year	Population
0	180
1	201
2	226
3	253
4	283
5	317
6	355

Figure 6.23.
Population data.

4. **Figures 6.21–6.23** show three data tables. Explain whether each represents an additive process.

5. When Paulo looked at the first two graphs in Item 3, he thought they looked like the same graph, except that the second one had been slid to the left. Algebraically, using parametric equations and their graphs, explain why he is correct.

You learned about simulation in Lesson 2. It was not particularly useful there, but now it becomes essential. Your answer to Item 10 of Activity 5 provides you with a mathematical form for the birth model. Since you have no data regarding the specific constants in your equation, you again have two choices. You can do research, or you make assumptions and do "what-if" simulations. In the interest of time, take the second option now. Results from research can then be used as an evaluation of your completed model.

Items 6–8 ask you to begin this process of doing what-if simulations and verifying that your new model is not just another additive model.

6. Select specific numbers to use for each unknown constant in the new recursive equation you proposed in Item 10 of Activity 5. Base your selections on your best guess, using common sense and what you know the numbers represent. Assume that the additional 100 moose are not brought into the park. Be sure to specify your initial value, for $p(1993)$, too. Then do a simulation of your model to predict values for $p(1994)$, $p(1995)$, $p(1996)$, and $p(1997)$.

7. Since you will eventually need to check $p(2013)$ with your model, assume that the additional 100 moose are added at the end of 1993, and repeat Item 6.

8. Use your simulation values from Item 7 to complete a few rows of a table with columns labeled year, current population, and yearly change in population. Then construct three graphs from your table: current population versus year, next population versus current population, and yearly change in population versus current population. Draw segments in the first graph to show the changes between years, and describe for each graph the features that indicate that the model from which it was constructed is not additive. (Hint: Look back at Items 3 and 4.)

In Activity 5, you developed a new model of the moose population. Your new model incorporates births of new moose instead of migration as the mechanism for growth of the herd. The difficulties with the additive model became apparent when you compared its predictions over a period of years to available information. The same kind of evaluation should be used on the new model. However, as you saw in Items 6–8 of Individual Work 5, predicting far into the future with the new model may not be easy if you are working by hand.

There are quite a few models that might have been proposed by you and your classmates in incorporating the birth assumptions into your work in Activity 5. In order to keep your class work manageable, your teacher will help you select one or more of those models for use in this activity.

One form that may be used in this activity is the recursive relation:

$$p_{next} = p_{current} + r \times p_{current}$$

Since $p_{next} - p_{current}$ = change in population, this recursive description really says that the change in population from one year to the next is proportional to the current population. If your model is not selected for use now, keep your work and check it later using the methods explored here.

1. a) Use your new model to predict $p(2013)$, where p represents population and the independent variable is year. If possible, carry out the necessary simulation both by using home screen iteration on a calculator and by entering appropriate formulas into a spreadsheet. Discuss the advantages and drawbacks of each tool.

SIMULATED MOOSE

b) If you needed to make a prediction for 70 years in the future, how easy would it be to use your recursive equation?

2. Keeping the form of the model as you have it, change the constants in your model until $p(2013)$ is as close to 1300 as you can get it. (This is another application of the general principle of evaluating and revising a model.) Again, comment on your choice of tools.

   Save your simulation data for all 20 years, 1994–2013, both for your best model and for a poor model. You will need these data in the remaining questions.

3. In Item 2, how easy was it to find the proper numbers to use in your equation in order to make the model fit the data reasonably well?

Your work in Items 1 and 2 should have given you a reasonable model for the growth of the moose population, at least based on what is known at this time. It includes appropriate assumptions about the actual situation and fits the given numerical information fairly well. However, you used a trial-and-error approach to tune your model's constants to the reported data. Such an approach is somewhat inefficient and time-consuming.

Think back to your experiences with linear models in a variety of contexts, from secret codes to scaling images, classifying bones, and creating animations. Some of your early work involved trial and error, but after you learned that all lines have equations of the same form and that the constants can be determined from two points, you could find the equations of the lines more quickly.

Linear functions are the closed-form representations for additive models. Their linearity is visible in a number of different kinds of graphs and representations. In finding equations for lines, "slope and intercept" or "slope and point" are the key ideas. So, finding quick ways to determine their slopes makes finding equations of lines easier. The slope can be calculated from the coordinates of two points. It can also be found directly from a recursive description.

ACTIVITY

SIMULATED MOOSE

6

In order to be able to work with models like the one you developed in Items 1 and 2, you need to know more about their properties. Perhaps there is something like slope that can help you move quickly to closed-form equations for easier long-term predictions. The following items ask you to consider some of the details of your new model.

4. a) Graph current population v. year for each of your saved simulation results from Item 2.

b) You knew when you built it that your model was not additive. Explain how the graphs make that clear. What geometric feature matters, and why?

c) In Unit 4, *Prediction*, you learned that any nonvertical line has an equation of the form $y = mx + b$. The m and b are control numbers, describing the line itself. Changing the values of m and b changed the line's tilt and location, but the graph remained a line. How are the graphs you drew in part (a) alike? What control number (or numbers) did you change in going from your poor graph to your good graph? What effects did they have on the graph?

d) What other number could have been treated as a control number? How is its value visible in your graphs?

5. a) Examine the similarity between the equations for the migration (additive) model and the birth model. In the migration model, the mathematical representation is $p_{next} = p_{current} + k$, where k is the net change in population due to migration (measured in moose). The birth model can be written as $p_{next} = p_{current} + \text{births}$, where births represent the net change in the herd size due to newborn calves (again, in units of moose). Rewriting that equation, $\text{births} = p_{next} - p_{current}$. Use your saved simulation results from Item 2 to add a births column to your data table.

ACTIVITY

SIMULATED MOOSE

6

b) Graph births versus $p_{current}$, and describe the graph as precisely as you can. If you can find an equation for the graph, do so. How does it relate to the original recursive model from which the simulation was run?

c) Interpret in words the equation you obtained for births in part (b). What does the equation tell you about the fundamental way the herd grows?

In Item 2 you used your recursive equation to predict population values beyond the initial year (1993). You probably used either a calculator or a computer for the actual computations. In Item 6, you will repeat those computations by hand to look for patterns that might save time in later work.

No calculators allowed!

6. a) Begin by relabeling 1993 as year 0. Then write the recursive equation for the population in 1994, that is, for $p(1)$, but do not complete any arithmetic you can not do in your head.

b) Use your answer from part (a) to write a formula for $p(2)$ in terms of $p(0)$ and constants. ($p(1)$ should not appear in your answer.) Again, do not carry out any arithmetic beyond what you can do mentally.

c) Extend your work in part (b) for the next three values, again using only $p(0)$ and constants. Then guess a general closed-form equation for the population in terms of year n.

d) Verify your final equation by checking $p(0)$ and $p(20)$ against the values for 1993 and 2013 obtained in your original simulation in Item 2.

e) Two constants describe your recursive equation. One is the initial population. The other has something to do with the number of births in the herd. How do these numbers appear in your final equation in part (c)?

INDIVIDUAL WORK 6

Everything's Relative

1. a) Simplify your recursive representation for the best birth model you obtained in Activity 6. Explain why it might be called a **multiplicative process** or model.

   b) Nina wrote her recursive equation as $p_{next} = (1 + r) \times p_{current}$ and then subtracted $p_{current}$ from both sides. What does the left side of her new equation represent?

   c) Interpret the meaning of the r in Nina's equation in simple language that anyone can understand.

In the last two activities, you have developed a pretty good model for the growth of the moose population, at least when judged by looking at the numbers. There are several ways of writing equations to describe the births-only model. Activity 6 suggested using an equation of the form $p_{next} = p_{current} + r \times p_{current}$. This equation says that the amount of change in the population from one year to the next is proportional to the population in the first of those two years. When working with a spreadsheet or doing home screen iteration on your calculator, it is an easy equation to use.

The number r in this equation is the constant of proportionality between population and change in population. It controls how fast the growth takes place. It is called the **relative rate of change** (or relative rate of growth) per year for this population. The rate is called a relative rate because it is given in relation to the current population—it is the growth as a fraction of what's already present. For example, if r were 0.5, then the population would be 50% larger from one year to the next.

The rate is per year because one year represents one step in the recursive process. The number r makes sense only in connection with a recursive representation of the growth. It is always the relative rate of change per recursive period.

2. Suppose the moose population in the park grows according to the equations:

$$p_{next} = p_{current} + 0.15 \times p_{current} , \ p(0) = 200.$$

This represents a relative rate of growth of 15% per year.

   a) Make a table of values of the population for the next 10 years.

   b) Check that the amount of growth from year 0 to year 1 actually is 15% of the initial population.

c) Show that the amount of growth from year 1 to year 2 is 15% of the population at the end of year 1.

d) Check the 15% growth rate for at least two other years.

e) Predict the percentage of growth that will take place in a 5-year period. Check your prediction using your table.

3. Write a recursive equation for a population having a 20% relative rate of growth per decade. (Hint: Measure time in decades, not years.)

4. a) In Activity 6, you were able to use your recursive representation to find a closed-form equation for growth of the moose population. Since a closed-form equation allows quick predictions of values far into the future, having such a representation is quite valuable. Use the same process you used in Item 6 of Activity 6 to find a closed-form equation for the recursive equation given in Item 2.

   b) Repeat part (a) for your poor model from Activity 6.

5. A possible model for one version of the moose population growth has a closed-form representation of $p = 127(1.123)^x$, where x is in years, with year 0 being 1993.

   a) Graph this equation in the window [0, 20] x [0, 1300]. Then write parametric equations that describe the same situation, and graph them on the same window to confirm that the graphs are the same.

   b) Translate the graph to the right by 93 spaces. (Be sure to adjust your window accordingly.) Show parametric and closed-form representations of the corresponding equation. Interpret what the graph and new closed-form equation mean.

Additive models have two control numbers: the amount added in each period and the initial amount. The graphs of quantity versus time for additive growth are lines, and the control numbers show up in the closed-form equation for lines as the slope and vertical intercept. That fact makes writing the closed-form equation for additive growth pretty easy. Look at the corresponding situation for the model describing the moose population.

6. a) In what ways are the closed-form equations from Items 4(a) and 4(b) and the equation from Item 6(d) in Activity 6 similar? If you can, write a generic form for these equations, using letters for any control numbers you need to include.

b) Write instructions that a friend could follow to go from a recursive equation of the form $p_{next} = p_{current} + r \times p_{current}$ to a closed-form equation, and comment on how the control numbers appear in the final equation (if they do).

After Item 6, you know that models having recursive equations of the form $y_{next} = y_{current} + r \times y_{current}$ for $r > 0$ must have closed-form equations that look like $y = a(1 + r)^x$. Here x represents the number of periods, y represents the population, and r represents the relative rate of growth per period. This new type of function joins the linear functions you have studied in earlier units as the second member of what might be called your tool kit of functions. Others will be added in later units. Each plays a significant role in being able to model our world.

The variables in this equation are written as x and y in order to match what your calculator expects, but you could use any letters you prefer. The other two letters, a and r, represent constants—control numbers—for the equation. They describe the behavior of the growth (much as m and b describe the behavior of growth in the linear model $y = mx + b$).

Because the response variable y is defined by an equation having the explanatory variable x in the exponent, it is called an **exponential function**. Any quantity that can be described with an equation of this form is said to grow exponentially. The quantity for which x is the exponent is called the **base** of the exponential function. Thus, $y = 100(1.01)^x$ is an exponential function with base 1.01. It is customary to restrict the definition of exponential functions to include only positive bases other than 1. Of course, letters other than x and y may be used; it is the fact that the explanatory variable is in the exponent that matters, not the letters you use.

Exponential behavior is quite common. If you pay attention, you are likely to hear (or read) in the news about exponential growth.

Notice that your calculator graphs equations such as $y = 100 \times 1.01^x$ for values of x other than integers. You may not yet know what a fractional exponent means, or you may remember from an earlier course. In either case, your work in this and later assignments will build your understanding of the properties of such expressions so that their meaning will be clearer.

7. a) Based on your answers to Item 4, explain the meaning of the number a in the closed-form equation $y = a(1 + r)^x$.

   b) Substitute $x = 0$ into your answers to Item 4(a). Assuming that your equation is valid, what is $(1.15)^0$?

c) Repeat part (b) for the equations you found in Item 4(b) and in Activity 6.

d) Summarize your findings as a general rule.

The fact that the time-series graphs of additive models are always lines means that all quantities that grow in an additive manner can be described by a closed-form equation of the form $y = mx + b$. You learned in Unit 4, *Prediction*, how to find the equation of a line from just two points; you don't need a recursive relation and its control numbers in order to find a closed-form representation. In fact, from just two points you were able to find the control numbers.

In Activity 6 and in Item 4, you started with recursive equations and found the corresponding closed forms. However, in Activity 6, getting the numbers correct in the recursive form took a lot of trial and error. Can something like the two-point calculations for lines apply to situations like moose population growth?

8. Assume that a particular population grows in such a way that at time 0 there are 100 members of the population and at time 10 the population has increased to 300.

a) Sketch a rough graph of population versus time assuming that the growth is additive.

b) Find a closed-form equation for this population versus time, assuming that its growth is additive. Explain your method carefully.

9. Assume that a particular population grows in such a way that at time 0 there are 100 members of the population and that at time 10 the population has increased to 300.

a) Sketch a rough graph of population versus time assuming that the growth is exponential.

b) Substitute the given (time, population) pairs (separately) into the closed-form exponential formula, and simplify each expression as far as possible.

c) Using just the two given points, algebraically find the numerical values of a and r without trial and error.

d) Verify that your equation is correct, first by again substituting the given (time, population) pairs and then by entering the equation into a graphing calculator and tracing the graph to the given points.

10. Use the same method you used in Item 9(c) to find a closed-form equation for the situation in Activity 6. Compare your result to that you obtained in Item 6 of the activity.

11. Comment on the amount of work needed in order to get closed-form equations using trial and error as in Activity 6 and using the method from Items 9 and 10.

12. Recall that the decision facing you as the commissioner is whether to bring in an extra 100 moose. Assume that the relative rate of growth you found in Activity 6 and Item 10 correctly describes the growth of moose herds. Compare the predicted herd sizes for the year 2013 with and without bringing in the extra 100 moose. You have already done at least one of these.

13. In addition to a traditional savings account, you can save money and earn interest by investing in savings certificates, which are called certificates of deposit (CDs). Some CDs earn interest monthly, not annually. Suppose that you open such a certificate with $500 and it grows to $526.89 in one year. What is the relative rate of growth per month for your CD?

14. Look back at the closed-form equation you developed for the moose population at the end of Activity 6. Recall that, in the equation, year 0 means 1993. If you were to use actual years, starting at 1993 instead of 0, the resulting graph would move right by 1993 units. Write a closed-form equation for that translated model.

15. In Items 12 and 13, you knew that some quantity grew exponentially. Explain how you think someone could tell whether the growth really was exponential if they were not told in advance.

PREPARATION READING

Do I Know You?

T he relative growth rate is one of the numbers that control the details of the moose population model. Another is the total number of moose initially in the park. You may have had other control numbers in your model. The recursive and closed-form equations, together with these control numbers, make up part of the structure of exponential behavior. Recall that part of the process of building a model and evaluating it involves moving out of the context of the original question and into the mathematical world. Making that move permits you to learn more about the structure of the system under study and to take advantage of it. For example, by being able to move easily from recursive to closed-form descriptions, you can make predictions much more quickly.

The moose population growth seems to be reasonably well described by an exponential model. This is the second major kind of function you have studied. You are already very familiar with linear functions from your work since Unit 2, *Secret Codes and the Power of Algebra*, and you know what to expect of them. What are the properties of exponential functions? And how could knowing properties of a kind of function help?

First, in order to know when you can apply your knowledge of exponential behavior, you need to be able to recognize such behavior when it hasn't already been labeled for you. Second, exponential functions describe a lot of situations, so if you know something about this kind of behavior, you actually know something about many contexts.

Note that the names of the kinds of growth refer to the closed-form equations. Thus, exponential functions have closed-form equations of the form $y = a(1 + r)^x$, and linear functions have closed-form equations of the form $y = mx + b$. As you have seen, exponential functions can have recursive forms that graph as lines.

I'M IN CONTROL

The exponential equation $y = a(1 + r)^x$ includes two constants, a and r. Use a spreadsheet or graphing calculator to carry out a detailed investigation to determine the main effects of each constant on the graph of the closed-form equation. If you use a spreadsheet, be sure to use the closed-form equation and to include values of x other than just integers. Remember, when investigating two quantities, it is best to investigate them separately so you will know what causes any changes you notice.

As you explore or after you finish, answer the following questions.

1. What is the primary feature controlled by r?

2. Two specific values of r are particularly interesting. In general, the graph is not very sensitive to changes in r. However, when r is near either of these special values, small changes have great effects. Find these two values, describe what happens to the graph as you change r past each of these numbers, and then explain from the population context why these effects are reasonable. Remember, values other than positive integers can make sense, too.

3. What is the primary feature controlled by a?

4. There is one particular value of a near which the function is sensitive. Find that value and describe what happens as a moves past it.

INDIVIDUAL WORK 7

Modeling Growth and Decay

1. Because of the unexponential behavior of the graph of $y = a(1 + r)^x$ when r is 0 or below -1, mathematicians do not use the term exponential for those cases, even though such equations formally have the exponential form.

   a) For $r = 0$ and $r = -1$, substitute for r in the equation and simplify the result. Explain, based only on the equation, why it should not be called exponential.

   b) Explain what $r = 0$ and $r = -1$ mean in the context of the birth model. Again, indicate why these lead to models that should not be called exponential.

In your investigations in class you should have noticed that not all exponential functions grow. However, for positive values of a, all exponential functions do have the same general shape, but some increase and some decrease. It is customary to refer to increasing exponential functions as **exponential growth**. Decreasing exponentials are referred to as **exponential decay**.

2. For each of the exponential graphs in **Figures 6.24–6.29**, identify the values of r and a as well as possible. Think of axis labels that seem reasonable for each graph (other than part (e)). If you are not able to find actual numbers, tell how the values are related to the special values discovered in Activity 7.

a)

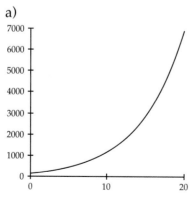

Figure 6.24. An exponential graph.

b)

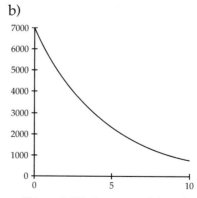

Figure 6.25. An exponential graph.

c)

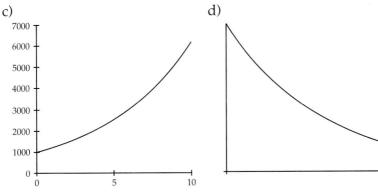

Figure 6.26. An exponential graph.   **Figure 6.27.** An exponential graph.

e)

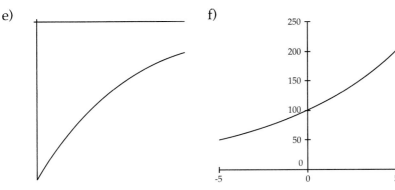

Figure 6.28. An exponential graph.   **Figure 6.29.** An exponential graph.

3. Suppose that a state buys a large forest to set aside as an animal refuge. Past population information does not exist for any animals, so the state wildlife manager begins keeping detailed records. At the time of the purchase, the population of one species is about 250; a year later it is about 280.

Assume for this item that the data are exactly correct and that the population grows exactly according to the recursive model $p_{next} = p_{current} + r \times p_{current}$.

a) Determine the relative rate of growth per year.

b) Draw an arrow diagram for this recursive process.

c) Use the arrow diagram to calculate the population of this species for the next year.

d) Use the arrow diagram to calculate the population of the species for the year prior to the purchase of the forest.

e) Now use the original two population values to find the closed-form equation for this growth. Then check your answer to part (d). (What will you need to use for x in your check?)

Item 3 points out one problem with the recursive form $p_{next} = p_{current} + r \times p_{current}$. It is easy to use to find values in the future recursively, but going backward recursively is difficult. Evaluating the closed-form equation with your calculator gets answers, but what is the calculator doing? You know that $(1 + 0.15)^3$ means to multiply $(1 + 0.15)$ by itself three times, but what does $(1 + 0.15)^{-1}$ mean? After all, that's the kind of expression you have to evaluate to go backward one time period. What you need is a recursive form that can be reversed more easily.

The model you used most recently was based on the recursion $p_{next} = p_{current} + r \times p_{current}$. It implies that the change in p is proportional to the current value of p. However, that may not be the form you came up with in your original work; other forms can do just as well. In fact, applying the distributive law to the right side, this equation may be rewritten as $p_{next} = (1 + r) \times p_{current}$. Since $(1 + r)$ is just some number, it can be replaced by a name of its own, say, b. Then the recursive equation becomes $p_{next} = b \times p_{current}$.

This says that the next p-value is proportional to the current p-value. In fact, these two proportionality conditions characterize exponential growth and decay. It is reasonable to think that the changes in a population (both births and deaths) are approximately proportional to the size of that population. The constants of proportionality reflect the general health of the population. In finance, the amount of interest is typically proportional to the amount of money invested (or owed); the constant is the APR, yield, or periodic rate. The amount of a radioactive element that decays in a particular time period is proportional to the amount present at the beginning; the constant is the decay rate. You may think of other contexts is which you find yourself saying something like "the amount of change is proportional to the amount present." That's exponential change.

> For recursively described exponential functions, both the amount of change in the response variable and the next value of the response variable are proportional to its current value. The constants of proportionality are the relative rate of change and the growth factor, respectively.

4. a) Write the recursive equation from Item 3 using this new form. What is the value of b for this situation?

   b) Make an arrow diagram for the new recursion equation.

   c) Use your arrow diagram to find the population of the species for the year prior to the purchase of the forest. How does this answer compare with the one you got in Item 3(e)?

d) Find the corresponding closed-form representation.

e) Based on parts (c) and (d), what does an exponent of –1 mean?

f) The number r is called the relative growth rate. It represents the amount of increase during one period, expressed as a fraction of the amount at the start of the period. What does the number b represent in your new equation?

The number b in the new form of the exponential function is called the **growth factor**. The growth factor is always 1 more than the relative growth rate for the same period. It gets its name from the recursive process; each step involves the multiplication of the prior value by this factor. As was the case with the relative growth rate, a growth factor has a direct interpretation only in the context of recursion. Thus you must always specify an interval to which it applies—the distance from the current value to the next value of the explanatory variable.

Recall from Lesson 3 that b is also known as the base of the exponential function. The base does not require a recursive interpretation in order to be used, but it does depend on the units in which the explanatory variable (usually time) is measured.

5. The September 30, 1996 issue of *U. S. News & World Report* stated that the buffalo population in Yellowstone National Park had grown from 25 to 4000 during the period from 1900 to 1996. Determine the annual growth factor for the buffalo population over that period. Explain in ordinary language what it means.

6. Get a large number of coins (at least 100). Count them. Use them as your population for generation 0. Now toss each coin once. Remove any coin that lands heads up, and put it away for the remainder of this experiment. After each coin has been tossed once, count the number remaining (not removed). Use those coins as your population for generation 1. Toss each member of generation 1 once, again removing any that land heads up. Record the number that remain as the population for generation 2. Continue until no coins remain. Based only on the rules of this process, do you expect it by be approximately exponential? Explain. Analyze your population-versus-generation data to verify your claim.

7. In Activity 7, you explored the effect of the relative growth rate on the graph of the corresponding function. You found that two particular values had special roles in that interpretation. Translate your earlier findings into statements about the growth factor, b.

8. Rework Item 2, this time using growth factors instead of relative growth rates.

9. It has been reported that satellite-based remote sensors (using microwave frequencies) measured a 6% decline in the area of the earth's surface that was covered in ice during the period 1978 to 1994.

   a) Find the corresponding annual relative rate of change in ice coverage. What is the annual growth factor? Give your answers to the nearest tenth of a percent.

   b) Using 1978 as year 0 and treating the extent of coverage as 100% in that year, write a closed-form equation for the extent of ice coverage remaining n years later.

   c) Graph your function over a large domain (several centuries). If this model holds far into the future, describe the long-term behavior of the amount of ice coverage.

10. Suppose a bank account currently has a balance of $300 and is earning interest at the rate of 3% per year. Assume that there have never been any service fees, deposits, or withdrawals.

   a) Write a recursive equation, using an annual growth factor, for the balance in this account. Based on your equation, explain how you could compute balances for prior years.

   b) Write a closed-form equation for the balance in this account (use the year with a $300 balance as year 0). Then graph it in the window [–50, 50] x [–100, 1000].

   c) Describe the overall shape of the graph of the account balance. In particular, what does the extreme left portion of the graph tell you about the balance? Use your answer to part (a) to explain why this makes sense.

11. Make up and graph the closed-form equations for several exponential functions. For each graph, use a variety of windows to look at the long-term behavior, both future and past, of such functions. Include both growth and decay functions in your investigation, but use only positive values for a. Write a short paragraph to someone who has never seen an exponential function in which you explain what their graphs look like.

ACTIVITY

CHANGING TIMES

8

The closed-form equation for any exponential growth or decay model includes two control constants—the initial value and either the relative rate of change or the growth factor. In your first encounter with this model, you used trial and error to evaluate the appropriate constants for the moose data. Now you can work more efficiently. You know what to do to find the initial value: find the value when the explanatory variable is 0. But both the relative rate and the factor were defined in terms of the recursive equations, so both depend on the recursive period.

There are two fundamental choices you must make in defining how to measure your explanatory variable (usually time). First, where is 0? Second, how long is one unit? How do these choices affect your ability to find a closed-form equation?

You have seen that working with exponential functions is really pretty easy. In general, if you know in advance that the model is exponential, then all you need in order to write a closed-form equation are coordinates for the initial point and one other point. Although the calculations are not extremely difficult, what happens when neither known point is the 0 point? And what if you are measuring in years and someone else is measuring in decades? Is one model easier to find? Do both methods lead to valid models? In the spirit of being efficient, what is the easiest way to find an exponential model from two given data points? That's the subject of this activity.

Here are some questions that you need to be able to answer in the quest for efficiency.

- For a particular exponential model, can recursion be used to get from some point to another point n steps away even if the first point is not at an integral time? If so, and the same recursion period is used, what changes are necessary in the relative rate and growth factor?

CHANGING TIMES

- For a particular exponential model, can recursion be used to get from some point to another point n steps away if the recursion period is changed? If so, what changes are necessary in the relative rate and growth factor?

- For particular data, are there good choices for the starting point and recursion period? (That is, do some choices make the calculations easier?)

According to published data, the world rate of crude oil production increased nearly exponentially from 1880 through 1970. Use 1930 as year 0. The rate at the end of 1930 was about 1412 million (1,412,000,000) barrels per year, and in 1964 it was about 10,310 million barrels per year. That is, had production suddenly been frozen, production would have been 10,310 million barrels for the next year. Another way to think about this idea is that production during the last month of 1964 was about 10,310/12 million barrels (1/12 of the yearly rate). Assume for the sake of this investigation that the rate of production grew continuously. For example, the rate of production at year 3.75 would be the rate three-fourths of the way through the fourth year, that is, at the end of September 1934.

1. Write a closed-form equation for world oil production, in millions of barrels, in terms of years. What is the corresponding recursive equation? Graph your equation and verify that it contains the two given data points.

2. Use the equation you found in Item 1 and a spreadsheet or calculator to find the rate of world oil production at some arbitrary noninteger time. Then use the same equation to find the world oil production rate for exactly one year later. Repeat for 2, 3, 4, and 5 years after your starting point. Does any recursive equation seem to govern these values? If so, what is it, and how are its relative growth rate and growth factor related to what you found in Item 1? If not, show how it fails the additive and multiplicative models. Compare your findings with those of other students in your class who used different starting years.

CHANGING TIMES

3. Use your closed-form equation from Item 1 to repeat the investigation of Item 2, this time using a recursion period other than one year.

4. a) Review the work you did in Item 1 to get your closed-form equation. What was the hardest part of the process? What would be impossible to evaluate without using a calculator or computer?

   b) How was the time scale involved in that calculation? If you could change the units of measure for time in this problem, what choice for "one unit" would make the difficult step you identified become much easier?

5. Combine what you have learned from your work in Items 2–4 to define a really convenient time scale for the world oil production data. Remember to define time 0 and the length of one unit. Write a closed-form equation using your time system, explaining how the time system makes your work easier. Then tell how to use your new system to predict the production rate for other years, such as 1970. Confirm your prediction for 1970 using the equation from Item 1.

6. Summarize your procedure from Item 5 by stating a general process that can be used to find a closed-form equation for exponential growth from two data points.

INDIVIDUAL WORK 8

Laws of Exponents

For any particular exponential function, its recursive description depends on the time scale you choose. Any time scale is valid, but your reader needs to know what you have chosen. One especially useful time scale assigns time 0 to the earliest observed value and assigns time 1 to the latest observed value. Then, by the definition of the growth factor, the growth factor for this function is just the ratio of the two observed values. Because of that, the closed-form equation can be found with essentially no work at all. Of course, the time scale may turn out to be in very unusual units!

1. a) Apply this quick method of finding closed-form representations for exponential functions to the original moose data: 127 moose in 1993 and 1300 in 2013. (These figures assume that the conservationists were allowed to bring in the extra 100 new moose in 1993.) Be sure to specify the time scale you use.

   b) The estimate of 127 moose in 1993 is based on adding 100 moose to a population of between 25 and 30 moose. Rework part (a) using 125 moose in 1993 and 1300 in 2013 and then with 130 moose in 1993 and 1300 in 2013. Discuss the sensitivity of the equation to the initial population.

2. The September 30, 1996, issue of *U. S. News & World Report* stated that the buffalo population in Yellowstone National Park had grown from 25 to 4000 during the period from 1900 to 1996. In an earlier item you found that the growth factor for the buffalo population over that period was 1.054 per year. Use the quick method from Activity 8 to write a new closed-form equation for this population. What is the corresponding growth factor (with units)? Use your new equation to predict the population for the year 2010.

   [Satchell 1996]

3. The earth's ice coverage decreased by 6% over a 16-year period beginning in 1978. Write a closed-form equation predicting the percentage of ice coverage for any time if this trend continues. Be sure to indicate how to use your equation for arbitrary times.

In trying to understand how to simplify the expression $(b^n)(b^m)$, think of 1 as being an initial value for some growth. Then $1 \times (b^n)$ represents n periods of growth with a factor of b per period. But that result could be used as the starting value for new growth having the same factor, this time for m periods. If so, the final amount would be $(b^n)(b^m)$. However, that is also just a total of $(n + m)$ periods of growth from the beginning, still using a factor of b per period. That means that $(b^n)(b^m) = b^{n+m}$.

4. Decide whether you think this reasoning is valid. Explain your conclusions.

5. a) Explain how to simplify $(b^n)/(b^m)$. Based on the reasoning above, how might you think about this question?

   b) Use the fact that $x = 0$ corresponds to the starting value for exponential growth to explain what b^0 is for positive values of b. (Note: 0^0 is undefined; it is not a number.)

   c) Combine the rules from parts (a) and (b) to give a rule for simplifying $(b^0)/(b^n)$. What does this new rule say about negative exponents?

6. Doug and Dave both wanted to model the moose data in Item 1. They decided to work in years instead of 20-year periods, and both wanted to use year labels calling 1993 year 93 (so 2013 would be 113). Doug substituted the points (93, 127) and (113, 1300) into the general exponential formula $y = ab^x$ to get two separate equations. He then divided one equation by the other to determine b and a. Dave used the points (0, 127) and (20, 1300) to find a closed-form equation and then translated the whole thing to the right 93 units. Recreate both solutions, check that each is valid (by graphing or substituting several values into both), and then relate your equations to the rule you found in Item 5.

7. When they got ready to investigate the case of the vanishing ice, Inês and Dana both decided to use 1978 as year 0. Inês wants to measure time in 16-year blocks in honor of her age, but Dana wants to measure time in years. The equations they came up with (with constants rounded off) are $c = 100(0.9400)^x$ and $c = 100(0.9961)^x$.

   a) Verify that these equations are correct by showing the exact value of each growth factor and where it came from. Indicate which equation belongs to whom.

   b) Show the calculations each person would need to carry out to predict the ice coverage for the year 2074. Use exact values, not the decimal approximations given above.

c) If you did things correctly and did not simplify, one of your answers in part (b) involves a power of a power, something like $(b^m)^n$. Look back at your reasoning before Item 4, and devise a similar explanation for how to simplify $(b^m)^n$. State the rule, explain how to see that it is correct, then apply it to the result in part (b) to show that Inês and Dana have the same prediction for 2074.

d) Use your work here to explain why the percentage growth in Item 2(e) of Individual Work 6 was not just 75%.

e) Use your rule from part (c) to explain what fractional exponents mean. That is, what is going on when you calculate $3^{1/4}$? Apply your rule to $3^{1/4}$ to verify your idea, and then give two additional examples of your rule in action.

8. In looking at a number of exponential graphs, including both growth and decay, the graphs all seem to have very similar shapes. In fact, every growth curve is matched with a corresponding decay curve in a very natural way. Consider the growth described by the equation $y = 100 \times 1.25^x$.

a) Graph this function on the window $[-10, 10] \times [0, 1000]$. Then write a recursive description for this function. Be sure to specify the period of your recursion.

b) Now think of making a decay curve by reflecting your given graph across the y-axis. It will still cross the axis at $y = 100$, so the equation of the new decay function will be something like $y = 100b^x$. Go backward on your recursive equation from part (a) to determine the value of b, and write the closed-form equation for this decay. Then graph your equation and verify that it really is the reflection of your answer in part (a). (Check pairs of points, such as $x = 2$ and $x = -2$ or $x = 8$ and $x = -8$.)

c) Graph the function $y = 100 \times 1.25^{-x}$. Use your work in part (b) and Item 5(c) to explain what you see.

9. In Item 1 you found a quick equation for the moose data. Start with that equation and explain how to use it to find the predicted moose population for any actual year (such as 1999). Make sure your rule is valid for any year, though, not just for 1999. Then convert your rule into a new, simple equation for population in terms of year. Verify your final equation by checking 1993, 2013, 1999, and one other year of your choice.

10. a) Remember that the original moose population model was intended to help describe the effects of adding 100 new moose to the park. Using the annual growth factor determined in Item 6, write the closed-form equations describing the moose population with the added 100 moose and without the new moose. As in Item 6, assume that the 1993 population before adding any moose is 27.

    b) Use the graph of your equation for the model of no new moose to determine how long it would take the moose population to reach 127 on their own. Explain your method.

    c) Graph the two models from part (a) on the window [–25, 40] x [–200, 1500]. The graphs should look as though one could be translated to form the other. Check on your graph to see whether that seems correct. If not, explain. If so, determine how large a translation is needed, write an appropriate translation equation, and explain the practical interpretation of this observation.

    d) The equations you found in part (a) differ only in their *a*-values. Use the translation equation from part (c) to explain how the *a*-values are related numerically to your results in parts (b) and (c).

11. Several of the previous items dealt with properties of the arithmetic of exponents. Look back at Items 4, 5 and 7. List all the rules for computation that you developed in those items. These rules are known as laws of exponents.

The change of one assumption about the growth of the moose population has led to a lot of new information. However, in all the examples since your development of the moose model, you have begun by assuming that the behavior was exponential. That is, you began knowing to use an equation of the form $y = ab^x$ or $y = a(1 + r)^x$.

For the moose population investigation, that form was not known and came from a recursive description; the recursive description reflected the assumption that the number of new births was proportional to the current herd size. In essence, then, that birth assumption was equivalent to assuming that the model would be exponential—you just did not know that was where it would lead you. If you begin by assuming that an exponential model describes a situation, then you need only two points to find appropriate equations.

In general, real situations described by exponential equations are of two main types. Bank accounts and loans are typical of the first kind. They really are exponential. They are governed exactly by the use of specific exponential equations. Thus, finding exponential descriptions for such situations gives exact results, except for rounding.

The other type includes situations such as wildlife populations and the assimilation of medicine by the body. Take the case of populations. Neither moose nor other animals sit down and do calculations with some formula in order to figure out how many calves they need next year! Any model of their population is only approximate. It works in describing the overall trend, but no one really expects actual population values to be exactly what the model predicts.

HOW WILL I KNOW IT'S YOU?

So how do you spot an exponential function from data? One method, of course, is to select two of the data points, use them to find an exponential equation, then see how well that equation matches all the data you ignored. However, it would be nice to be able to examine data and decide up front whether an exponential description even makes sense. Developing one or more methods to solve that problem is your task in this activity. To begin that search it is helpful to start with data that you know are exponential and examine their properties closely. Start, then, with a model for the moose population based on the data (1993, 127) and (2013, 1300). Assume that population figures are for the last day of the given year.

A spreadsheet will be extremely helpful in this investigation.

1. As noted before, the recursive and closed-form equations for exponential functions are closely related. Every exponential can be written in recursive form as $p_{next} = p_{current} + r \times p_{current}$ or $p_{next} = b \times p_{current}$. Write recursive and closed-form equations for the moose growth described above. Use one-year periods for your time scale, and include equations based on the relative rate of growth, r, and on the growth factor, b. Also write an equation for the number of new births.

2. Use your equations to build a year-to-year table of values showing each of the following quantities: year, population at end of current year, births in next year, and population at end of next year. If you are using a spreadsheet, build the year and current population columns from your defining equations first, then build the next population and births columns using relative references to those first two columns only.

HOW WILL I KNOW IT'S YOU?

3. Any pair of columns from your table can be used to make a graph. The graph of population versus year is the one you're trying to check as exponential. Based on your answers to Item 1, which other pairs of columns would produce graphs that would be easy to use to prove that the population versus year is exponential? Explain.

4. Graph births versus year, births versus current population, next population versus current population, and any others you would like to examine. If you are using a spreadsheet, you can have all your graphs defined at the same time so you can look from one to another. Which graphs are simple enough to recognize reliably (even if there are only 5 or 6 points instead of 21)? Find equations for any of these simple graphs for which equations can be found, then relate those equations to your work in Item 1.

5. Add columns to your table to show values for births/(current population) and for (next population)/(current population). Comment on the results.

6. Now suppose you are given just the first two columns of data: the year and the current values. Explain what additional columns you would like to have, how you could construct them from the two columns you are given, and what you would do with them after they were constructed in order to convince yourself that the data were (or were not) exponential.

7. If you used a spreadsheet to carry out your work in this exploration, leave its graphs defined as they are, but replace the year and current values with a new function that you know is exponential. The graphs and other columns should reflect the changed data automatically. Do your methods in Item 6 appear to identify correctly the new function as exponential?

If you have used a spreadsheet for this work, you may wish to save your worksheet so that it can be used to test other data in the future.

Seeking the *r* and *b*'s

E very exponential can be written in recursive form as $p_{next} = p_{current} + r \times p_{current}$ or $p_{next} = b \times p_{current}$. Thus exponential growth is multiplicative, whereas linear growth is additive. Each of these equations is linear, with variables $p_{current}$ and p_{next} and constants *r* and *b*. Notice that each has no constant term. That means that their graphs are straight lines that cross the axes at (0, 0) and that their slopes provide information about *r* and *b*.

These same equations may be solved algebraically for *r* and *b*. Then, given population values, you might elect to calculate the corresponding *r* or *b* for a number of observations and check that they remain constant.

Both graphical and tabular methods are good ways to verify that data are exponential. Remember, though, that real data are never perfect—the moose don't have equations and calculators!

1. Money in a savings account earns interest. Suppose an account earns interest once a month at a 0.25% per month.

a) Based on this description, how often are new balances calculated?

b) Write an appropriate recursive representation for the growth of the money in the account.

c) Make a table of the balance in the account at the end of each month for a year, beginning with $1000 and making no deposits or withdrawals.

d) Add a column to your table for the amount of interest paid at the end of each period.

e) Use your table to make graphs of the balance versus time and of interest versus balance. Describe the main features of these graphs. In particular, write the slope from point to point in each graph.

2. Data tables from Item 4 of Individual Work 5 (Lesson 3) are repeated in **Figures 6.30–6.32**. Sketch change-versus-population graphs for each table. Then, decide which of the tables describe exponential behavior. If the graphs are not exponential, identify what they are, if possible.

a)

Century	Population
0	120
1	124.7
2	129.4
3	134.1
4	138.8
5	143.5
6	148.2

Figure 6.30.
Population data.

b)

Day	Population
0	1
1	5
2	13
3	25
4	41
5	61
6	85

Figure 6.31.
Population data.

c)

Year	Population
0	180
1	201
2	226
3	253
4	283
5	317
6	355

Figure 6.32.
Population data.

3. Repeat your checks of the data in Items 2(a–c) using at least one other method of your choice. Explain your process.

4. a) Decide whether each of the graphs in **Figures 6.33–6.38** could be the recursive graph of an exponential process. Explain by writing an appropriate recursive equation.

   b) Decide whether each of the graphs in Figures 6.33–6.38 could be the closed-form graph of an exponential process. Explain carefully, using the definition of "multiplicative."

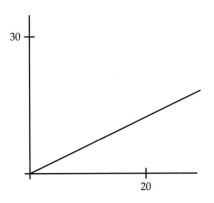

Figure 6.33. A line.

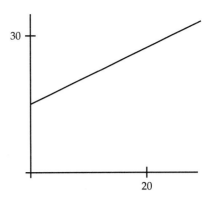

Figure 6.34. A line.

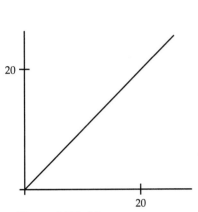

Figure 6.35. A line.

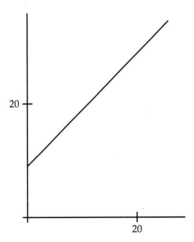

Figure 6.36. A line.

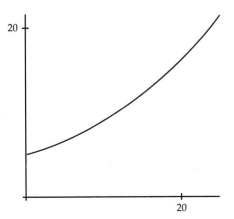

Figure 6.37. A curve.

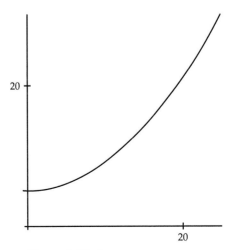

Figure 6.38. A curve.

No. of sheets	Light intensity
0	0.803
1	0.596
2	0.452
3	0.357
4	0.260
5	0.210
6	0.139

Figure 6.39.
Experimental light intensity results.

5. Several students noticed that some cars have tinted windows. Some cars appear to have darker tints than others. In an attempt to determine just how much light the tinting material lets into a car, the students decided to do an experiment. They obtained several sheets of the tinting material, a flashlight, and a light-intensity meter. **Figure 6.39** shows their data. Determine first whether the data appear to be exponential, and justify your decision. If they are exponential, find a reasonable closed-form equation, explain your method, state the relative rate of decay, and tell what it means in plain language. If they are not, explain how you know.

Time (days)	Concentration (mg/1)
0	0.500
1	0.345
2	0.238
3	0.164
4	0.113
5	0.078
6	0.054
7	0.037
8	0.026

Figure 6.40.
Digoxin decay.

6. Digoxin is a medicine given to some heart patients. In order to prescribe proper doses, medical personnel need to know how rapidly the digoxin leaves the body. **Figure 6.40** shows data for one patient. Determine whether the decay of the concentration is linear, exponential, or neither. If it is linear or exponential, find a reasonable closed-form equation.

7. A hot temperature probe was allowed to cool in air for about 2 min. Its temperature (in degrees Celsius) was recorded every 10 sec. The table in **Figure 6.41** shows the data.

58.5	45.69	38.08	33.01	29.27	26.83	24.98	23.81	22.99	22.29	21.82	21.36

Figure 6.41.
Temperature data from cooling experiment.

Newton's law of cooling states that the difference between the temperature of an object and the temperature of its surroundings should be an exponential function. Room temperature during this experiment was 21.35°C. Use that fact to create a table of values for the difference in temperatures between the probe and the air and determine whether Newton's law of cooling applies to these data. Explain your work.

8. The equation $y = 100 \times 1.1^x$ describes exponential growth.

a) Select eight x-values that are equally spaced (for example, 0, 1, 2, 3, 4, 5, 6, and 7, but be more creative than that). Find the corresponding y-values from the equation. Then graph y_{next} versus $y_{current}$ for your selected values. Describe the graph, and relate what you see to a recursive description of the growth.

b) Select eight x-values that are not equally spaced. Find the corresponding y-values from the equation. Then graph y_{next} versus $y_{current}$ for your selected values. Describe the graph, and relate what you see to a recursive description.

c) Formalize your observations in parts (a) and (b) to the general case of examining data to determine whether they are exponential or not. Explain, based on your knowledge of the recursive representations of exponential functions.

LESSON FIVE

Refining the Models

KEY CONCEPTS

The mathematical modeling process

Simulation

Identifying exponential data

Probabilistic models

PREPARATION READING

Give Me the Details

*I*n the model you developed in Lesson 3, you made some general assumptions about the entire population of moose in Adirondack State Park. Those assumptions probably involved differing levels of detail for different students. For example, someone might have suggested, "In a large herd, about the same fraction of moose will have calves each year." That leads directly to a model based on the equation

$$births = r \times p_{current}.$$

Someone else may have said something like, "In the long run, we would expect about half the moose to be female, and some fixed fraction of those females will have calves during any given year." That assumption means that the number of new moose each year will be equal to the total population multiplied by 0.5 times the fraction of females that have calves: births = $p_{current}$ x $0.5c$. So, one way to interpret what you did in Lessons 3 and 4 is that you estimated (indirectly) c, the fraction of females having calves, since r and $0.5c$ play the same role in the models.

Both these descriptions, though, are assumptions about the entire population. They do not take into account the differences among moose. Are such broad, population-wide assumptions reasonable? Can a population actually act that way? An alternative approach is to make assumptions about individual moose or smaller groups of moose. For example, the fraction of females that have calves might depend on the ages of the females. What if the herd has few females, something that seems likely if no additional moose are brought in? Simulation then allows you to play out those assumptions and see whether they lead to new models. Again, this falls under the general heading of evaluating your model, this time by focusing on the assumptions more than on the numbers.

How would you would use detailed information about the moose if it were available? What would you do to create a better simulation?

ACTIVITY

10

MOOSE DEMOGRAPHICS

Listed below are some sample assumptions to use during your investigation. However, you may wish to do some research and replace one or more of these assumptions with better information. Remember that all the numbers are assumptions, based on the observations that were reported. You should investigate the sensitivity of the resulting model to all these assumptions. Determine whether small changes in a particular assumption result in large changes in the model. If so, that assumption had better be verified carefully.

KNOWN INFORMATION:

- The 1993 population is estimated at 25 to 30.

- Moose that migrate into the park are almost always male and are almost always young (3–5 years old).

- Females were not observed in the park prior to 1988.

- The average life of a moose is 15 to 23 years, independent of sex.

ASSUMPTIONS:

- There were 22 males and 5 females in the population in 1993.

- Female moose can first have calves at age 4 and remain fertile through age 9.

- Each female that can bear young will have exactly one calf each year.

- Exactly half of all newborn moose are female.

- Every moose dies at age 20, no exceptions.

- When new moose are brought into the park, they will include equal numbers of each gender.

- When new moose are brought into the park, they will include equal numbers of moose of each age from 2–11 years.

MOOSE DEMOGRAPHICS

ACTIVITY

10

THE MODEL:

1. Begin with these assumptions or those you have made based on your own research, and complete the population table for the year 1993 on Handout H6.4. Discuss your table with your group, and agree on a reasonable setup. For now, do not include the additional 100 moose for 1993, so ignore the last two assumptions listed here.

2. Sharing the work with your group, transfer your 1993 population distribution to Handout H6.5, and complete each year's entries based on the assumptions listed here and on information from the prior year.

3. Use a computer spreadsheet to extend your work from Item 2 through the year 2013. (Hint: Use one column for each year (1993, etc.), with the first 20 rows for females, by age. Then use the next 20 rows for males, by age.)

4. Use the results of your simulation to make a table of: year, current total moose population, and change in total moose population.

5. Graph current total population versus year, change in total population versus current total population, and next total population versus current total population. Based on these graphs, decide whether these assumptions lead to an essentially exponential model.

INDIVIDUAL WORK 10

Something Old, Something New

1. a) Use your graph of change in total population versus current total population from Item 5 of Activity 10 to determine reasonable recursive and closed-form equations for your simulated population. Graph your equations with your data and comment on how well they fit.

    b) Repeat part (a) using your graph of next total population versus current total population.

    c) Compare the results of parts (a) and (b).

2. Comment on the reasonableness of each assumption listed at the start of Activity 10. For any that seem unlikely to be correct, suggest specific improvements that could be tested in a simulation.

3. Return to your model and implement the final two assumptions, related to bringing in an additional 100 moose. How well does the model appear to fit with the numerical assumptions that you used throughout Lessons 3 and 4?

4. In your models (from earlier lessons) involving the whole population, you had direct control over the population growth. To make it grow faster, you could increase r or b; to slow it down, you could decrease them. In the model you developed in Activity 10, however, the direct control is not obvious. What can you do? Describe ways that you could modify your model to produce a lower overall growth.

5. Nadine read the assumption that all moose died at age 20 and suggested that an alternative way of saying the same thing is to say that 1/20 of the moose die each year. Is she exactly correct? If so, why? If not, is her method at least approximately the same? Under what conditions is it best?

6. Roger is holding a leaky bucket while Emily pours water into it at a constant rate of 0.5 cups/sec. The leak removes 5% of the water in the bucket each second. However, the water level in the bucket is not changing.

   a) How fast is water leaking from the bucket, in cups per second?

   b) How long does an average water molecule stay in the bucket?

   c) How much water is in the bucket?

7. Many calculators have a "random" command. When used, it produces a number between 0 and 1. Repeating the command gives a list of numbers with no apparent pattern. Some are large, and some are small. Some pairs are close together; others are far apart.

   a) Use your calculator to create a list of ten randomly generated numbers.

   b) Suppose that a weather forecaster says that for each of the next ten days the chance of rain on that day is exactly 30%. Explain how to use your ten random numbers to simulate the rain/no rain situation for those ten days.

   c) If you repeated parts (a) and (b) to simulate the weather for another ten days, would you expect the same sequence of values? Would you expect the same days to be rainy? The same number of rainy days? Explain.

   d) Explain how random numbers could be used to improve the simulation of the moose population growth.

RANDOM MOOSE

Looking back at your work in this unit, it is clear that you have gone through the modeling cycle several times. Early models were very simple, based on a few assumptions. Since there were not many assumptions, they were also easy to check. Control was pretty easy, too, but the models did not do a very good job of matching reality. You have added detail in later models. They have a better chance of describing what is really going on, but they require more research because they are built on many more assumptions.

To this point, however, each model treated the moose herd as if everything were controlled by you. The assumptions forced the population growth to do the same thing no matter who ran the simulation or how often it was run. Moving closer to reality, it is not likely that two different but identical herds of moose left alone for 20 years would end up being identical. Many features of the growth of a herd can be measured or assumed, but those features are generally going to be averages. They describe typical behavior, but any particular herd will have its differences from those average behaviors.

This activity gives you a chance to examine and modify a model in which randomness is added. In particular, the survival of each moose from one year to the next, whether each female has a calf, and the gender of all calves will be determined randomly. Models in which features are randomly generated are referred to as **probabilistic models**.

The table in **Figure 6.42** shows the probabilities of survival to the next year for both male and female moose, as well as the probabilities that females of particular ages will have calves by the next year. Note that some of the assumptions for this model differ from those used in Activity 10, but this model also assumes that all moose die following their 20th year. This table presents the main assumptions for this model; they become the control points for tuning its behavior and the objects of further research for more accuracy.

RANDOM MOOSE

Age	Female Survival %	Female Birth %	Male Survival %	Age	Female Survival %	Female Birth %	Male Survival %
0	75%	0%	75%	10	93%	50%	93%
1	90%	0%	90%	11	90%	40%	90%
2	80%	90%	95%	12	85%	30%	85%
3	95%	90%	95%	13	80%	0%	80%
4	95%	90%	95%	14	80%	0%	80%
5	95%	85%	95%	15	80%	0%	80%
6	95%	80%	95%	16	80%	0%	80%
7	95%	75%	95%	17	80%	0%	80%
8	95%	70%	95%	18	80%	0%	80%
9	95%	60%	95%	19	80%	0%	80%

Figure 6.42.
Survival and calving probabilities.

Use a clean copy of Handout H6.4, and enter the same initial population for 1993 as you used in Activity 10. However, this time use tally marks instead of numbers to show the population. The first few rows of an example are shown in **Figures 6.43 and 6.44**.

Females

Age	1993	1994
0	I	
1		
2		
3	II	
4	I	

Figure 6.43.
Possible partial age-distribution tally for 1993 female moose.

Males

Age	1993	1994
0	I	
1		
2	I	
3	III	
4	IIII	

Figure 6.44.
Possible partial age-distribution tally for 1993 male moose.

As suggested by Figure 6.42, you must determine for each moose whether or not it survives to the next year. In addition, if the moose is female, you must determine whether or not she has a calf and, if so, whether the calf is male or female. Each of these events must be determined separately, as you did in the rain/no rain determination in Item 7 of Individual Work 10.

1. Write a rule for using the random command on your calculator or on a spreadsheet to determine whether a two-year-old male moose survives to be three years old.

2. Generalize your rule in Item 1 to cover the determination of the survival of any male moose.

3. Repeat Item 2 for female moose survival.

4. Write a rule to determine whether a particular female has a calf during the next year. (Assume that you already know that she will survive to next year.)

5. Apply your rules to complete Handout H6.5 for each moose in the current herd through 1998. (You may wish to circle each moose (tally mark) after you have determined whether it survives so that you will not lose track of which moose have already been considered.)

What Did You Expect?

1. Examine your data from the 1993–1998 simulations you did in Activity 11 to determine whether they seem to indicate that growth under this model is still exponential. Explain.

2. After working by hand with five years of the data, you can understand why people use calculators or computers to do this kind of work. Use a program on a computer or calculator to simulate the full 20 years of data. Then repeat Item 1 with the new data.

3. One feature of probabilistic simulations is that they rarely give exactly the same results when repeated. Explain how you could use the information from several simulations (all starting with the same assumptions, of course) to get a better idea of the general behavior than just by looking at one.

4. Recall that this unit began by seeking to understand the effects of adding 100 new moose to the existing few in Adirondack State Park. Repeat your Item 2 analysis on a 127-moose simulation.

Wrapping Up Unit Six

1. If you are given a table of ordered pairs, how can you tell whether the table represents linear growth?

2. a) If you are given a table of ordered pairs, how can you tell whether the table represents exponential growth?

   b) Apply your method to these data:

x	10	20	30	40	50	60	70	80	90	100
y	100	170	290	500	830	1400	2400	4100	7000	12,000

3. a) Assume that the city of Troy was approximately 120 ft. above sea level in 1500 B.C. and that its elevation increased an average of 4.7 ft. per century. Write a recursive formula that describes the elevation of Troy. Be sure to specify units for your values and the period for your recursion.

   b) What type of growth is this?

   c) Suppose you were told that the elevation of Troy was 167 ft. above sea level in 500 B.C. Is that value consistent with the information in part (a)? Explain how you computed your answer.

   d) Assume that Troy survived and that it continued to rise at the same rate. Use home screen iteration to predict its elevation in the year 2000.

   e) Write a closed-form equation for Troy's elevation. Be sure to state what each of your variables means. Then use your formula to check your answer to part (d) and to predict when its elevation would reach 500 ft. Explain your method and why you chose it.

4. Hiroshi told his classmates that every (nonvertical) straight-line graph represents an additive situation. In fact, he claimed that for every closed-form equation of a straight line they could give him, he could produce two different recursive representations, both starting at the same initial point but with different quantities added. Can you do what he claims to be able to do? If so, tell how (exactly). If you think you could stump Hiroshi, write the equation you would give him, and explain what difficulties he would have.

5. a) Elena opened a savings account with an initial deposit of $50, and she deposits $25 per month from her earnings each month. The bank pays her interest at the rate of 0.25% per month. That is, each month the bank adds 0.0025 × (current balance) to the account and then Elena adds $25 of her own money. Write a recursive equation describing her balance under this process.

   b) Use home screen iteration or a spreadsheet program to simulate the growth of her account using this saving scheme.

   c) Does using this method result in an additive model for the balance? Is it an exponential model? Justify your answers.

6. A population grew exponentially from 1000 to 17,000 in 25 years. Find a closed-form equation for this population, sketch its graph, and use the equation to approximate the population at the end of 18 years.

7. a) A worker plans to retire in 20 years and expects inflation to average about 3% per year between now and then. If she were at retirement age now, she would want $1500 per month in income. How much income will she want to have in 20 years in order to maintain that same standard of living?

   b) The worker can secure an investment that is safe and will pay 1% per month. She wants to be able to use the interest from this investment as her sole income during retirement. (She will remove each monthly interest payment as soon as it is put into the account; the principal will never be touched.) How much does she need to have in the account when she retires to meet the goal described in part (a)?

8. a) The front page article of the Durham, NC, *Morning Herald* on Sunday, February 7, 1988, stated: "From 1960 to 1980 the population of Durham, Orange, and Wake counties increased 60 percent, from 324,047 to 521,167." Based on the paper's statement, what was the annual relative growth rate for the population of these three counties during the 1960s and 1970s?

   b) Predict the population of these counties for the year 2000. Explain your method(s).

9. a) The population of a small town grew from 4000 to 5500 between 1950 and 1970. Assuming that the growth was exponential, what was the yearly relative rate of growth?

b) Use your answer to part (a) to write a closed-form equation for population versus time, in years. Think of 1950 as year 0, 1951 as year 1, etc. Graph your equation, and use trace or table to verify that it goes where it should.

c) Wright wants to have a closed-form equation in which the actual year is used. That is, instead of having to remember that 1950 is year 0 (easy) or that 2011 is year 61 (harder), he wants to be able to substitute 1950 or 2011 directly. Without actually finding an equation for him, describe how a graph of his equation would compare to the graph you found in part (b).

d) Using what you know about translating graphs and their equations, write an equation for Wright.

10. Suppose a quantity grows according to the recursive relation $p_{next} = p_{current} + r \times p_{current}$, with $r = 0.07$. Assume that each time period is one year. This situation represents 7% growth per year. Use 100 as the initial population.

a) Write a closed-form equation for this population versus time, with time measured in years.

b) Trace on the graph of your equation to determine the relative growth rate for two-year periods. Does the two-year relative growth rate seem constant?

c) Determine the relative growth rate for five-year periods. Does it seem constant?

d) Determine the relative growth rate for 20-year periods. Does it seem constant?

e) Explain how your observations in parts (b–d) could be used to find a quick prediction of the population for the three North Carolina counties in Item 8 for the year 2000.

11. Over recent history, the population of Bangladesh has grown approximately according to the equation $p = 104.1 \times 1.0955^x$, where p is in millions and x is time measured in decades (with 1985 as year 0). Rewrite this equation in a form in which the explanatory variable is the actual year number, check your new equation for 1985, then use your equation to predict the year in which the population will reach 300 million.

12. a) An object that is warmer than its surroundings tends to cool off; an object cooler than its surroundings tends to warm up. In fact, the cooling or warming is such that the difference between the temperatures of the object and the surroundings changes exponentially. Medical examiners can use this information to help determine the approximate time of death in some cases. Of course, every case is different, depending on the local air temperature, clothing on the deceased, etc. Assuming that the exponential model is exactly correct, explain how you would determine the time of death. (What measurements would you take and what would you do with them?)

   b) What assumptions did you need to make in order to answer part (a)?

13. Suppose that a friend has an old cassette tape player. It has a counter that keeps track of how many times the take-up reel has gone around; it does not measure time directly. List assumptions that you think would be useful in building a mathematical model that could be used to convert the reel counts into elapsed time. You do not need to formulate a model.

14. There once was a company in Tennessee that made pianos. As do many manufacturers, they stamped serial numbers on their pianos, counting 1, 2, 3, . . . throughout their history. You know that the factory opened in 1817 and you have piano 5100, made in 1850. A friend finds piano 945 at a flea market and wants to know how old it is (there is no date). Discuss and construct models that might be used to help your friend estimate the age of the piano.

15. In Unit 4, *Prediction*, Individual Work 3, you played the game *Target*. **Figure 6.45** is the same as Figure 4.24 on page 310. It shows one possible way to score the game. Write a closed-form equation for the rule that governs each round's score.

Dots on the line	1	2	3	4	5	6	7	8	9	. . .
Score	1	2	4	8	16	32	64	128	256	. . .

Figure 6.45.
Table assigning scores to the number of dots on the line.

Mathematical Summary

*I*n this unit you have studied two major mathematical ideas: exponential functions and mathematical modeling. You also revisited linear functions, parametric equations, and the mathematics of translation and scale change, all encountered in earlier units.

Mathematical modeling is an active process. In its simplest form, mathematical modeling is whatever people do to use quantitative information to understand something in real life. More formally, it consists of several stages, each of which is vital, although not necessarily mathematical in nature. These main stages are as follows:

Step 1.

Identify a situation: Notice something that you wish to understand, and pose a well-defined question indicating exactly what you wish to know.

Step 2.

Simplify the situation: List the key features (and relationships among those features) that you wish to include for consideration. These are the assumptions on which your model will rest. Also note features and relationships you choose to ignore for now.

Step 3.

Build the model: Interpret in mathematical terms the features and relations you have chosen. (Define variables, write equations, draw shapes, gather data, measure objects, calculate probabilities, etc.)

Step 4.

Evaluate and revise the model: Go back to the original situation and see if the results of your mathematical work make sense. If so, use the model until new information becomes available or assumptions change. If not, reconsider the assumptions you made in step 2 and revise them to be more realistic.

Some other ideas help guide this process. Among them are these rules of thumb: Keep it simple. If you are looking for a relationship between quantities then treat them as variables. Change only one thing at a time. Such guiding principles are based on the objective of understanding more about a particular situation.

Linear functions exhibit steady change at a constant rate. Therefore, they may always be described by recursive equations of the form $y_{next} = y_{current} + c$, where c is the constant amount of change in one recursive step. The value of c depends on the size of the recursive step used. If one recursive step is 1 unit for the explanatory variable, then c also turns out to be the slope of the closed-form description of the linear function. Because of the form of recursive equations for linear functions, such functions are also referred to as additive functions.

In contrast to linear functions, which are additive, exponential functions can be described as multiplicative. Their recursive descriptions all have the form $y_{next} = b \times y_{current}$. In that form, b represents the growth factor, the factor by which the function grows in one recursive step. An equivalent form, more easily compared to the additive form of linear functions, is $y_{next} = y_{current} + r \times y_{current}$. This form stresses the fact that exponential growth may also be thought of as adding a constant percentage to the function in each recursive step. The quantity r is called the relative growth rate. As with linear functions, the constants b and r for exponential growth depend on the size of the recursive step used.

The constants b and r control the overall behavior of the exponential function. In particular, they are what distinguish between growth and decay curves. All exponential functions have closed-form graphs that look very much alike. Decay curves are the same shape as growth curves but are reflected across the vertical axis. Both growth and decay graphs have one end very near the horizontal axis and the other end rising more and more steeply at greater distances from the vertical axis.

Because each of the recursive equations for linear and exponential functions is itself linear, the graphs of these functions are straight lines. Straight lines are much more easily identified visually than are any other shapes. Thus, the recursive equations lead to several simple graphical tests to determine whether data can be described by linear or exponential functions. Helpful graphs include y_{next} versus $y_{current}$ and change versus $y_{current}$. According to the recursive equations given before, each of these should be linear and contain $(0, 0)$ for exponential systems. Of course, since linear functions are already linear, such tests are not frequently applied to them.

Parametric equations may be used to graph any function. They also provide one way to understand the mathematics of translation and scale change. For example, since $x_1 = t$, $y_1 = 100(1.1)^t$ describes an exponential function that starts at $(0, 100)$, the new parametric pair $x_2 = x_1 + 5$, $y_2 = 100(1.1)^t$ describes a similar function that has been moved five units to the right, containing $(5, 100)$. But that means that $x_1 = x_2 - 5$, so substitution relates y_2 directly to x_2 and gives $y = 100(1.1)^{(x-5)}$ as another (not parametric) representation of the second function. In general, substituting $(x - h)$ for x in an equation has the effect of translating the original graph h units to the right.

In a similar manner, scale changes may be accomplished by substituting and expression such as cx for x. The effect is to cause the original graph to shrink by a factor of c back toward the vertical axis (horizontally).

Glossary

ADDITIVE PROCESS:
Any process described by a recursive equation of the form $Q(n + 1) = Q(n) + k$, where k is a constant.

BASE:
The number that is raised to a power in an exponential expression. For example, in the expression b^x, b is the base.

EXPONENTIAL DECAY:
Behavior described by an exponential function having a growth factor between 0 and 1 or a relative rate of change between –1 and 0.

EXPONENTIAL FUNCTION:
A function having a closed-form description of the form $y = a(1 + r)^x$ or $y = ab^x$.

EXPONENTIAL GROWTH:
Behavior described by an exponential function having a growth factor greater than 1 or a relative rate of change greater than 0.

FUNCTION NOTATION:
A way to show the value of one quantity whose value is determined by the value of another quantity. For example, $c(1) = 25$ indicates that the value of c is 25 when the other quantity is 1.

GROWTH FACTOR:
The amount by which an exponential quantity is multiplied in one recursive period.

INEQUALITY NOTATION:
A mathematical statement of the relationship between two numbers that are not necessarily equal. For example, 1 is less than 4 is written as $1 < 4$; x is between 0 and 2 is written as $0 < x < 2$; and y is greater than or equal to 7 is written as $y \geq 7$.

MATHEMATICAL MODELING:
The process of beginning with a situation and gaining understanding about that situation through the use of mathematics.

MULTIPLICATIVE PROCESS:
Any process described by a recursive equation of the form $Q(n + 1) = kQ(n)$, where k is a constant.

ORDERED PAIR:
Two numbers designating the values of two quantities, where the order in which the numbers appear indicates their meaning. For example, (4, 16) indicates that 4 is the value of the first quantity and 16 is the value of the second.

PROBABILISTIC MODEL:
A model that includes randomly occurring events.

RELATIVE RATE OF CHANGE:
The change in a quantity during a given period, expressed as a fraction of its value at the start of the period; $\frac{\Delta y}{y}$.

SCALE CHANGE TRANSFORMATION:
A transformation in which all first coordinates (or second coordinates) are multiplied by the same constant, resulting in a horizontal (or vertical) stretch of the corresponding graph.

SENSITIVITY ANALYSIS:
The investigation of the effects of making small changes in specific assumptions in a model. If small changes lead to large effects, the model is said to be sensitive to that change.

SIMULATION:
Acting out the details of a situation you are modeling. Simulation may use equations, graphs, technology, physical objects, etc.

UNIT

7

Imperfect Testing

In earlier units you have seen the importance of probability in dealing with problems as diverse as cracking codes and predicting wildlife populations. In each case, however, the probability information was determined "perfectly." The probabilities you used did not depend on any additional information in any obvious way. On reflection, though, you may have used models in Unit 6, *Wildlife*, for which survival rates depended on gender, and surely the probability of having a calf depended on whether the particular moose in question was male or female!

In this unit you focus your attention on methods of measuring probabilities, primarily in the field of medical tests, that are not perfect. How can reliable information be obtained from such tests? Your primary tools will be the mathematics of conditional probability and the algebra of functions.

PERFORMANCE-ENHANCING DRUGS

The use of drugs to enhance athletic performance is not a new phenomenon. For example, in the 1904 St. Louis marathon, Tom Hicks was able to restore his dwindling energy near the end of the race by taking doses of strychnine and brandy. He won the race, and, at that time, no one objected.

However, times have changed. It is now known that, in addition to giving an athlete an unfair advantage, the use of performance-enhancing drugs can cause a variety of serious health problems. Today, major amateur and professional organizations such as the International Olympic Committee (IOC), the National Football League (NFL), and the National Collegiate Athletic Association (NCAA) have instituted drug testing of athletes. Now, when it is discovered that an athlete has been able to win not because of his or her own natural abilities and hard work, but because the athlete has taken performance-enhancing drugs, it makes the headlines. A well-known example is Ben Johnson. In 1988, the Canadian sprinter won the 100-meter gold medal in the Olympics in Seoul, South Korea and set a world record of 9.79 seconds. However, after testing positive for anabolic steroids, he was stripped of his title. The story ran in newspapers throughout the country and the world.

How widespread is the use of performance-enhancing drugs among athletes? If an athlete has not taken a performance-enhancing drug, could a drug test turn out to be positive? How likely is it that an athlete who has taken performance-enhancing drugs can "beat" a drug test and get a negative result?

In this unit you will learn some mathematics that can be used to answer these questions.

A Sporting Chance

KEY CONCEPTS

Drug testing

Sample percentages

Variability of samples

Probability

Mutually exclusive events

Complementary events

The Image Bank

PREPARATION READING

Estimating the Use of Performance-Enhancing Drugs

How widespread is the use of performance-enhancing drugs? The first drug-testing program at a major sporting event took place at a bicycle race in France in 1955. Over 20% of the competitors tested positive for the use of banned substances. Five years later, Danish cyclist Knut Jensen died during a road race in Rome. He had been using drugs to enhance his performance. His death prompted the International Olympic Committee (IOC) to institute its drug-testing policy.

More recently, a study on substance use and abuse was conducted at various colleges and universities in the United States. Of the 2505 athletes surveyed, 1719 were males and 786 were females. Only 2.5% of the athletes responded "Yes" when asked whether they had used anabolic steroids sometime during the previous year. Of those, 40% reported that they began steroid use while in middle school or high school. This report was presented in 1993 to the National Collegiate Athletic Association (NCAA).

The examples above refer to two different methods for estimating the percentage of a population that uses performance-enhancing drugs. The first example relies on results from drug testing, and the second relies on results from a sample survey. In this lesson, you'll explore information from sample surveys. In Lesson 3, you'll investigate situations that involve testing.

CONSIDER:

1. What percentage of the cyclists did not test positive for banned substances?

2. Do you think that any of the cyclists who tested negative might have been able to "beat" the test; in other words, they tested negative even though they had taken one of the banned substances?

3. How many of the athletes in the national study admitted that they had taken steroids? How many of these students began taking steroids in middle school or high school? Do you think that use of anabolic steroids is a problem in your school?

4. Do you know of any reasons, other than to enhance performance in sports, that a student might take steroids?

ACTIVITY

1

IN MY ESTIMATION

In the NCAA survey cited in the preparation reading, the athletes were asked:

QUESTION A:

> Have you used anabolic steroids sometime during the past 12 months? Yes or No.

Those conducting the study worried that student athletes might not answer this question truthfully if they thought that college officials or their coaches might see their responses. Students were assured that they could keep their answers confidential by sending their completed surveys directly to Michigan State University for processing.

This lesson deals with ways of using survey data to estimate a percentage. You need to survey the students in your class. Do you think that everyone in your class would answer Question A truthfully? Because this is a sensitive question and because students themselves would be interested in each other's answers, you probably could not guarantee to keep all responses confidential. Instead, ask the following, less-sensitive question:

QUESTION B:

> Do you participate in your school's sports program?
> Yes or No.

You might agree that students are more likely to respond truthfully to Question B than to Question A.

1. Within your group and before doing a survey, estimate the percentage of students in your school who would answer Yes to Question B. Explain how you arrived at your answer.

   If you gave Question B to each student in your school and got an honest Yes or No answer from every student, you would be able to calculate the exact percentage of students who participate in your school's sports program. However, it is unlikely that everyone would return the survey.

IN MY ESTIMATION

2. The NCAA study received replies from 78% of those who were eligible to participate. This is considered a good response to a survey.

   a) Given that 2505 student athletes completed surveys, how many students were eligible to be in this study?

   b) What percentage of the eligible students chose not to participate in the study?

Instead of trying to ask every student in your school, you might select a **sample**, a smaller portion of the students and administer the survey to the students in the sample. If they are representative of the students in your school, then their percentages of Yes or No will be a good estimate of the school-wide percentage. A **representative sample** is similar in composition to the **population**—the entire group—being studied.

3. Suppose you want to estimate the percentage of students in your school who are avid readers. Three samples are described below. For each, discuss whether the sample would be representative of the students in your school (at least in terms of being avid readers). Include reasons for your answer.

   a) Ten students are selected at random from an honors English class.

   b) All students' names are written on paper and placed in a hat. The papers are mixed, and then 10 names are drawn from the hat.

   c) The sample consists of the first 10 players who show up for basketball practice during basketball season.

4. Return to the problem of estimating the percentage of students who participate in your school's sports program. The most convenient sample consists of the students present in your math class today. Is there any reason why your class, as a group, might not be representative of the students in the rest of the school, in terms of participation in your school's sports program? Explain.

ACTIVITY

1

IN MY ESTIMATION

5. Now, it's time to collect some data. Each person in the class should write his or her answer to Question B on a piece of paper and the papers should be collected.

a) Complete the entries in **Figure 7.1** based on the class responses to Question B.

Figure 7.1.
Class data.

Number of students in class who participate in school's sports program	
Number of students present in class	

b) What portion (fraction or decimal) of the students present in class participate in the school's sports program? What percentage of the students participate?

c) What portion of the students from this class do not participate in the school's sports program? What percentage of the students do not participate?

d) Suppose that your class is not representative of the school and that the percentage of students in your class who participate in the school's sports program is higher than the school-wide percentage. If you use the percentage from part (c) to estimate the school-wide percentage of students who do not participate in your school's sports program, will your estimate be too high or too low? Explain.

Frequently, area diagrams are used to represent a population's breakdown in terms of a particular trait. To construct an area diagram, you begin by drawing a shape whose area represents the population. Then you divide the shape into two regions, one

ACTIVITY

IN MY ESTIMATION

1

representing the portion of the population that has the trait and the other the portion that does not have the trait. In Items 6 and 7, you will display the results of your class survey using two common area diagrams, percentage bars and pie charts.

6. A percentage bar looks very much like a thermometer. On a thermometer, the height of the mercury represents temperature; on percentage bars, the height of the bottom region (which is frequently shaded) represents the percentage of the population that has a particular trait.

   a) Draw a rectangle similar to the one in **Figure 7.2** to represent the students in your class. Starting at the bottom, shade a region that represents the percentage of students in your class who participate in the school's sports program.

   b) Explain how you determined what portion of the rectangle you would shade.

   c) What does the unshaded region represent?

7. In a **pie chart**, a circular region represents the entire population. A wedge, resembling a slice of a pie, is shaded to represent the portion of the population that has the trait of interest. (See Figure 7.2.)

   a) Draw a circle to represent the students in your class. Then shade a wedge that represents the students who participate in your school's sports program.

   b) How did you determine the size of the wedge?

   c) Which display do you like better, the percentage bar or the pie chart? Why?

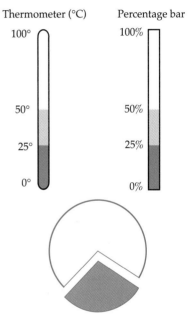

Figure 7.2.
A percentage bar, which resembles a thermometer, and a pie chart.

ACTIVITY

1

IN MY ESTIMATION

8. The relationship between the portion of students who participate in the school's sports program and the portion of students who do not participate can be generalized for any particular trait.

   a) Given any population, what is the relationship between
      the portion of the population "with the trait" and the por-
      tion "without the trait"? Make a graphic display that illus-
      trates your answer.

   b) What is the relationship between the percentage of the
      population "with the trait" and the percentage "without
      the trait"?

In this activity, you used the percentage of students in your class
who participate in your school's sports program to estimate the
school-wide percentage of students who participate. Provided
your class is representative of the students in your school, your
estimate for the students who participate in your school's sports
program should be fairly close to the school-wide percentage.
However, because your class is a convenient sample and perhaps
not a representative sample, your estimate may be too high or
too low.

Remember that you are learning about the estimation process
using the class responses to Question B, even though you really
would like their responses to Question A. In Lesson 2, you will
study a method that masks individual responses to sensitive
questions, which may allow you to ask Question A.

INDIVIDUAL WORK 1

Test Your Taste in Movies

1. Do you like horror movies?

2. What percentage of the students in your school do you think like horror movies?

3. If you randomly select a student from your school, how likely is it that this student likes horror movies?

   The answer to Item 1 is easy—you either said Yes or No. However, you could probably improve your answers to Items 2 and 3 if you had some additional information.

4. Suppose, when 30 students from your school are polled, 10 claim that they like to watch horror movies. Assume that this sample is representative of students in your school.

   a) What does it mean to assume that this sample is representative of the students in your school?

   b) Based on the results of the sample poll, estimate the percentage of students in your school who like horror movies.

   c) What percentage of the students do not like horror movies?

   d) Make an area diagram to represent the breakdown of the students in your school into those who like horror movies and those who do not. Explain how you determined the size of the region that represents "students who like horror movies."

   e) Based on the results of the sample poll, if a student is randomly selected from your school, is the student more likely to like horror movies or not like horror movies? Justify your answer.

5. Suppose that the movie industry hires a polling organization to investigate the breakdown by age of moviegoers who like horror movies. A partial summary of the results is given in **Figure 7.3**.

   a) Create a table similar to the one in Figure 7.3 and complete the entries.

Age (years)	Percentage who like horror movies	Percentage who do not like horror movies
13–18	30	
19–34		88
35–50	15	

Figure 7.3.
Breakdown of those who like and do not like horror movies.

b) Use area diagrams to represent the data in your table. If your area diagrams were part of a newspaper article about the popularity of horror movies, describe what you would want a reader to learn from your diagrams.

c) Suppose you randomly selected two people and find that the first is 15 years old and the second is 45 years old. Which person do you think is more likely to enjoy horror movies? Justify your answer using results from the poll.

6. In a study of senior citizens, suppose it was found that four-fifths of the senior citizens did not like horror movies.

a) What fraction of the senior citizens admitted to liking horror movies? What percentage is this?

b) Suppose the study of senior citizens was based on interviews with 25 men and 75 women and that 16% of the women said they enjoyed horror movies. What percentage of the men liked horror movies? Explain how you arrived at your answer.

c) Imagine randomly selecting a woman and a man from the senior citizens in the study. Is it more likely that the man or the woman would say that they like horror movies? Justify your answer using results from this study.

Item 6(c) asks you to use percentages to determine which of two possibilities is more likely. In the next activity, you will use probability to assess how likely it is that certain outcomes occur.

SAMPLE EXAMPLES

At a major cycling race in France in 1955, approximately 20% of the cyclists tested positive for banned substances. All of the cyclists at this event were tested, hence the exact percentage of positive tests could be calculated. However, testing for performance-enhancing drugs takes time and costs money. So, at major sporting events such as the Summer Olympic Games, only a portion of the athletes are tested. Although you can use the percentage of positive tests in the sample to estimate the percentage of Olympians who would test positive, you need to be aware that sample information generally varies from sample to sample, even when the samples are taken from the same group.

In this activity, you will simulate taking samples and have an opportunity to observe how much the sample percentage can vary from sample to sample and how far off the sample percentage can be from the group-wide percentage.

For example, suppose that 20% of the students from Adams High School are wearing white socks today.

1. a) Imagine that the students present today in your class are a representative sample of students from Adams High School. Based on this assumption, if you were to look down at their feet, how many pairs of white socks would you expect to see? Explain how you arrived at your answer.

   b) Suppose that, as part of a data-gathering assignment, you randomly select a student from Adams High, record the color of his socks, let him go back to class, then randomly select a second student, record the color of her socks, and so forth. After repeating this process ten times, how many pairs of white socks would you expect to observe?

   c) Suppose you repeated the process in part (b) two more times. (You will have observed a total of 30 students.) How many pairs of white socks would you expect to observe?

SAMPLE EXAMPLES

If you were selecting actual samples of students from Adams High School, you would need to consider the following questions:

- Suppose that you randomly select five samples of ten students from Adams High School and, for each sample, record the number of times white socks are observed. How much variability might you notice among the sample percentages? What if you select five samples of 500 students?

- Suppose you didn't know that 20% of the students at Adams High were wearing white socks and had to estimate the percentage of students who were wearing white socks based only on information from a sample. Would you get a better estimate from a sample of 10 students, 30 students, or 500 students?

You can't actually answer these questions by going to Adams High School, selecting samples of students, and observing whether or not they are wearing white socks. However, you will simulate the outcomes of such experiments by first using a simple random device such as a container with two colors of beads (in Item 2) and then using a random-number generator (in this case, a calculator, in Item 4). The simulation experiments in Items 2 and 4 provide enough background for you to reconsider the above questions when you work on Item 5.

2. **Simulation Experiment with Beads**

This experiment should be performed in small groups. Each group needs a container of beads or chips, identical except that 20% should be white. (You may substitute another color.) The white beads represent the students who are wearing white socks. Afterward, you will combine your group's results with those from the other groups.

Step 1:

a) Mix the beads thoroughly. Then, without looking into the container, draw out a bead. Record the "sock" color, and

SAMPLE EXAMPLES

return the bead to the container. Repeat this process another nine times (for a total of 10 draws).

b) Record your results, the number of draws and the number of white socks, in the first row of Figure 1 on Handout H7.2.

c) Next, compute and record in Figure 1 the percentage of times that white socks were observed in the 10 draws.

Step 2:

Continue drawing and replacing another 10 times as you did in Step 1. Do not start your count over; continue by adding the results of 10 new draws to your 10 previous draws. Then compute and record in Figure 1 the percentage of times that white socks were observed in 20 draws.

Step 3:

Continue to draw and replace another 10 times as you did in Step 2. Continue adding the results of the 10 new draws to your 20 previous draws. Record your results on Handout H7.2 in the third row of Figure 1.

3. **Class Results of Simulation Experiment**

a) Record in Figure 2 (Handout H7.2) each group's results from the first 10 draws and then from all 30 draws. Check that each group has computed their percentages correctly.

Look down the column "Percentage of times white socks were observed." You should notice that even though you are drawing samples from a mixture containing exactly 20% white beads, your samples seldom contain exactly 20% white beads. Instead, the sample percentages fluctuate around the 20% mark.

b) Do the percentages based on 10 draws or 30 draws stay closer to 20%?

ACTIVITY

2

SAMPLE EXAMPLES

c) Next, complete the entries in the first two columns of Figure 3 (Handout H7.2) as follows:

• Enter your group's results from your 30 draws into row 1 of Figure 3 on the handout.

• To get the entries for the first two columns of row 2, add the next group's results from their 30 draws to your group's results.

• To get the entries for the first two columns of row 3, add another group's results from their 30 draws to the combined results of your first two groups.

• Continue this process until you have included the results from each group in the class.

d) Complete the third column in Figure 3 by computing the percentages of white socks for each row.

e) Make a scatter plot of the percentages versus the number of draws. (Be sure to choose appropriate scales for the axes and label your axes.) Then, connect the dot corresponding to 30 draws to the dot corresponding to 60 draws, and then the dot corresponding to 60 draws to the dot corresponding to 90 draws, and so forth. Describe any trends that you see.

Do you think that you would get the same results if you repeated the bead-drawing experiment 500 times? Fortunately, you don't need to do this by hand! Instead, you can use a graphing calculator and its random-number generator to simulate this experiment.

4. **Calculator Simulation Experiment**

a) Each group will use a calculator to simulate the results of 500 draws from a 20% mixture of white and non-white socks. After all groups have completed the simulation, compile the group results (as you did in Item 3), and then complete the entries in Figure 4 (Handout H7.2).

b) Make a scatter plot of the sample percentages versus the number of draws (in groups of 500). Use the same scale on the vertical axis that you used in Item 3(e). Compare the amount of fluctuation in the percentages within your two graphs.

5. Below you see the list of questions that preceded the simulation experiments. Write a brief paragraph to answer these questions. Base your answers on information from Handout H7.2.

• Suppose that you randomly select five samples of ten students from Adams High School and, for each sample, record the number of times white socks are observed. How much variability might you notice among the sample percentages? What if you select five samples of 500 students?

• Suppose you didn't know that 20% of the students at Adams High were wearing white socks and had to estimate the percentage of students who were wearing white socks based only on information from a sample. Would you get a better estimate from a sample of 10 students, 30 students, or 500 students?

For many experiments, some outcomes are more likely than others. An **event** is any outcome or collection of outcomes. For example, if you are interested in tomorrow's weather, possible events could be (1) it will rain, (2) it will snow, or (3) it will precipitate (rain, snow, sleet, or hail). While the events (1) and (2) are single outcomes, event (3) consists of a collection of 4 outcomes. For any event, you can assign a number between 0 and 1 that indicates how likely that particular event is to occur. This number is the **probability** of the event. The higher the probability, the more likely it is that the event will occur.

For example, if you randomly select a student from Adams High School, the student is wearing white socks or not wearing white socks. Each of these outcomes, "wearing white

ACTIVITY

2

SAMPLE EXAMPLES

socks" and "not wearing white socks," can be assigned a probability. That number is based on how frequently the outcome occurs when the experiment is performed repeatedly (as in your simulations). Look at the percentages in Figure 4 (Handout H7.2). As the number of repeats of the experiment gets larger, the percentage of "white socks" gets very close to 20%. So, in this case, it makes sense to say that the probability that a randomly selected student from Adams High will be wearing white socks is .20.

6. What number would you assign as the probability that a student is not wearing white socks? Justify your answer.

7. A weather forecaster predicted an 80% chance of rain on three different days. It turned out that it rained on the first day, did not rain on the second, and rained on the third. Based on the fact that it only rained on two of the three days, do you think the weather forecaster made a terrible series of predictions? (Use information observed during the simulations to support your answer.)

In your simulation experiments, you were told that the percentage of students who had a particular trait (wearing white socks) was 20%. In Item 7, you were told that there was an 80% chance of rain. Generally, you won't know how likely it is that a certain outcome occurs. However, in your simulations, notice that the percentage having a particular trait found after a high number of trials is very close to the stated probability for that trait. Percentages found through using large sample sizes are good approximations for probabilities.

8. Take, for example, flipping a tack. If you flip a tack it can either land point up or point down (**Figure 7.4**). Let P(point up) and P(point down) represent the probabilities of the tack's landing point up or point down, respectively.

SAMPLE EXAMPLES

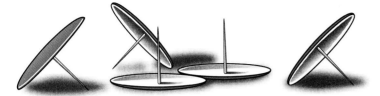

Figure 7.4.
Outcomes of tossing a tack.

a) Do you think it is more likely that the tack will land point up or point down? Explain.

b) Suppose a tack is flipped 1000 times. A graph of the percentage of times that the tack lands point up versus the number of times the tack is tossed appears in **Figure 7.5**. Estimate *P*(point up) based on information from the graph. (Pay attention to the scale on the vertical axis.)

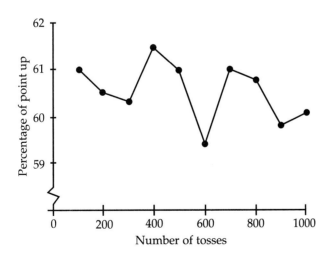

Figure 7.5.
Percentage of point up v. number of tosses, by groups of 100.

c) What is your estimate for *P*(point down)?

ACTIVITY

SAMPLE EXAMPLES

2

Select one of the experiments described in Items 9–11. Repeat your selected experiment at least 100 times. Report the results of your experiment. Then estimate the requested probabilities.

9. Experiment: Flip a coin.

   a) Record how many times the coin lands heads up and how many times the coin lands tails up.

   b) Estimate P(coin lands heads up).

   c) What is your estimate for P(coin lands tails up)?

10. Experiment: Roll a die.

   a) Record the number of times the die lands with 1 or 6 facing up and the number of times that it does not.

   b) Estimate P(either 1 or 6 faces up).

   c) What is your estimate for P(either 2, 3, 4, or 5 faces up)?

11. Experiment: You will need a container with two colors of beads, Color A and Color B. Mix the beads. Without looking into the container select a bead.

   a) Record the number of times Color A is observed and the number of times Color B is observed.

   b) Estimate the probability of drawing Color A.

   c) What is your estimate of the probability of drawing Color B?

INDIVIDUAL WORK 2

Driving Ourselves to Distraction

*I*n this assignment, you will have an opportunity to review what you have learned so far in this lesson. In addition, you will learn some new vocabulary commonly used in probability.

1. Suppose that a school has 200 students, 75 of whom are seniors.

   a) If you select a student at random, what is the probability that the student is a senior? Interpret the meaning of this probability.

   b) If you select a student at random, what is the probability that the student is not a senior?

   c) How are the following probabilities, *P*(the student is a senior) and *P*(the student is not a senior), related?

Notice that the students in the school can be divided into two non-overlapping groups: students who are seniors and students who are not seniors. No student can be in both groups, so the outcomes "senior" and "not a senior" are **mutually exclusive events**. When two mutually exclusive events combined comprise the entire population, they are called **complementary events**. When you combine the seniors with the non-seniors you get all the students in the school. This means that the events "senior" and "not senior" are complementary events.

2. Suppose that the population consists of the students in your school. Determine which of the pairs of events below are mutually exclusive events. Then determine which of the pairs of events are complementary. Explain your answers.

   a) male; female

   b) member of the football team; member of the baseball team

   c) students in grades 11 and 12; students in grades 11 or lower

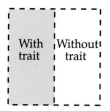

Figure 7.6.
Square representation.

Suppose that the square in **Figure 7.6** represents a school's student population. The shaded region within the square represents that part of the school's population having a particular trait or characteristic (such as students with blue eyes or students who regularly eat school lunches).

Since half of the square is shaded, the probability of randomly selecting a student with this trait is .5 or 50%.

3. Suppose the trait you are looking for is "student has after-school job." Information from four schools is shown in **Figure 7.7**. The shaded region corresponds to students having this trait.

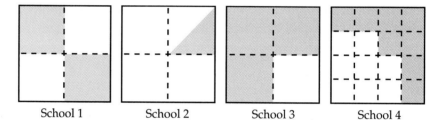

Figure 7.7.
Four schools.

School	Probability that a randomly selected student has an after-school job	Probability that a randomly selected student does not have an after-school job
School 1		
School 2		
School 3		
School 4		

Figure 7.8.
Probabilities of holding after-school jobs.

a) Based on the corresponding diagram, complete the entries in **Figure 7.8**.

b) How are the probabilities in the last two columns of Figure 7.8 related?

4. Suppose that 200 people have applied for ten job openings at a fast-food restaurant.

a) What percentage of the applicants can get jobs?

b) What is the probability that a randomly chosen applicant will get a job, assuming that all are equally qualified?

c) What is the complementary event to "applicant gets a job"? How likely is it that this event occurs?

5. a) The survey referred to in **Figure 7.9** reported that 6% of those interviewed admitted that they read while driving. Assuming that results
were rounded to the nearest whole percent, what is the range of possible percentages that would give rise to the reported 6%?

Figure 7.9.
Driving ourselves to destruction? Source: *USA Today* Website:
www.usatoday.com. ©1997 USA Today, a division of Gannett Co., Inc.

b) Estimate the number of people surveyed who admitted that they
read while driving. Estimate how many people responded that
they did not read while driving.

c) What fraction of the respondents admitted that they fixed their hair while driving? What fraction of the respondents said that they never fixed their hair while driving?

Assume that the sample used in the magazine's telephone survey is representative of drivers in general.

d) If a driver is selected at random, what is the probability that the driver does not eat while driving? Interpret the meaning of this probability.

e) Are drivers more likely to talk on the telephone or fix their hair while driving? Explain.

f) Do you agree with the statement "about one in five drivers talks on the telephone while driving"? Support your answer.

There is little additional information that can be extracted from the graph in Figure 7.9. However, if you had access to the original surveys, you could determine considerably more information about people's distracting driving habits than shown in the figure.

For example, suppose that you wanted to know if people who eat while driving are more likely to talk on the telephone while driving than people who do not eat while driving. This question involves two traits, drivers' eating habits and drivers' phoning habits. You would be able to answer this question if you had a completed two-way table similar to that in **Figure 7.10**.

	Talks on phone while driving	Does not talk on phone while driving	Total
Eats while driving			833
Does not eat while driving			429
Total	227	1035	1262

Figure 7.10.
Two-way table on people's phoning and eating habits while driving.

Two-way tables are used to break a group down by two traits. In Figure 7.10, the row trait relates to drivers' eating habits and the column trait to drivers' phoning habits. There are two rows because the survey allowed for only two distinct eating habits: "eats while driving" and "does not eat while driving." Similarly, there are two columns because the survey allowed for two distinct phoning habits: "talks on the phone while driving" and "does not talk on the phone while driving."

6. a) In Figure 7.10, the numbers to the right of the rows and at the bottom of the columns are estimates of the numbers of people in each of the row and column categories, respectively. Explain how the estimates (833, 429, 227, and 1035) were computed.

   b) The blank cell in the upper-left corner of Figure 7.10 should contain the number of people who said that they both ate and talked on the phone while driving. What should the blank cell in the upper-right corner contain?

   c) Because you do not have the questionnaires from *Prevention*'s telephone survey, create hypothetical data for the blank cells that are consistent with the row and column totals.

   d) Imagine that the data you created are actual data from the survey. Based on these data, is it more likely that a person who eats while driving talks on the phone while driving than a person who does not eat while driving? Explain how you arrived at your answer.

According to *Prevention* magazine's survey, 18% of drivers talk on the phone while driving. To answer Item 6(d) you estimated this percentage again, but this time a condition was added, namely, that the person eats while driving. When you took into account the condition, you probably changed your estimate of the percentage of drivers who talk on the phone. In the next lesson, you will learn more about how knowledge of conditions can change probabilities.

LESSON TWO

On One Condition

KEY CONCEPTS

Randomized-response
technique

Tables

Conditional
probability

Independence

Tree diagrams

The Image Bank

PREPARATION READING

Who is More Likely to Use Performance-Enhancing Drugs?

How widespread is the use of anabolic steroids among male athletes? Is steroid use any less prevalent among female athletes? Is a randomly chosen female athlete more or less likely to use performance-enhancing drugs than a randomly chosen male athlete?

Each of these questions involves estimating a percentage or a probability for a specific subgroup of athletes who share a certain trait or condition. Knowledge of the condition, which in this case is gender, frequently changes your estimate of a percentage or probability of another trait (here, steroid use).

These questions may be interesting to researchers, but you can't ask a question about use of anabolic steroids in a classroom setting and expect that all students will answer honestly. In this lesson, you will learn about one model called the Warner model that is specifically designed to gain participants' cooperation in answering "sensitive" questions honestly. In addition, you will discover that mathematical methods used to answer the questions in the first paragraph of this reading can be applied when you analyze results from the Warner model.

CONSIDER:

1. Suppose the setting is a large high school. Which do you think is most likely: a randomly selected student has used anabolic steroids during the past year; a randomly selected member of the school's football team has used anabolic steroids during the past year; or a randomly selected senior has used anabolic steroids during the past year?

2. Each of the statements above relates to a randomly selected high school student. In the last two statements a condition has been added. What are the two conditions?

THE RANDOMIZED-RESPONSE TECHNIQUE

The problem of getting people to answer sensitive questions on surveys honestly is a serious matter. If people respond untruthfully to a sensitive question or fail to respond to the question, the results of the survey could be very misleading. To get people to be more forthright on surveys, researchers developed different methods of asking questions. One of these is called the **randomized-response technique**.

In Activity 1, you saw this sensitive question:

Have you taken anabolic steroids in the past 12 months?
Yes or No.

Imagine that your class is participating in a survey and this is the first question. Would you wonder how students sitting near you might answer? Could you keep your eyes on your own paper and not look over to see how they responded?

Next, imagine that you have been using anabolic steroids. If the principal finds out, you could be ineligible for many school activities. The student sitting next to you is clearly trying to look at everyone's answers. Will you chance answering this question truthfully? Suppose that your response to the survey question could be masked in such a way that other students could not tell from your answer whether or not you used anabolic steroids. Then would you be more willing to answer honestly?

The Warner Model

The Warner model, introduced in 1965, is an example of a randomized-response technique. This doesn't mean that you respond randomly to a question. Instead, you get to make a random selection of the question you will answer. Because nobody knows your question, the answer you give is of no interest to anyone else. Here are the details about using the Warner model.

THE RANDOMIZED-RESPONSE TECHNIQUE

Step 1:

Transform your sensitive question into two statements opposite in meaning.

Statement A:
I have taken anabolic steroids during the last 12 months.

Statement B:
I have not taken anabolic steroids during the last 12 months.

Step 2:

Each person who is participating in the survey is given a copy of the two statements. Each participant individually uses a randomizing device to select one of the two statements. However, the two statements must not be equally likely to be selected.

Step 3:

After participants have selected their statements they answer the question:

Do you agree with this statement? Yes or No.

1. Imagine that each person in your class has followed Steps 1–3. You see the student sitting next to you answers Yes. What does his answer tell you about his use of anabolic steroids?

   After you collect the data from all those participating in the survey, you can use the percentage of Yes responses to estimate the percentage of those who have used anabolic steroids. You are probably skeptical that any meaningful results could be obtained, so we'll demonstrate the method by using a question you have already tested. For example, use the "nonsensitive" question about school sports participation in Activity 1 (Lesson 1).

ACTIVITY

THE RANDOMIZED-RESPONSE TECHNIQUE

3

Testing the Model

In Items 2–5, apply the Warner model to estimate the percentage of students who would respond Yes to the following question:

Do you participate in your school's sports program?
Yes or No.

2. **Step 1:** Translate the preceding question into two statements with opposite meanings. What are your statements?

3. **Step 2:** Each student must use a randomizing device designed to select Statement A one-fourth of the time and Statement B three-fourths of the time. Describe some possible randomizing devices that can do this.

   Then, use your randomizing device to select the statement you will answer. (Different group members may choose different methods.) Do not tell anyone, not even the members of your group, which statement you will answer!

4. **Step 3:** It's time to collect some data from the class.

   a) Record on paper your own answer to the following question:

      Do you agree with the statement that you have selected? Yes or No.

      After all students have responded to their randomly selected statement, their responses should be collected and tabulated.

   b) Record the class results in a table similar to **Figure 7.11**.

Number of Yes responses	
Number of No responses	

Figure 7.11.
Class responses.

THE RANDOMIZED-RESPONSE TECHNIQUE

5. Based on the class data, what is your estimate of the percentage of students who participate in your school's sports program? Explain how you arrived at your estimate. (If you are unable to arrive at an estimate that you can justify mathematically, describe some of the ideas that your group discussed.)

Don't worry if you were unable to determine a mathematically justifiable estimate in Item 5. You will return to Item 5 at the end of this lesson. However, don't give up until you have gained some insight into the breakdown of the Yes responses into two mutually exclusive groups.

6. Suppose that a student is randomly selected from your class. The selected student tells the class that she responded Yes to the question "Do you agree with the statement you have selected?" Is it more likely that the student does or does not participate in your school's sports program? Explain.

7. If Steps 2 and 3 were repeated with your class, do you think the second data set would be exactly the same data as the data in 4(b)? Explain why or why not.

Warner's model is an example of a two-stage process. First, you select the statement, A or B; that's Stage 1. Then, depending on the statement that you have selected, you respond Yes or No; that's Stage 2. At the end of Activity 2, based on a two-way table containing hypothetical data, you answered the question "Is it more likely that a person who eats while driving talks on the phone while driving than a person who does not eat while driving?" On the surface, this question may seem to have nothing to do with a two-stage process. However, you can analyze this situation as a two-stage process. For Stage 1, divide the participants of the survey into two groups, those who eat while driving and those who do not; for Stage 2, divide each group into two groups, those who phone while driving and those who do not. In the next activity, you will examine how probabilities change in a two-stage process such as the one just described.

ACTIVITY

4

WELL CONDITIONED

Suppose that 1000 students attend Hampton High School and 250 of these students participate in the sports program.

1. What is the probability that a randomly selected student participates in the sports program?

2. Suppose that the randomly selected student is a girl (Stage 1). Do you think that the probability that she participates in Hampton's sports program (Stage 2) is the same as the probability you found in Item 1? Why?

For Items 1 and 2, you determined the probability that a randomly selected student participated in Hampton's sports program. However, the second probability takes into account the condition that "the student is a girl" (the outcome of Stage 1). To differentiate the **conditional probability** from the unconditional probability, mathematicians use the following notation and language:

Unconditional probability:

P(student participates in Hampton's sports program). Read this as "the probability that a student participates in Hampton's sports program."

Conditional probability:

P(student participates in Hampton's sports program | student is a girl). Read this as "the probability that a student participates in Hampton's sports program given that the student is a girl."

To calculate the first probability, determine the fraction of students from Hampton High who participate in their school's sports program.

$$\frac{\text{(number of students in sports program)}}{\text{(number of students at Hampton High)}}$$

WELL CONDITIONED

To calculate the conditional probability, determine the fraction of girls from Hampton High who participate in their school's sports program.

$$\frac{(\text{number of girls in the sports program})}{(\text{number of girls at Hampton High})}.$$

Probabilities can be expressed as a decimal between 0 and 1, as a percentage, or as a fraction.

Now apply what you have learned about conditional and unconditional probability (or percentages) to analyze the results of two-question surveys.

3. Suppose that you are going to distribute the two-question survey in **Figure 7.12** to the students in your school. Assume steps have been taken to keep responses confidential, so that there is no need to use the randomized-response technique introduced in Activity 3.

Survey Questionnaire

Please circle your answers.

1. Gender Male Female

2. Have you used anabolic steroids during the past 12 months?

Yes No

Figure 7.12.
Survey on use of anabolic steroids.

a) Assume that the questionnaires have been returned. Make a list of at least eight questions that you could answer based on the survey data. Your questions should satisfy these criteria:

(1) Your questions should involve percentages and/or probabilities.

(2) At least four of your questions should contain conditional information.

(3) The answers to your questions should be of interest to students, parents, and officials of the school and community.

ACTIVITY

4

WELL CONDITIONED

b) Identify the questions that include a condition and state the condition (or conditions). Then explain how you would compute the fractions that would be involved if you were to compute the answers to these questions.

Now it's time to collect data from your class so that you can practice answering questions similar to the ones that you posed in Item 3. However, substitute this less-sensitive question for Question 2:

Substitute Question 2:
Do you participate in your school's sports program?
Yes or No.

4. On a separate piece of paper, create a survey questionnaire that consists of Question 1 from Figure 7.12 and substitute Question 2. Then, respond to the survey. Turn the completed surveys in to your teacher for tabulation.

Sports/Gender Data
_____

1. Number of
   students in class: _____

2. Number of students
   who participate in the
   school's sports program: _____

3. Number of girls in class: _____

4. Number of girls who
   participate in the
   school's sports program: _____

a) Enter the class results in a form similar to **Figure 7.13**.

b) Notice that the two traits of interest, "sports participation" and "gender," divide the students in your class into four non-overlapping subgroups or categories. What are these categories?

Figure 7.13.
Sports/gender data form.

ACTIVITY

WELL CONDITIONED

4

c) There are many ways to present the same data. The form in (a) is one way. The two-way table in **Figure 7.14** is another. Use your data to complete entries in a table similar to Figure 7.14.

	Participates in school's sports programs	Does not participate in school's sports programs
Girls		
Boys		

Figure 7.14.
Two-way table of sports/gender data.

d) Which is the row trait in your table, "gender" or "sports participation"? Which is the column trait?

e) Refer to your list of eight questions (your answer to Item 3). Modify each question as follows. Replace references to "uses anabolic steroids" with "participates in the school's sports program." Replace any reference to "does not use anabolic steroids" with "does not participate in the school's sports program." Use the data from your two-way table to answer each of the modified questions.

5. Answer parts (a–c) based on the data in your two-way table from Item 4(c).

a) What fraction of students who participate in the school's sports program are boys ?

b) What fraction of boys participate in the school's sports program?

c) What is the difference between finding the fraction of students who are boys participating in the school's sports program and finding the fraction of boys who are participating in the school's sports program?

6. Recall from Individual Work 2 (page 621) the *Prevention* magazine telephone survey on driving distractions. Focus on two questions from this survey:

Do you talk on the phone while driving? Yes or No.

Do you eat while driving? Yes or No.

WELL CONDITIONED

	Talks on phone while driving	Does not talk on phone while driving	Total
Eats while driving	100	733	833
Does not eat while driving	127	302	429
Total	227	1035	1262

Figure 7.15.
Hypothetical data from
Prevention's telephone survey.

A hypothetical breakdown of the results from these two questions is presented in **Figure 7.15**.

Answer parts (a–f) based on the information in Figure 7.15.

a) What fraction of the drivers talk on the phone while driving?

b) What fraction of the drivers both eat and talk on the phone while driving?

c) What fraction of the drivers who eat while driving also talk on the phone while driving?

d) What is the difference between finding the fraction in (b) and the fraction in (c)?

Next, assume that the drivers interviewed in *Prevention's* telephone survey are representative of drivers in general. Use the results presented in Figure 7.15 to estimate the pairs of probabilities in parts (e) and (f).

e) P(person eats while driving) and
P(person eats while driving | person talks on phone while driving).

Did knowing that a person talks on the phone while driving increase, decrease, or leave unchanged the probability that the person eats while driving?

f) P(person talks on the phone while driving) and
P(person talks on the phone while driving | person does not eat while driving).

WELL CONDITIONED

In Item 6(e) you first calculated the probability that a person eats while driving given no conditional information, and then you calculated the same probability given the condition that the person talks on the phone while driving. These two probabilities did not turn out to be the same. When this happens, the two events involved in the conditional probability are said to be **dependent events**. In this case "eats while driving" and "talks on the phone while driving" are examples of dependent events.

g) Are the events "person talks on the phone while driving" and "person does not eat while driving" dependent events? Explain?

In Figure 7.15, the **row totals**, the numbers to the right of the table, and the **column totals**, the numbers at the bottom of the table, are estimates based on actual percentage results of *Prevention*'s magazine survey. Unfortunately, the graph in Figure 7.9 did not provide enough information to estimate the four entries in the two-way table. The table in Figure 7.15 shows only hypothetical data consistent with the estimated row and column totals. The actual data might have been quite different.

7. Create a new set of hypothetical data for Figure 7.15 that is consistent with the estimated row and column totals.

a) Change the numbers in the two-way table in such a way that your answers to Items 6(a) and 6(c) are equal (at least to the nearest percent). What is your two-way table? Check that the row and column totals for your table are the same as for Figure 7.15.

8. Use your table from Item 7 to estimate the following pairs of probabilities. Round your answers to the nearest percent—the same precision used in the graph in Figure 7.9.

a) *P*(person eats while driving) and
P(person eats while driving | person talks on phone while driving).

ACTIVITY

WELL CONDITIONED

4

b) *P*(person does not talk on phone while driving) and *P*(person does not talk on phone while driving | person eats while driving).

c) In parts (a) and (b), how did the addition of a condition affect the probability?

If you answered Item 7 correctly, you noticed that the conditional and unconditional probability pairs turned out to be the same. For example, using your hypothetical data, you should have found that (at least to the nearest percent)

P(person eats while driving) = *P*(person eats while driving | person talks on phone while driving).

When the unconditional probability and the conditional probability are equal, the two events involved in the conditional probability (in this case,) are said to be **independent events.** So, in this case "person eats while driving" and "person talks on the phone while driving" are examples of independent events.

In Item 3, you listed eight questions about the use of steroids. You may have asked questions such as "Is a higher percentage of male athletes or female athletes using steroids?" Researchers are frequently interested in how percentages differ depending on a given condition. For example, in the 1993 NCAA study discussed in the preparation reading in Lesson 1, the researchers wanted to know much more than simply the percentage of users. Here are some questions that researchers might like to ask.

Did the percentage of athletes using steroids depend on:

- gender? • the particular sport? • geographic location?
- whether the athlete played on a Division I, II, or III team?

In this activity, your analysis was limited to two-question, Yes-No surveys. Individual Work 3 will show you how to use a new tool, called a "tree diagram," to analyze two-stage situations. Then, in Activity 5 you will work with tables and charts published in the NCAA report. Your group will become a team of researchers. In this role, you will find answers to some of the preceding questions and pose questions of your own.

INDIVIDUAL WORK 3

Representing Information on Two Traits

When you analyze data from a survey, you find that interesting questions often involve responses to two questions. In Activity 4 you organized your data into a two-way table. In this assignment, you will learn about two other methods for representing data on two traits, a **tree diagram** and a **glyph display**.

Gridville High

Two traits of students at Gridville High School are pictured in the glyph face display in **Figure 7.16**. The hair and mouths of the faces indicate whether the student is male or female and whether the student has a driver's license.

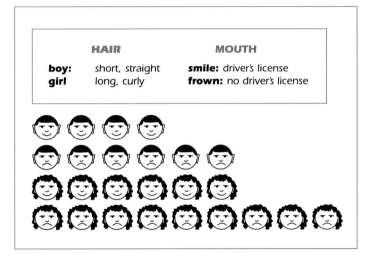

Figure 7.16.
Characteristics of students at Gridville High. Each face in the display represents ten students.

1. What is the probability that a randomly selected student from Gridville High School has a driver's license? What is the probability that a randomly selected student does not have a driver's license?

2. Suppose that the randomly selected student is a girl. What is the probability that she has her license? Suppose that the randomly selected student is a boy. What is the probability that he has his license?

3. Does knowing the student's gender affect the probability of having a driver's license? (Be sure to reduce fractional probabilities to lowest terms or write probabilities in decimal form when justifying your answer to this question.)

4. Are "having a driver's license" and "being a boy" independent events? Why or why not?

5. The glyph face display in Figure 7.16 presents, in a picture, information on numbers of students in four mutually exclusive categories: boys who have their licenses, boys who do not have their licenses, girls who have their licenses, and girls who do not have their licenses. Use information from this display to add the number of students in each category to the table in **Figure 7.17**. Then, answer parts (a–c) based on the data in your table.

Figure 7.17.
Two-way table based on glyph face display.

	Boy	Girl
License	🙂	🙂
No license	🙁	🙁

a) What fraction of the students are girls?

b) What fraction of the students are girls who have driver's licenses?

c) What is the difference between finding the fraction of students who are girls that have their licenses, and finding the fraction of girls who have their licenses?

6. The glyph faces in Figure 7.16 and the table that you completed for Item 5 provide two methods of representing information. Another method ideal for representing two-stage processes is a tree diagram. For Stage 1, you select the first factor, generally the trait that provides conditional information; then for Stage 2, you select the second factor, or in this case, the remaining trait.

a) Make a copy of **Figure 7.18** and then follow steps 1–3 to complete your tree diagram.

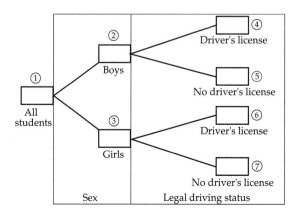

Figure 7.18.
Incomplete tree diagram.

Step 1:

Enter the total number of students in Box 1.

Step 2:

The first factor in the tree diagram splits the students in Box 1 into two groups: boys and girls. Record the number of boys in Box 2 and the number of girls in Box 3.

Step 3:

The second factor splits the boys (Box 2) and girls (Box 3) each into two groups: those who have their driver's licenses and those who do not. Record the number of boys with driver's licenses in Box 4, boys without driver's licenses in Box 5, girls with driver's licenses in Box 6, and girls without driver's licenses in Box 7.

b) What is the relationship between the numbers in Boxes 1–7 of your completed tree diagram and the numbers in the table that you completed for Item 5?

c) Determine ratios of numbers from two boxes in your tree diagram that correspond to the following probabilities. Then add each of these probabilities to your tree diagram by writing the probability above the appropriate branch connecting two boxes.

$P(\text{license} \mid \text{girl})$

$P(\text{no license} \mid \text{girl})$

$P(\text{license} \mid \text{boy})$

$P(\text{no license} \mid \text{boy})$

$P(\text{boy})$

$P(\text{girl})$

d) Next, draw a second tree diagram. This time use "driving status" as factor 1 and "gender" as factor 2. Label each branch with its corresponding probability.

Camp Grenada

During one session of Camp Grenada there were 140 boys and girls attending the camp. At the beginning of the summer everyone was given a swimming test. **Figure 7.19** shows the breakdown of the campers by gender and swimming ability.

	Swimmers	Non-swimmers
Girls	45	10
Boys	60	25

Figure 7.19.
Breakdown of Grenada campers by gender and swimming ability.

7. Base your answers to parts (a–d) on the data in Figure 7.19.

a) How many boys attended camp?

b) How many girls attended camp?

c) How many of the campers were non-swimmers?

d) How many of the campers were swimmers?

8. a) If a camper is chosen at random, what is the probability that the camper is a swimmer?

b) If a randomly chosen camper is a boy, what is the probability that he is a swimmer?

c) What is the probability that a randomly chosen camper is a boy who swims?

d) What is different about finding the probabilities in (b) and (c)?

9. Are girl campers at Camp Granada more likely or less likely to be swimmers than boy campers? Explain.

10. Explain in writing the difference in meaning between $P(\text{swimmer} \mid \text{boy})$ and $P(\text{boy} \mid \text{swimmer})$.

11. a) Organize the information on campers into two different tree diagrams. Begin one by dividing the campers by gender. For the other, begin by dividing campers by swimming ability. For each diagram, label the characteristics of the students in each box. Above each branch write the corresponding probability.

   b) Which tree diagram is helpful in determining $P(\text{swimmer} \mid \text{male})$ and $P(\text{non-swimmer} \mid \text{female})$? Should "gender" or "swimming ability" be factor 1 of this tree diagram? Why?

   c) What are the values of $P(\text{swimmer} \mid \text{male})$ and $P(\text{non-swimmer} \mid \text{female})$? How can you get these probabilities from your tree diagram?

   d) The other tree diagram should be helpful in determining $P(\text{male} \mid \text{swimmer})$ and $P(\text{female} \mid \text{non-swimmer})$. Should "gender" or "swimming ability" be factor 1 of this tree diagram? Why?

   e) What are the values of $P(\text{male} \mid \text{swimmer})$ and $P(\text{female} \mid \text{non-swimmer})$? How can you get these probabilities from your tree diagram?

12. Use your tree diagrams from Item 11 to answer the following questions.

   a) What is the probability that a camper chosen at random is a girl who can swim? How could you get this probability from probabilities corresponding to the branches of one of your tree diagrams?

   b) What is the probability that a camper chosen at random is a boy who cannot swim? How could you get this probability from probabilities corresponding to the branches of one of your tree diagrams?

13. Camp Wilder has separate swimming areas for beginners, intermediates, and experts. There are 180 campers at Camp Wilder, twice as many boys as girls. Half of the boys and one-fourth of the girls are classified as beginners. Half of the girls and one-fourth of the boys are classified as intermediates. All of the remaining campers are classified as experts.

   a) Draw a tree diagram that represents this situation.

   b) Determine the number of swimmers in each swimming classification.

   c) Arrange the information into a two-way table. What are the row and column traits?

ACTIVITY

5

FINDINGS FROM THE 1993 NCAA REPORT

The researchers at Michigan State University who wrote the 1993 NCAA report cited in Lesson 1's preparation reading had to do a great deal of planning prior to their study. They had to decide:

- what information they wanted to learn.

- how to word the questions.

- who would take the survey.

- who would administer the survey.

- where students would meet to take the surveys.

- when the survey should be given.

Respondent Characteristics
2505 athletes (78% response rate)
1719 male athletes and 786 female athletes

NCAA Division	Percentage of the respondents
Division I	57%
Division II	27%
Division III	16%

NCAA Region	Percentage of the respondents
East	21%
South	19%
Midwest	42%
West	19%

Figure 7.20.
Characteristics of respondents by gender/division and region.

After the students completed the survey and returned their responses to Michigan State University, the researchers had to organize, analyze, and, most importantly, interpret the data.

In this activity, imagine that you are one of the researchers on the project. Partial results from this study are shown in **Figures 7.20–7.23**. Your job is to answer questions related to these results and to pose questions of your own. Take a few minutes to review those results.

Figure 7.20 presents information on the number of respondents by gender. In addition, you will find the percentage breakdown of the respondents by NCAA Division and NCAA Region.

ACTIVITY

FINDINGS FROM THE 1993 NCAA REPORT

5

The respondents in the 1993 NCAA study participated in at least one of five men's sports (baseball, basketball, football, tennis, and track/field) or women's sports (basketball, softball, swimming/diving, tennis, and track/field). Gender and participation in specific sports are frequently thought to influence the likelihood that an athlete uses performance-enhancing drugs. Figure 7.21 summarizes the use of two types of performance-enhancing drugs, anabolic steroids and major pain medicines, within each of the five men's and five women's sports.

Researchers were interested to learn if use of anabolic steroids differed within NCAA Divisions or NCAA Regions. Figures 7.22 and 7.23 present these findings.

Performance-enhancing-drug use, by sport		
	Anabolic steroids	Major pain medicines
MEN'S SPORTS		
Baseball	0.7%	32.9%
Basketball	2.6%	26.9%
Football	5.0%	34.0%
Tennis	0.0%	23.1%
Track/Field	0.0%	23.7%
WOMEN'S SPORTS		
Basketball	1.5%	34.1%
Softball	1.7%	28.9%
Swimming	0.6%	30.5%
Tennis	2.7%	20.0%
Track/Field	2.7%	26.7%

Figure 7.21.
Percentage of performance-enhancing-drug users within each gender/sport category.

Location	Percentage of users
East	4.0
South	1.7
Midwest	2.3
West	2.6

Figure 7.22.
Anabolic-steroid use within NCAA divisions.

Division	Percentage of users
I	1.9
II	4.3
III	1.9

Figure 7.23.
Anabolic-steroid use within NCAA regions.

ACTIVITY

5

FINDINGS FROM THE 1993 NCAA REPORT

The researchers also looked for evidence that could link use of anabolic steroids and five other ergogenic substances (performance-enhancing drugs). Their findings are displayed in **Figure 7.24.**

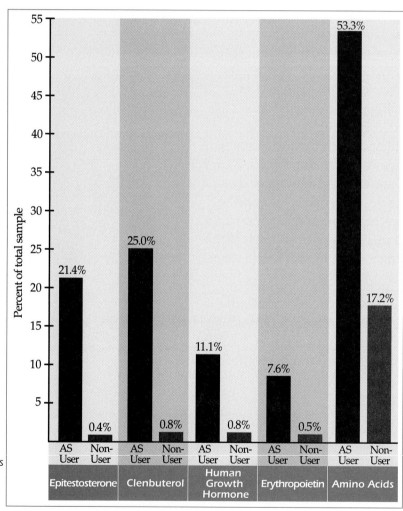

Figure 7.24.
Reported use of other ergogenic substances by users and non-users of anabolic steroids. (Anderson, Albrecht, and McKeag 1993)

FINDINGS FROM THE 1993 NCAA REPORT

1. Make a list of five questions that you think would be of interest to researchers. Then answer as many of your questions as you can. (At the end of this activity, you can return to any questions that you can't answer at this point.)

In Item 1 you had the opportunity to pose and answer questions of your own. You may find that some of your questions are repeated in Items 2–6. If so, then you have already answered some of the remaining questions in this activity.

2. Estimate the number of student participants who played in Division I sports, Division II sports, and Division III sports.

3. Are athletes in Division I, Division II, or Division III the most likely to use anabolic steroids? Support your answer using percentages.

4. a) The preparation reading reported that 2.5% of the students who responded to this survey reported using anabolic steroids. Is the percentage of anabolic-steroid use dependent or independent of the NCAA Region? Explain.

   b) If your answer to (a) was "dependent," how would Figure 7.22 need to change in order for your answer to be "independent?" On the other hand, if your answer to (a) was "independent," how would Figure 7.22 need to change in order for your answer to be "dependent?"

5. Refer to Figure 7.21 for this question.

   a) Are male or female track and field athletes more likely to use anabolic steroids? Support your answer by stating the relevant percentages.

   b) Which sport among the men's sports has the highest percentage of anabolic steroid users? What about the women's sports? Does either answer surprise you? Explain why or why not.

   c) Make up three additional questions that you can answer based on the data in Figure 7.21. (You may use a question posed for Item 1 if it is based on Figure 7.21.) Then, answer your questions.

ACTIVITY

FINDINGS FROM THE 1993 NCAA REPORT

5

6. One of the questions that the researchers wanted to answer was whether the use of five other ergogenic substances (epitestosterone, clenbuterol, human growth hormone, erythropoietin, and amino acids) was dependent on anabolic steroid use. Refer to the histogram in Figure 7.24. Based on this histogram, write a brief paragraph for the researchers. In your paragraph, state whether use of any of these five ergogenic substances appears dependent on the athletes' use of anabolic steroids. If you find that it does, explain whether steroid use increases or decreases the likelihood that an athlete uses this particular ergogenic substance. Report any other findings that interest you.

7. a) It is clear from Figure 7.24 that a substantial percentage of both anabolic steroid users and non-users take amino acids. Draw a tree diagram using "anabolic steroids" and "amino acids" as the two factors. Use your tree diagram to estimate the number of respondents in the NCAA study who said they used amino acids. What percentage of the respondents is this?

   When you construct your tree diagram, round the numbers in Boxes 1–7 to the nearest whole number. Check that the sum of the numbers in Boxes 4–7 equals the number in Box 1 (the number of respondents in this study).

   b) Create similar tree diagrams for anabolic steroids and each of the remaining four ergogenic substances. For each substance, estimate the number of respondents who said that they used the substance and then calculate that percentage. Group members should divide up this work.

   c) Summarize your findings from (a) and (b) in a two-way table similar to **Figure 7.25**. Then, describe the use of each of these ergogenic substances among college student athletes.

FINDINGS FROM THE 1993 NCAA REPORT

Anabolic steroids	Percentage of users: five other performance-enhancing drugs				
	Epitestosteron	Clenbuterol	Human growth hormone	Erythropoietin	Amino acids
User					
Non-user					

Figure 7.25.
Percentages of users and non-users of each of five ergogenic substances.

8. If you were unable to answer any of your questions from Item 1, go back and see if you can answer them now. If you are still unable to answer one of your questions, state what additional information you would need in order to answer it.

In this activity, you have acted as researcher and have used unconditional and conditional probabilities or percentages to analyze data from the 1993 NCAA national survey. Because of the sensitive nature of the questions in the NCAA survey, you couldn't ask the students in your class to respond to the same questions.

In the next activity, you'll return to Warner's model, a method for dealing with sensitive questions. This time, you'll use tree diagrams to represent Warner's model and estimate the percentage of students who would answer Yes to a sensitive question.

INDIVIDUAL WORK 4

Conditional Comparisons

Up to this point, you have been calculating conditional probabilities. In this assignment you must decide which of the unconditional or conditional probabilities is larger.

1. For each probability pair below, state whether you think the conditional probability is higher than, lower than, or the same as the unconditional probability. In each case, write a sentence explaining why you think as you do. There are no clearly right answers in all cases; the explanation supporting your choice is what is most important.

   a) P(high school student has driver's license)
      P(high school student has driver's license | student is in ninth grade)

   b) P(person likes coffee)
      P(person likes coffee | person is under ten years old)

   c) P(person's favorite ice cream is chocolate)
      P(person's favorite ice cream is chocolate | person is male)

   d) P(student watches Monday Night Football)
      P(student watches Monday Night Football | student participates in school sports)

2. a) Based on your answer to 1(a) are the events "high school student has driver's license" and "student is in ninth grade" dependent or independent events?

   b) Based on your answer to 1(b) are the events "person likes coffee" and "person is under ten years old" dependent or independent events?

   c) Based on your answer to 1(c) are the events "person's favorite ice cream is chocolate" and "person is male" dependent or independent events?

   d) Based on your answer to 1(d) are the events "student watches Monday Night Football" and "student participates in school sports" dependent or independent events?

3. Next, you must contend with three probabilities, one unconditional and two conditional. Order the probabilities from smallest to largest. (If you think that two probabilities are the same, then you can declare a tie.) Write a sentence that supports your ordering. Again, there may not be clear right answer in all cases; the explanation supporting your choice is what is most important.

   a) P(man is tall)
      P(man is tall | man is a professional basketball player)
      P(man is a professional basketball player | man is tall)

   b) P(person has blue eyes)
      P(person has blue eyes | person is female)
      P(person is female | person has blue eyes)

4. Now it's your turn to create your own probability pairs. Make up three pairs of conditional and unconditional probabilities. You should have a pair that you think satisfies each of the requirements below:

   a) The unconditional probability is greater than the conditional probability.

   b) The conditional probability is greater than the unconditional probability.

   c) The conditional probability is the same as the unconditional probability.

5. Make up a pair of dependent events, call them A and B, and compare conditional probabilities: $P(A | B)$ to $P(B | A)$. State which of the conditional probabilities you think is largest and explain why.

ACTIVITY

6

WARNER'S MODEL REVISTED

Researchers know that people may ignore or be untruthful about sensitive questions. Suppose that your class was asked these questions:

(1) Have you ever cheated on an exam?

(2) Have you ever lied to a teacher?

(3) Do you like Disney movies?

You might not want other students to know your honest answers to one or more of the questions. If the researchers used the Warner model, introduced in Activity 3, you might be more likely to answer honestly because the randomized-response technique used in the Warner model makes it impossible for someone to look at your answers and know your answers to the sensitive questions. Although individual responses to Questions 1–3 would not be known, researchers could use the results to estimate the percentages for the class.

In this activity, you will use the Warner model with the question, "Do you like Disney movies?" Then you will analyze the results with a tree diagram.

1. First, you'll need some data. Refer to Activity 3 and modify Steps 1–3 for the Disney question. After the data are collected, record the results in Figure 1 on Handout H7.5.

2. Let p represent the fraction of students in class who like Disney movies. (Eventually, you will be asked to estimate the value of p.)

   a) Draw a tree diagram of the two-stage process used in Warner's Model (Stage 1: select a statement; Stage 2: respond Yes or No). Label what you can. Assign the variable p for the appropriate branches. Then complete the tree in terms of that variable.

WARNER'S MODEL REVISTED

b) Use your tree diagram to write an expression involving p that represents the total number of Yes responses you would expect from the class.

c) Equate your answer to (b) with the number of Yes responses actually obtained from the class. Use your equation to determine a value for p. What is your estimate of the percentage of students in class who like Disney movies?

Now that you have an estimate for the percentage of students who admit that they like Disney movies, how good is your estimate? Your teacher could ask the students who like Disney movies to raise their hands so that you could calculate the exact percentage. But do you think everyone who likes Disney movies will admit it in front of other students? Also, your class is a very small sample. Most surveys consist of many more participants. So, you should test it using more participants than the number of students in your class, in order to be get a realistic result.

Role Playing

Next, you will test Warner's model using four times as many participants as there are students in your class. To do this, each student will play the role of four different people. For each of your four people, you will be handed a card. The card will be marked Yes if the person likes Disney movies and No otherwise.

3. Place your four cards in front of you. Use a randomizing device to select a statement for each of your four people. Then, for each of them, answer the question "Do you agree with the selected statement?" Keep track of this information in Figure 2 on Handout H7.5.

ACTIVITY

WARNER'S MODEL REVISTED

4. Next, record the number of Yes responses to the question "Do you agree with this statement?" After results from the class have been collected and tabulated, record the class results in Figure 3 on Handout H7.5.

5. Using the results in Figure 3, estimate the percentage of participants (represented by the cards) who like Disney movies. Compare your estimate to the actual percentage of cards that had Yesses. How good was your estimate?

6. In response to concern from town officials that many students at the high school were using anabolic steroids, the math teacher agreed to conduct a study. She randomly selected 90 students and gave each student two statements.

   Statement A:
   I have used anabolic steroids in the past 3 months.

   Statement B:
   I have not used anabolic steroids in the past 3 months.

   Each student rolled a die. If the die landed with 1 or 2 facing up, they selected Statement A; otherwise, they selected Statement B.

   After the math teacher collected the responses, she recorded 50 Yes responses. In her report, should she indicate that there is reason for alarm or that concerns are unfounded? Explain your answer by estimating the percentage of students using anabolic steroids.

WARNER'S MODEL REVISTED

7. Suppose the randomized response process in Item 6 is changed to the following: Each student flips a coin. If the coin lands heads up, select Statement A; otherwise, select Statement B. Repeat your analysis of the data in Item 6 (50 Yes responses). What makes this situation different from that in Item 6 and your class test of Warner's model in Items 3–5?

In Lessons 1 and 2 you have used information obtained from surveys to estimate the prevalence of a particular trait in a group. However, even when the responses to a question are masked, as with Warner's model, you can't be certain that people will answer honestly. In Lesson 3, you will estimate the prevalence of a particular trait based on objective tests. You will find that objective tests don't yield perfect results either; however, you will be able to apply what you know about probability to estimate the percentage of the population having the trait for which the test was designed.

In the remaining lessons of this unit, you will find tree diagrams and tables useful to represent conditional information and to aid calculation of various conditional and unconditional probabilities.

LESSON THREE

Nobody's Perfect

KEY CONCEPTS

Conditional probability

Tree diagrams

Linear model

Domain of a model

Inverses of functions

The Image Bank

PREPARATION READING

Dealing with Imperfect Information

You have learned in Lessons 1 and 2, that some traits (for example, uses performance-enhancing drugs, has a driver's license, eats while driving) can be studied through simple surveys. However, surveys have two major limitations. The first is that you can't get reliable information from questions to which the respondent doesn't know the answer. The second is that no matter how carefully the anonymity of the respondents is protected, some of the respondents will not answer sensitive questions truthfully.

An alternative to surveys for gathering information about a particular population is to use some form of an objective test. You've probably had experience with situations in which tests were used to determine if you had a particular trait. For example, standardized tests are used to assess the percentage of students who read at or above grade level. Blood tests on

ACTIVITY

THE RESULTS ARE IN

7

b) What is $P(-\mid \text{user})$? What does this mean in words?

c) Finally, let U represent the fraction of athletes who use performance-enhancing drugs. What does $1 - U$ represent?

d) Make a copy of your tree diagram from Item 1. Place U, $1 - U$, and the values for $P(+\mid \text{user})$, $P(+\mid \text{non-user})$, $P(-\mid \text{user})$, and $P(-\mid \text{non-user})$ on the appropriate branches of the tree diagram.

3. Suppose that a test with the characteristics described in Item 2 is administered to a random sample of 225 athletes. Of the 225 test results, 44 are positive and 181 are negative.

   a) What percentage of the tests are positive?

   b) Is the percentage of positive tests a good estimate of the percentage of athletes using performance-enhancing drugs? Explain why or why not.

You probably decided that the percentage of positive tests is not a good estimate of the percentage of performance-enhancing-drug users. If not, you'll understand why it isn't a good estimate after completing part (c). In part (c), you examine a situation for which you know the percentage of users. This allows you to compare the percentage of positive tests to the percentage of users.

c) Suppose that this test is given to 225 football players and that 1/9 of these players actually use one or more of the banned substances. Complete the tree diagram in **Figure 7.26**.

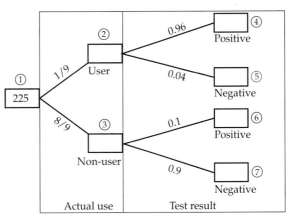

Figure 7.26.
Tree diagram for drug tests given to a sample of football players.

THE RESULTS ARE IN

d) How many of the football players' tests would you expect to be positive? Explain how you arrived at your answer.

e) What fraction of the test results are positive? Compare this fraction to the fraction of users, 1/9. Is the fraction of positive tests a good estimate of the fraction of users? How do these results compare when converted to percentages?

You have just seen one case in which the fraction of users differs from the fraction of positive tests. See if the same holds true for a test with a different set of characteristics.

4. Suppose that a test for anabolic steroids correctly reports a positive result for a user 90% of the time and correctly reports a negative result for a non-user 80% of the time. A sample of 120 athletes, one-third of whom are using anabolic steroids, is tested.

a) Draw a tree diagram for this situation.

b) Calculate the total number of positive test results you would expect if you tested all the athletes. Estimate the fraction of test results that would be positive. Compare this fraction with the fraction of actual users, 1/3. How different are these fractions when converted to percentages?

By now you should be convinced that the percentage of positive test results is not a good estimate of the percentage of drug users. So, are the test results useless? No! In the next activity, you'll find that there is a relationship between the quantity you know, the percentage of positive test results, and the quantity you want to estimate, the percentage of actual users.

INDIVIDUAL WORK 5

Calculating Test Results

*I*n this assignment you calculate the number of positive results that you would expect in a given testing situation.

1. A performance-enhancing-drug test correctly detects the presence of drugs with probability $p = .98$, and correctly reports the absence of drugs with probability $r = .92$. Assume that you test 250 athletes, of whom 150 are actually using drugs. Answer the following questions. (A tree diagram may be useful.)

   a) A **true positive** means that the test results are positive and the person tested has used a performance-enhancing drug. How many true positives should you expect this test to produce?

   b) If you actually tested 250 athletes, do you think that the results of your tests would match exactly your answer in (a)? Why or why not?

   c) A **false positive** means that the test results are positive but the individual has not used performance-enhancing drugs. How many false positives should you expect the test to produce?

   d) Estimate the fraction of all the tests that will be positive. What percentage of the tests is this?

   e) Is the actual fraction of users less than, equal to, or more than the fraction of positive test results?

   f) A **true negative** means that the test results are negative and the individual has not used performance-enhancing drugs. How many true negatives should you expect the test to produce?

   g) A **false negative** means that the test results are negative but the individual has used performance-enhancing drugs. How many false negatives should you expect the test to produce?

   h) Estimate the fraction of all tests that will be negative. What is the relationship between the fraction of tests that are negative and the fraction of tests that are positive?

2. A certain performance-enhancing-drug test correctly detects the presence of drugs with probability $p = .9$ and correctly reports the absence of drugs with probability $r = .8$. Assume that you randomly select 400 athletes for testing from a population that is estimated to

contain 2.5% users of these drugs.

a) Estimate the fraction of tests that will be positive. Express this number as a percentage.

b) How does the fraction of positive test results compare to the actual fraction of drug users? Is the percentage of positive test results a good estimate for the percentage of users in this population?

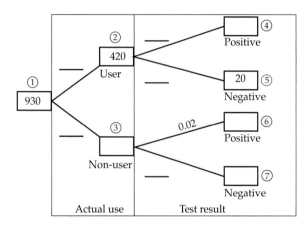

3. Copy the tree diagrams in **Figure 7.27**. Some of the values are given and others are missing. Fill in as many of the missing values as you can. Write a short paragraph interpreting all of the values in at least one of your tree diagrams.

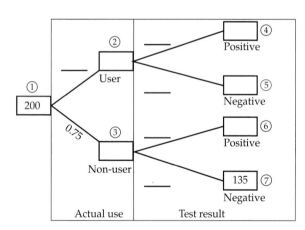

4. Make up a problem for a two-stage probability tree. Use the categories of "in school band" and "not in school band" for the first stage, and the categories of "saw school play" and "didn't see school play" for the second stage. Provide enough basic information about the probabilities and/or the numbers of students involved so that a reader can answer the question(s) in your problem. Then write a solution for your problem.

Figure 7.27.
Two incomplete tree diagrams.

BUILDING A RELATIONSHIP

CONSIDER:

1. How can you calculate most efficiently the expected percentage of positive test results given the following information about the test and the composition of the sample?

   Test characteristics:

   p = probability that the test correctly detects the trait = .9

   r = probability that the test correctly reports the absence of the trait = .8

   Number in sample = 120

   Percentage of sample with trait = 25%

In Activity 7 you discovered that the percentage of positive test results is not always close to the percentage of the population that has the trait. However, they are related even though they are not equal. The goal of this activity is to design a method that lets you use these test results. The first step toward this goal is to gather some data and then use the data to describe a relationship between the fraction of actual users and the fraction of positive drug tests.

1. Divide data sets A–D among the members of your group. For each data set, calculate the fraction of positive test results for the three scenarios. Express the fraction as a decimal. Round your answers as little as possible so as not to mask any relationships among the fractions. For each scenario, write your answers as an ordered pair: (actual fraction of users, fraction of positive test results). Check your calculations with others in your group.

   Note: In this problem you will be working in a "theoretical world" and not the "real world." So, in your calculations, assume that 20% means exactly 20%, not 19.5% to 20.5%.

ACTIVITY

8

BUILDING A RELATIONSHIP

DATA SET A

SCENARIO 1

$p = .95, r = .88$
Number of athletes = 120
Percentage of actual users = 5%

SCENARIO 2

$p = .95, r = .88$
Number of athletes = 120
Percentage of actual users = 10%

SCENARIO 3

$p = .95, r = .88$
Number of athletes = 120
Percentage of actual users = 90%

DATA SET B

SCENARIO 1

$p = .95, r = .88$
Number of athletes = 120
Percentage of actual users = 10%

SCENARIO 2

$p = .95, r = .88$
Number of athletes = 120
Percentage of actual users = 20%

SCENARIO 3

$p = .95, r = .88$
Number of athletes = 120
Percentage of actual users = 25%

DATA SET C

SCENARIO 1

$p = .95, r = .88$
Number of athletes = 120
Percentage of actual users = 25%

SCENARIO 2

$p = .95, r = .88$
Number of athletes = 120
Percentage of actual users = 40%

SCENARIO 3

$p = .95, r = .88$
Number of athletes = 120
Percentage of actual users = 60%

DATA SET D

SCENARIO 1

$p = .95, r = .88$
Number of athletes = 120
Percentage of actual users = 60%

SCENARIO 2

$p = .95, r = .88$
Number of athletes = 120
Percentage of actual users = 75%

SCENARIO 3

$p = .95, r = .88$
Number of athletes = 120
Percentage of actual users = 90%

ACTIVITY

BUILDING A RELATIONSHIP

8

2. After your group has analyzed all the data, record your results in a table similar to that in **Figure 7.28**.

3. Graph the eight ordered pairs in your table. (If you graph the ordered pairs by hand, you will need to round the decimals.) Label the vertical axis "Fraction of positive tests" and the horizontal axis "Fraction of actual users."

4. Describe the relationship between the fraction of users and the fraction of positive tests. Draw a smooth curve or line through your data points. If possible, determine an equation that describes the relationship between the fraction of actual users, U, and the fraction of positive tests, T. (If you wish, use your calculator to graph the data and then check that your equation describes the relationship.)

Fraction of actual users U	Fraction of positive tests T
0.05	
0.10	
0.20	
0.25	
0.40	
0.60	
0.75	
0.90	

Figure 7.28.
Data from tests with $p = .95$, $r = .88$.

5. Use your graph from Item 3 to answer the following questions concerning this testing situation. (Notice that your graph deals with the fraction of users and positive tests, but the questions below deal with percentages.)

a) Suppose that 25% of the tests were positive. What percentage of the population has been using performance-enhancing drugs?

b) What if only 15% of the tests were positive?

c) What if 80% of the tests were positive?

d) What if only 2% of the tests were positive?

ACTIVITY

8

BUILDING A RELATIONSHIP

6. a) Finally, determine a model that you could use to estimate the fraction of actual users if you are told the fraction of positive tests. (To do this, you need to write an equation that gives U in terms of T.)

   b) Test your equation. Rework your answers to Item 5 using your equation from (a). Then compare your results with the approximate answers that you obtained from your graph.

   c) Using your equation, estimate the percentage of performance-enhancing-drug users if the test results are 10% positive. What can you learn from this answer?

   d) What is a reasonable domain (possible values for T) for your equation from part (a)?

In Item 4, you wrote a function that describes T in terms of U. In Item 6, you reversed that function; you found the inverse function, which describes U in in terms of T. Because testers generally do know T but do not know U, the inverse function is actually the more useful of the two. Due to the imperfect nature of the test, the domain of this function is not the set of all conceivable test results 0.00–1.00, but some smaller interval. False positives and negatives prevent having "no positive tests" or "all positive tests" as a result.

The end result of this activity is a model (an equation and a domain over which it produces useful information) that you can use to estimate the actual fraction of performance-enhancing-drug users given the results of a group of tests. However, there is one other limitation on this model. It only holds for tests with characteristics $p = .95$ and $r = .88$. If a test with different characteristics is used, you need to repeat the process to determine a model for that test.

INDIVIDUAL WORK 6

A New Relationship

This assignment will give you a chance to check your understanding of the work you did in Activity 8.

1. A survey on sports participation is given to 80 boys and 60 girls. According to the survey results, 20 boys and 40 girls belong to a swim team. Suppose that one of the students who participated in the survey is chosen at random. Find the following probabilities.

   a) What is the probability that the student is a boy?

   b) If the student is a girl, what is the probability that she swims on a team?

   c) If the student swims on a team, what is the probability that the student is a girl? How does this question differ from (b)?

   d) Are the events "belongs to a swim team" and "boy" independent events? Why or why not? Support your answer mathematically.

In Activity 8, your group gathered data on a test described by the following characteristics: $p = .95$ and $r = .88$. When you plotted your ordered pairs of data, all your points fell on a straight line. Will this be true for tests with different characteristics?

2. Suppose, due to technological advances, there is a new test for detecting the use of performance-enhancing drugs. The characteristics of this test are as follows: $p = .97$ and $r = .92$.

   a) Use the new test to complete a data table similar to that in **Figure 7.29**.

   b) Plot the points in your data table. Describe the graph and write its equation. (If your points do not lie on a line, check your calculations.)

   c) If you test the athletes and find that one-fourth of the tests turn out positive, what would you estimate the fraction of drug users to be?

   d) Finally, determine a model that gives U, the quantity that you want to estimate, in terms of T, the quantity that you know from the drug tests. State a reasonable domain for your model.

Fraction of actual users U	Fraction of positive tests T
0.15	
0.30	
0.45	
0.60	

Figure 7.29.
Data table for test with characteristics $p = .97$ and $r = .92$.

e) If you test the athletes and find that 100% of the tests are positive, what would you estimate the percentage of drug users to be? Is your answer reasonable? Explain.

f) If two-fifths of the drug tests turn out to be positive, estimate the fraction of the athletes who are using performance-enhancing drugs.

g) If you know that none of the athletes use drugs, what fraction of the athletes would you expect to test positive?

3. Write a paragraph describing the process that you used in Item 2 and in Activity 8 to find a model that could be used to estimate the fraction of the population that has a particular trait given the fraction of positive tests.

4. In both Activity 8 and Item 2, the relationship between actual drug use and positive tests was linear. What number (or numbers) do you think control the line? That is, what determines exactly what line describes the relationship in a given situation? Can you find a way to predict the equation of the line without having to "start from scratch" every time?

GENERALLY SPEAKING

In Activity 8 and Individual Work 6 you solved one basic modeling question for this lesson: How can you use the test results to estimate the prevalence of the trait for which you are testing? For two specific tests for performance-enhancing drugs, you determined models that could be used to estimate the percentage of performance-enhancing-drug users given the percentage of positive test results. The only problem is that each time the characteristics of the test are changed, you have to repeat the process of determining a new model all over again!

The goal for this activity is to develop a general model, one that can adjust to changes in the test characteristics. In Item 4 of Individual Work 6 you may have noticed a relationship between the numbers in your individual models and the characteristics of the tests. You may actually have discovered the model that you will develop here. However, instead of a data-gathering approach based on a few specific values for U and V, you will use a tree-diagram approach based on general variables.

First, you'll try the "tree-diagram/general-variables" approach using the same test that you used in Activity 8. If the new approach is successful, then you can generalize your results.

1. a) Draw a tree diagram for the test with characteristics $p = .95$ and $r = .88$. Represent the fraction of performance-enhancing-drug users by U and the number of people in the study by N.

   b) Based on your tree diagram, estimate the number of positive tests with an expression involving N and U.

   c) Let T represent the fraction of all tests that turned out to be positive. What is your expression for T?

ACTIVITY

9

GENERALLY SPEAKING

d) After completing (c) you should have a relationship between T, the fraction of positive tests, and U, the fraction of drug users. Now, solve your equation for U in terms of T and you have your model. How does this model compare with the one that you determined in Activity 8?

2. Now change the characteristics of the test to $p = .98$ and $r = .91$.

a) Why might you prefer using this test instead of the test in Item 1?

b) Use the tree-diagram approach to determine a model that can be used to estimate U in terms of T.

c) Re-express your relationship in (b) in the form $U = mT + b$.

d) Suppose two groups of athletes were tested for the use of performance-enhancing drugs. The fraction of positive tests between the two groups differed by 0.05. By how much did the estimates of the fraction of drug users differ? What would this value be in terms of percentages? Explain how you determined your answer.

e) Does the value of b have any meaning in the context of testing for performance-enhancing drugs?

ACTIVITY

GENERALLY SPEAKING

9

3. You've worked through the process of determining a model given specific values for the characteristics of the test (p and r). Now go back and repeat the process using the letters p and r in place of numeric values for the test characteristics. In the end, you want an equation that tells you how to determine U when you are given the values for r, p, and T.

a) Complete a tree diagram for the general situation using the variables U, N, p, and r. Then use your diagram to help you determine a model for U in terms of p, r, and T.

b) What is a reasonable domain for your model?

c) To show that U and T are linearly related, express your model in the form $U = mT + b$. What quantities control m and b in your linear equation?

d) If the value of p is increased, will the values for m and b in your linear equation increase, decrease, or remain the same? What happens if the value of r is increased? Explain.

Now that you have a model for estimating the actual fraction from the fraction of positive test results, you can use your model even when the test characteristics and the contexts change.

INDIVIDUAL WORK 7

Get the Lead Out

*I*n Activity 9 you developed a general model that could be used to estimate the value of U given the value of T and the characteristics of the test, p and r. Here is one possible model. Yours should be algebraically equivalent to this one.

Let:

U = the fraction of the population with a particular trait
T = the fraction of positive test results after testing for the trait
p = the probability that the test correctly identifies the trait
r = the probability that the test correctly reports the absence of the trait.

(Note that the values for U, T, p, and r are between 0 and 1.)

Then you can estimate the value of U given values for T, p, and r by using the following model:

$$U = \frac{T - (1 - r)}{p + r - 1}$$

Use this model to complete the remainder of this assignment.

The presence of lead in homes (in the paint, window blinds, etc.) poses a risk to children and pets. To test for lead you can purchase special sticks that change color when rubbed on a surface containing lead. However, these home lead tests are not perfect.

1. Suppose that one particular test correctly reports the presence of lead 90% of the time. This test also correctly reports the absence of lead 85% of the time. A neighborhood wants to determine the percentage of homes with lead problems.

   a) Use the values of the test characteristics to find a model for predicting the fraction of homes with a lead problem given the results of lead testing. Be sure to specify a reasonable domain of your model.

   b) Of the 250 homes tested, 30% tested positive. Use your model to estimate the fraction of homes with lead problems. What is this number as a percentage? Should the neighborhood be concerned?

   c) What can you say for sure about the homes that tested positive for lead? Do they all have a lead problem? What advice would you give to one of the residents of these homes?

2. Suppose you have designed a new test to detect the presence of lead.

   a) What are the characteristics of your test? (Make up your own set of characteristics by selecting appropriate numbers for p and r between 0 and 1.)

   b) Use the values of your test's characteristics to determine a model for predicting the fraction of homes with a lead problem, given the results of lead testing.

   c) Suppose you test 175 homes in another neighborhood for lead using the test in part (a). Of these tests, 15 turn out to be positive. What percentage of the homes in this neighborhood do you estimate have a problem with lead?

   d) Suppose that in part (a), Clarice chooses $p = .4$ and $r = .5$. Is this a reasonable test? Explain why or why not?

   e) Use the characteristics of Clarice's test to determine a model for predicting the fraction of homes with a lead problem given the results of lead testing. Is this model reasonable? Explain why or why not. If not, what is wrong?

3. a) Suppose the model for a particular test is given by the equation:

$$U = \frac{T - 0.15}{0.72.}$$

   What are the characteristics for this test? How did you determine your answer?

   b) Suppose the model for a particular test is given by the equation:

$$U = 1.4T - 0.26$$

   What are the approximate characteristics for this test? How did you determine your answer?

In this lesson, you developed a model that you could use to estimate the percentage of the population that had a particular trait from the percentage of the population that tested positive for that trait. The next lesson poses this question: If a person tests positive for a particular trait, how likely is it that the person actually has the trait? The answer may surprise you.

LESSON FOUR

But I Didn't Do It! Really!

KEY CONCEPTS

Test effectiveness

Rational functions

The Image Bank

PREPARATION READING

Focus on the Individual

*I*n previous lessons, you used survey data or test results to determine the prevalence of a particular trait in a population. For example, given survey data or test results for several different groups of athletes, you estimated the percentage of the athletes who used performance-enhancing drugs. Although it is important for competitive sports organizations to monitor the prevalence of the use of performance-enhancing drugs among its athlete members, each drug test affects the life of an individual athlete.

How likely is it that an athlete who tests positive for performance-enhancing drugs actually has used performance-enhancing drugs ? As you discovered in Lesson 3, drug tests given at athletic events are not perfect. Because of the severe consequences of a positive test result, athletes who test positive are frequently retested before any disciplinary action is taken. In the first round of the 1994 Soccer World Cup, for example, Argentine star Diego Maradona tested positive for performance-enhancing drugs. After failing the second test, he was banned from competition for life!

Even though you may never take a drug test, you probably have already been given some sort of medical test. Have you ever known someone whose medical test indicated that he had a particular disease, and after additional testing, he learned that he did not have the disease?

In this lesson, you will determine the probability that a positive test result is a false positive. In the process, you will discover that this probability depends not only on the characteristics of the test, but also on the prevalence of the trait being tested.

CONSIDER

1. Suppose you are given a TB tine test for tuberculosis and the test turns out positive. Do you think you have tuberculosis? Why or why not?

2. How prevalent is tuberculosis in the United States?

TESTING HERSHEY'S KISSES®

In previous activities of this unit, you have focused on the whole population that has a particular trait. Now the focus is on the effects of test results on individuals.

Step 1:
The Population Composition

The trait of interest is almonds; some chocolate Hershey's Kisses® have almonds and some don't.

1. Record the number of Kisses with and without almonds.

2. How prevalent is the trait of having an almond within your group of Kisses? Write a statement about the composition of the population.

Step 2:
The Test

Suppose that the characteristics of the Kiss test are that the probability of correctly detecting an almond from an almond Kiss is .95 and the probability of correctly detecting the absence of an almond from a plain Kiss is .90.

3. a) How many of the almond Kisses would you expect to test positive?

b) How many of the plain Kisses would you expect to test positive?

c) What percentage of the total Kiss population would you expect to test positive?

Now determine which of the Kisses will test positive as follows. Select someone in the class to be the "Kiss tester." From the almond Kisses, the designated tester should randomly select the number of Kisses in Item 3(a); then, from the plain Kisses randomly select the number of Kisses in Item 3(b). Together they are the Kisses that have tested positive. Wrap each of these positive Kisses in aluminum foil so that you can't tell which are plain and which have almonds. Then put them in a container and mix well. Set the "negative" Kisses aside.

TESTING HERSHEY'S KISSES®

Step 3:
Identification of the Almond Trait

The designated tester should select one of the positive Kisses at random.

4. a) What is the probability that this Kiss is plain?

   b) What is the probability that this is an almond Kiss?

Distribute the Kisses from the container among the class. Each person holding a Kiss has tested positive. As a penalty, each will be banned from all school activities for the rest of the year. Now, unwrap your Kiss.

5. Imagine you are holding one of the Kisses. When you unwrap it, you see that your Kiss is plain. How do you feel about being banned from all school activities for the rest of the year? How would you feel, in general, if you were punished for something that you didn't do?

6. Now, look back at the Kisses that tested negative. Among that group there is at least one almond Kiss that avoided detection and escaped the penalty. How would you feel if you were able to "beat the test" in this way? How do you think plain-Kiss holders would feel if they knew?

7. What does the Kiss activity have in common with testing for performance-enhancing drugs or a medical test such as the TB tine test?

8. Suppose that in the Kiss activity, the Kisses with almonds represent individuals who use performance-enhancing drugs and those without almonds represent those who do not.

   a) Draw and label a tree diagram that represents this situation.

   b) What percentage of the positive drug tests was from non-users? What is the significance of this result to individuals taking this test?

   c) What percentage of the negative test results came from users? How does this differ from the percentage of users who test negative?

INDIVIDUAL WORK 8

Impossible Failures

Imagine that you are part of a group selected to take a drug test before joining a sports team. Afterwards, you discover that one of your friends tested positive. What is the probability that your friend is using drugs? The answer to this question may surprise you.

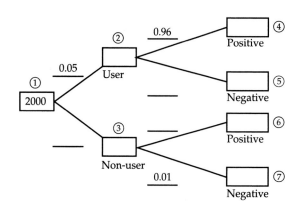

Figure 7.30.
Tree diagram for a drug test.

1. Suppose that a random sample of students is being tested for use of performance-enhancing drugs. The tree diagram in **Figure 7.30** represents this situation.

   a) Complete the tree diagram in Figure 7.30.

   Use your completed tree diagram to help you answer the following questions.

   b) How many people would you expect to correctly test positive for drug use?

   c) How many false negative results would you expect?

   d) How many people would you expect to be identified correctly as non-users?

   e) How many people would you expect to be identified incorrectly as drug users?

   f) How is it possible that more people were incorrectly identified as users than were correctly identified as users?

   g) What is the probability that a person who tests positive is actually a user? Explain how you determined your answer.

   h) What percentage of the positive test results are false positives? How is this answer related to your answer in part (g)?

2. A certain drug test correctly detects the presence of performance-enhancing drugs with a probability of $p = .98$ and correctly reports the absence of drugs with a probability of $r = .92$. Assume that only 1% of the athletes at a major international sports festival are actually using such drugs.

a) If all athletes at the festival are tested, what percentage of the drug tests would you expect to be positive? Would your answer change if only a random sample of the athletes were tested? Explain why or why not.

b) What is the probability that an athlete who tests positive is a drug user? What percentage of the positive test results are true positives?

c) What is the probability that an athlete who tests positive is not a drug user? What percentage of positive test results are false positives?

3. Suppose the test in Item 2 is given to a different group of athletes. Twenty percent of the athletes in this group are using performance-enhancing drugs.

a) If the athletes in this group are tested, what percentage of the drug tests would you expect to be positive?

b) What is the probability that an athlete who tests positive is a drug user? What percentage of the positive test results are true positives?

c) What is the probability that an athlete who tests positive is not a drug user? What percentage of positive test results are false positives?

4. Compare your answers to Items 2(b) and 3(b). Does this test do a better job in identifying actual users in the situation described in Item 2 or in Item 3? What accounts for the difference?

5. Assume that a test has been developed to aid in the diagnosis of a common childhood illness. Doctors give this test only if a patient has certain symptoms. Only 40% of the patients with these symptoms actually have this disease. (Other illnesses have the same symptoms.) The results from this test can be positive (the disease is detected), negative (no indication of the disease), or inconclusive (the disease may or may not be present). If the disease is present, the probability that test results will be positive is .90 and inconclusive .06. If the disease is not present, the probability that the test results will be negative is .85 and inconclusive .07.

a) Draw a tree diagram that represents this situation. Label all branches with the corresponding probabilities.

b) What percentage of the tests would you expect to be positive?

c) What percentage of the positive tests come from people who have the disease?

d) What percentage of the positive tests come from people who do not have the disease?

e) What percentage of the tests would you expect to be inconclusive?

f) What percentage of the inconclusive tests would you expect to come from people who have the disease?

g) What percentage of the negative tests come from people who have the disease?

6. Rework your answer to Item 5. This time assume that only 10% of the people who have the symptoms actually have the disease.

7. What impact did changing the percentage of those who had the disease from 40% to 10% have on your answers to parts (b–d) in Items 5 and 6?

By now, you should have discovered at least two things about tests as they relate to the individuals being tested.

- If you test positive on a test, you do not necessarily have the trait the test was intended to detect.

- You may not be able to distinguish the false positives from the true positives without further testing.

In the next activity, you will see that **population composition**, which is the make-up of the population (the percentage of the population with the trait), is related to the percentage of positive tests that are true positives.

ACTIVITY

TRUE FAILURES

11

CONSIDER:

1. How does the population composition affect the percentage of positive tests that are true positives?

Suppose that a particular test for performance-enhancing drugs correctly detects drug users 90% of the time and correctly reports the absence of drug use 80% of the time. (In other words, this test can be characterized by $p = .90$ and $r = .80$.) Your goal in this activity is to establish a relationship between the fraction of positive tests that identify actual drug users and the fraction of the population that uses drugs (population composition). Stated in terms of probabilities, this means that you will try to determine a relationship between the probability that a positive test actually identifies a drug user, $P(\text{user} \mid +)$, and the probability that a randomly selected individual from this group is actually a drug user, $P(\text{user})$.

In order to study this relationship, first you will need to gather some data just as you did in Activity 8. Record your data in Figure 1 of Handout H7.6. Before you begin, work through the Check below.

Check:
Using the test with characteristics described above, if 40% of the people in the population use performance-enhancing drugs, then 75% of the positive drug tests will come from drug users. In this case, the ordered pair (fraction of users, fraction of positive drug tests that are true positives) is (0.40, 0.75).

If you agree with the statement in the check, then enter the value 0.75 in the second column of the table to the right of 0.40. If you disagree with this statement, then recheck your calculations.

ACTIVITY

TRUE FAILURES

11

1. Distribute Data Sets A–D among the members of your group.

a) For each data set, you are given three scenarios. Each scenario has a different population composition (fraction of actual users in the population). Based on the population composition, calculate the fraction of positive tests that are true positives (in other words, determine $P(\text{user} \mid +)$). Enter your results in Figure 1 on Handout H7.6.

DATA SET A

SCENARIO 1

Population composition:
$P(\text{user}) = .05$

SCENARIO 2

Population composition:
$P(\text{user}) = .10$

SCENARIO 3

Population composition:
$P(\text{user}) = .95$

DATA SET C

SCENARIO 1

Population composition:
$P(\text{user}) = .25$

SCENARIO 2

Population composition:
$P(\text{user}) = .80$

SCENARIO 3

Population composition:
$P(\text{user}) = .85$

DATA SET B

SCENARIO 1

Population composition:
$P(\text{user}) = .15$

SCENARIO 2

Population composition:
$P(\text{user}) = .20$

SCENARIO 3

Population composition:
$P(\text{user}) = .90$

DATA SET D

SCENARIO 1

Population composition:
$P(\text{user}) = .30$

SCENARIO 2

Population composition:
$P(\text{user}) = .70$

SCENARIO 3

Population composition:
$P(\text{user}) = .75$

TRUE FAILURES

ACTIVITY

11

b) After your group has calculated all its data, carefully graph the 13 ordered pairs from your table on the grid provided in Figure 2 on Handout H7.6. After you have plotted your points, draw a smooth curve or line through the points.

c) Stop for a moment and study the relationship shown by the table and graph. What do they tell you about the conditions under which this test is most useful in identifying users? Under what conditions is it least useful?

2. Answer these questions based on your table (Figure 1) and graph (Figure 2) on Handout H7.6.

a) If one in ten members of an athletic team are using performance-enhancing drugs, what fraction of those who took and failed a drug test (in other words, the test results were positive) are users?

b) If 20% of the members of a swim team are using performance-enhancing drugs, what percentage of those who failed their drug test are users?

c) If 23% of the members of a team are using performance-enhancing drugs, what percentage of those who test positive on a drug test are users? What percentage of those who test positive are non-users? How are these percentages related?

d) If 60% of those who test positive on a random drug test are actually users, what percentage of the population that you have sampled is using drugs? How is this question different from the basic question asked in (a–c)?

e) In a perfect test, 100% of those who test positive would be users and 100% of those who test negative would be non-users. The test that you used to generate your data in Figure 1 (Handout H7.6) is not perfect. In order for this test to give results that you would consider reasonable, what percentage of the group being tested would have to

ACTIVITY

TRUE FAILURES

11

be users? Be prepared to explain your answer in class. (For your answer, focus only on the effectiveness of a positive test as an indicator that a person is using drugs.)

3. a) What is the relationship between $P(\text{non-user} \mid +)$ and $P(\text{user} \mid +)$?

   b) Use your relationship in part (a) to sketch a graph of $P(\text{non-user} \mid +)$ versus $P(\text{user})$. What does this graph tell you about the usefulness of the test?

4. The characteristics for the TB tine test (the values for p and r) will differ from the characteristics of the drug test used in this activity. Assume, however, that the graph of the relationship between the fraction of positive tests that correctly identify people with the disease, $P(\text{TB} \mid +)$, and the fraction of people with the disease, $P(\text{TB})$, has the same shape as the graph that you drew for Figure 2 on Handout H7.6.

   a) Suppose that you have just gotten your first full-time job. Before you can start your job, you are required to have a TB tine test. You are told that your test results are positive. What does the shape of the graph of $P(\text{TB} \mid +)$ versus $P(\text{TB})$ tell you about how you should react to this news?

   b) Frequently, people whose TB tine tests are positive do not have active TB. Given this situation, why do you think that the TB tine test is such a popular test?

In the next activity, you will use tree diagrams to determine an equation that describes the relationship in your data table and graph (Figures 1 and 2 on Handout H7.6). Later, you will use this relationship to assess when it makes sense to institute a program of "blanket testing," testing everyone in a large population for a particular trait.

INDIVIDUAL WORK 9

Hats off to the Machines Gone Wild

1. The table in **Figure 7.31** shows some information about members of the 9th grade class at one high school. You are to pick a student at random from this group to talk to your math class about dress codes. Use the information in the table to answer the questions that follow.

	Had a hat this morning	Did not have a hat this morning
Male	31	29
Female	13	27

Figure 7.31.
Characteristics of students in a high school.

a) What is the probability that the selected student is male?

b) What is the probability that the selected student had a hat this morning?

c) Assuming you selected a male student, what is the probability that he had a hat this morning?

d) Given that a student with a hat was selected, what is the probability that the student was male?

e) What, if anything, does this problem have in common with the drug-testing situations you have analyzed lately?

2. You and your partner have started a company to make spirit buttons for all the schools in your area. The buttons are produced in two locations. Bob and his friends make 2/3 of your buttons at his house, and the rest are made at Suzy's. Each group places an identifying mark on the buttons they make. You have found that 5% of Suzy's buttons are defective. Bob claims that 10% of his are also defective. You have never checked to see if Bob is correct.

After receiving one shipment, your partner counts all the defective buttons and notices that 8 of every 10 come from Bob's group. Your partner is very upset about this. After all, if Bob was telling the truth that only 10% of his buttons are defective, why are so many of the defective ones coming from him?

a) Draw a tree diagram that represents this situation.

b) Based on Bob's claim that only 10% of his buttons are defective, what percentage of the defective buttons would you expect to have come from Bob's house? Should you question Bob's claim that only 10% of his buttons are unusable?

c) Suppose that in another shipment 1/2 of the defective buttons come from Suzy's house. Your partner thinks that everything is fine. Do you agree with your partner, or do you think that something has changed? Explain.

3. You and your group have decided that you did not like the results of the drug test in Activity 11. (Recall $p = .90$ and $r = .80$.) You have requested two drug-testing companies to create new tests. They each present their tests. Now you need to decide which you think is better. With the new test from company A, the probability that a user will test positive is increased to .95, but the probability that a non-users will test positive is increased to .25. The corresponding probabilities for the new test from company B are .85 and .01.

a) What would be the benefits of switching from the test in Activity 11 to Company A's test? What would be the disadvantages?

b) What would be the benefits of switching from the test in Activity 11 to Company B's test?

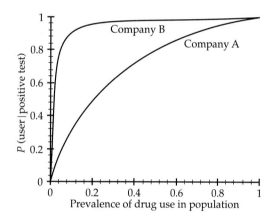

Figure 7.32.
P(user|+) versus P(user) for tests from two companies.

4. **Figure 7.32** presents two graphs of the relationship "test effectiveness" versus "population composition," one for Company A's test and the other for Company B's test. **Test effectiveness** is the fraction of positive tests that are true positives or $P(\text{user} \mid +)$, and population composition is the fraction of the population who are users or $P(\text{user})$.

a) Based on these graphs and information about the characteristics of the tests, under what conditions would you prefer Company A's test? Company B's test? Explain.

b) If you use Company A's test, how high does the percentage of actual users need to be for at least 80% of the positive tests to come from users?

c) If you use Company B's test, how high does the percentage of actual users need to be in order for at least 80% of the positive tests to come from users?

In the final activity of this lesson, you will develop equations that describe graphs such as those shown in Figure 7.32. From your equations and graphs you will be able to make policy decisions about when it makes sense to screen (or test) an entire population to identify individuals with a particular trait (such as drug use or the presence of the HIV virus).

TEST EFFECTIVENESS V. POPULATION COMPOSITION

12

In Individual Work 9, you compared performance-enhancing-drug tests made by two companies, Company A and Company B. Your comparison was based on graphs of the test effectiveness v. population composition. In this activity, you will determine equations for the graphs in Figure 7.32. Then, you will generalize your results and use them to model the relationship between "population composition" and "test effectiveness" for tests with characteristics p and r.

Recall the test characteristics for the tests manufactured by Companies A and B:

Company A: $p = .95, r = .75$

Company B: $p = .85, r = .99$

Let:

N represent the number of tests performed;

U represent the population composition, the fraction of the population that uses performance-enhancing drugs;

E represent the effectiveness of the test, the fraction of positive tests that come from drug users.

1. a) Draw and label a tree diagram representing the test for Company A.

   b) How many tests would you expect to be true positives? (Your answer should be in terms of N and U.)

   c) How many positive results would you expect?

   d) Write an equation for E, the fraction of positive tests that are true positives.

2. Repeat Item 1 for Company B.

3. a) Specify reasonable domains for each of your models in Items 1 and 2.

   b) What can you say about the effectiveness of either Company A's or B's tests if nobody in the population being tested is using performance-enhancing drugs? Does your answer make sense?

   c) What can you say about the effectiveness of either Company A's or B's tests if everybody in the population being tested is using performance-enhancing drugs ? Does your answer make sense?

4. Using a graphing calculator, graph your models from Items 1 and 2 in the window [0, 1] x [0, 1]. Why does it make sense to use this window? Compare your graphs to those in Figure 7.32.

5. For each of the tests from these two companies, determine how large the population composition must be before the test effectiveness is at least 50%. Explain how you could answer this question using your graphs. Then, show the steps you would use to solve the problem using algebra.

6. a) Now, generalize the process you used to get the equations for the tests manufactured by Companies A and B. Use the variables p and r for the test characteristics. Assume that p and r could have any values between 0.5 and 1.0. What is your model (equation) for the relationship E versus U?

   b) What is a reasonable domain for the general model?

TEST EFFECTIVENESS V. POPULATION COMPOSITION

12

7. a) Specify the test characteristics for two different tests of your choosing. (However, choose tests with characteristics that differ from the ones for the tests manufactured by Companies A and B.) Then, use your model from Item 6 to determine equations for the relationship E versus U corresponding to each of your tests.

   b) Graph each of your models in the viewing window $[0, 1] \times [0, 1]$.

   c) Based on the graphs you have observed, describe the shape of the graph of the relationship E versus U for any given test.

   d) Based on your two graphs in part (b), under what conditions would you prefer your first test? Your second test? Explain.

8. What do the test characteristics p and r control on the graph of E versus U?

   a) Design an exploration that will help you answer this question. Describe your design.

   b) Carry out your exploration. Based on what you have learned, how do the test characteristics p and r control the graph of E versus U? Describe anything else that you found interesting.

INDIVIDUAL WORK 10

They're Totally Rational

The models that you used in Activity 12 all had a common form:

$$E = \frac{pU}{(p + r - 1)U + (1 - r)}$$

where:

p and r are the characteristics of the test;

U is the fraction of the population having the trait being tested;

E is the fraction of the positive tests from people (or whatever is being tested) having the trait being tested.

1. Suppose that a competitive sports organization is planning to implement mandatory drug testing for its athletes.

   a) What fraction of the drug users among these athletes would test positive with a test where the relationship between U and E is given by the equation

   $$E = \frac{0.98U}{0.88U + 0.10}?$$

   b) What fraction of non-users would test positive with the same test?

   c) Suppose that 15% of this group of athletes are taking performance-enhancing drugs. If the athletes are given the drug test in part (a), what percentage of the positive tests would you expect to come from drug users?

   d) Suppose that when another group of athletes is given this test, only 40% of the positive tests are from drug users. Estimate the percentage of this group that are using performance-enhancing drugs. Show the algebraic steps needed to solve this problem.

2. A blood test is used to detect the presence of the HIV virus. Suppose for one particular test, when HIV antibodies are present, the HIV test is positive with a probability of .997. When HIV antibodies are not present, the HIV test gives a negative reading with a probability of .985.

a) Should all high school students who participate in sports be tested for HIV? Without doing any computations, how would you answer this question?

b) Assume that less than 1% of the high school student population is infected with HIV. If mandatory testing of high school students for HIV is implemented, what percentage of the positive tests would you expect to be false positives?

c) Write a model that describes the relationship between E and U for this situation.

d) What percentage of the high school population would need to be infected with HIV before you would expect that 70% of the positive tests were from students infected with HIV?

The models for E versus U in Items 1(a) and 2(c) are members of the $y = (ax + b)/(cx + d)$ family, the quotient of two linear expressions involving the explanatory variable. For the remainder of this activity, you will work with equations from this family. All members of the $y = (ax + b)/(cx + d)$ family belong to an even larger family called the **rational functions**. Just as rational numbers are quotients of two integers, rational functions are quotients of two algebraic expressions. (You will learn a more precise definition for rational functions in a later unit.)

3. Let $y = 2x/(x + 1)$.

Plot1 Plot2 Plot3
\Y₁=2X/X+1
\Y₂=
\Y₃=
\Y₄=
\Y₅=
\Y₆=
\Y₇=

Figure 7.33.
Calculator screen for Item 3(b).

a) What value for x is not in the domain of this relationship? Explain why not.

b) Daiva wanted to graph this relationship using her calculator. **Figure 7.33** shows the expression that she entered. What is the equation of the relationship that she actually entered? What was Daiva's error?

c) Graph this relationship in the window $[-5, 5] \times [-5, 5]$. Make a sketch of your graph on graph paper. After graphing this relationship on paper, draw a dotted vertical line at the x-value in part (a). (Remember, there is no point on the graph of this relationship that corresponds to the x-value in part (a). So, this dotted line is not part of the relationship between x and y.)

d) Explain how you could approximate the solution to $4 = 2x/(x + 1)$ using your graph from (c). What is your approximate solution?

e) Now, use algebra to get an exact solution to the equation in part (d). (In solving equations such as this one, you may find it helpful to multiply both sides of the equation by the expression in the denominator, $x + 1$.) After you have found a solution, remember to check that it is a solution to the original equation.

4. Let $y = (x + 3)/(2x - 4)$

a) What value for x is not in the domain of this relationship?

b) Graph this relationship in a window that shows the key features of its graph. Sketch the graph on a piece of graph paper.

c) Approximate the solution to $(x + 3)/(2x - 4) = -1$ using your hand-drawn graph. What is your approximate solution?

d) What is the exact solution? How did you determine the exact solution?

5. How are Items 3(d), 3(e), 4(c), and 4(d) like Items 1(d) and 2(d)?

6. Here are two more members of the $y = (ax + b)/(cx + d)$ family:

$$y = (2x - 1)/(3x + 2) \text{ and } y = (-2x + 1)/(3x + 2).$$

a) Graph each of these equations and then make sketches of their graphs.

b) How are these graphs alike and how are they different? What accounts for the difference?

IMPERFECT TESTING

Wrapping Up Unit 7

The term "polygraph" literally means "many writings." The name refers to the manner in which several physiological activities, such as breathing and pulse rate, are simultaneously recorded. Polygraph techniques are commonly referred to as "lie-detector" testing.

The polygraph technique is widely used by the law-enforcement community in two ways: as part of the pre-employment screening for police candidates and as a tool in investigating criminal offenses.

The APA Research Center at Michigan State University conducted a survey of police executives in the United States to determine the extent of polygraph testing for pre-employment screening.

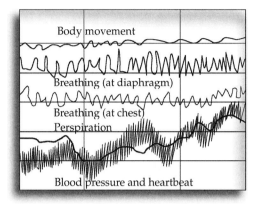

1. Of the 699 largest police agencies surveyed, 626 agencies responded. What percentage of the contacted agencies responded?

2. The survey found that 388 of these agencies had an active polygraph screening program, 194 did not, and 44 had discontinued a previously active program (usually because of laws prohibiting the practice). Create an area model to represent the percentage of agencies that fall into each of these three categories.

Suppose that a lie-detector test correctly identifies deceptive responses with a probability of .88, and correctly identifies truthful responses with a probability of .86.

UNIT SUMMARY

3. The results of the survey of police executives indicate that a high percentage of law-enforcement agencies conduct polygraph pre-employment screenings of police-officer applicants.

   a) Identify the traits, or factors, that are relevant to this situation and draw a tree diagram of it. Use variables for quantities that are unknown. Define each of your variables.

   b) Results from the APA Research Center survey also indicate that approximately 25% of those tested are disqualified from police employment based on the results of their polygraph tests. Estimate the percentage of candidates who were being deceptive. Explain how you arrived at your answer. Does your answer seem reasonable?

c) What is the difference between finding the fraction of positive tests among deceptive applicants and finding the fraction of deceptive applicants among those who tested positive? (Note: A positive test result here means that the test indicates deception.)

d) For this particular lie-detector test, what is the relationship between U, the fraction of applicants who were being deceptive, and E, the fraction of positive tests from applicants who were being deceptive?

e) Estimate the percentage of those who tested positive who were being deceptive.

f) Estimate the percentage of those who tested positive who were being truthful. Does this seem fair?

4. Imagine that during a hearing on the validity of polygraph results, police argue that 99% of the people who have tested positive during criminal investigations were, in fact, being deceptive. What does this tell you about the population that the police are testing? Justify your answer.

5. Sketch the graph of the relationship of E versus U from Item 3(d). What does this graph tell you about the lie-detector test's effectiveness for the following two groups of people: the police-officer candidates and the alleged criminals?

6. Polygraph tests can be positive, negative, or inconclusive. Suppose that when a person is being deceptive, the lie-detector test is positive with probability .88, and negative with probability .05. When a person is being truthful, the test is negative with probability .86 and positive with probability .06.

a) Draw a tree diagram that represents this situation. Use variables for any unknown quantities. Be sure to define your variables and label all branches with their corresponding probabilities.

b) Suppose that 10% of the police candidates being screened for jobs are being deceptive. Approximately what percentage of their polygraph tests would you expect to be positive? Negative? Inconclusive?

c) Assume that 10% of the candidates were being deceptive. Would you expect a higher percentage of the inconclusive results to be from deceptive candidates or truthful candidates? Justify your answers by computing the percentages.

Mathematical Summary

Probability represents the long-term relative frequency of some event's occurring. Relative frequency refers to the ratio of the number of occasions on which the event does occur to the number of all observed occasions. Thus, one way to estimate a probability is to count the number of successes in a set of trials. This view of probability implies that there is some experiment, something (or someone) that is being observed. That is, in order to count successes in a set of trials, you have to know what a "trial" is.

Sometimes different events in an experiment are related in some way. For example, think of some event. Then for any particular observation, that event can either happen or not happen. That means that "not happen" is also an event, and it is clear that these two are related in two ways. First, "happen" and "not happen" cannot both occur on the same observation. Second, one of them surely will occur on each observation. These are complementary events. Complementary events represent one possible relationship among events. Events may also be independent or mutually exclusive (which are not the same things).

Just as events may be related, so may probabilities. Each of the relationships among events translates into a computational rule about probabilities. In addition, a variety of representations prove useful in describing and working with related probabilities. Among these representations are tree diagrams, tables, and glyphs. Remember that every probability shown along a branch segment in a tree diagram is a conditional probability. At times that makes trees very useful, but it sometimes makes them difficult to construct, too.

You used one other major idea in this unit that was not new, but its use with probability may make it seem new: two quantities may be related in such a way that changes in one quantity will cause predictable changes in the other quantity. That is, a function is present. In the context of interpreting imperfect tests, many desired quantities are functions of a probability. That means that changing a particular probability associated with the test changes the interpretation of the outcome of that test. Knowing to look for a function allows for it to be studied, which in turn allows for "perfect" interpretation of the imperfect test results. Tree diagrams and tables prove useful in examining these functions.

Glossary

COLUMN TOTALS:
in a two-way table, the sum of the entries in each column.

COMPLEMENTARY EVENTS:
events that have no outcomes in common but when combined include all outcomes.

CONDITIONAL PROBABILITY:
a number between 0 and 1 that assesses the likelihood of a chance outcome based on the knowledge that another chance outcome (the condition) has occurred.

DEPENDENT EVENTS:
Events A and B are called dependent events if $P(A) \neq P(A \mid B)$. In other words, the probability that A occurs is changed by knowing that B has occurred.

EVENT:
an outcome or collection of outcomes of a chance experiment.

FALSE NEGATIVE:
the test results for a particular trait are negative, and the person tested has the trait.

FALSE POSITIVE:
the test results for a particular trait are positive, and the person tested does not have the trait.

GLYPH DISPLAY:
a graphical display in which features of individual elements of the display represent characteristics of objects in the population being represented.

INDEPENDENT EVENTS:
Events A and B are called independent events if $P(A) = P(A \mid B)$. In other words, A and B are independent events if the probability that A occurs is unchanged by knowing that B has occurred.

MUTUALLY EXCLUSIVE EVENTS:
events that consist of non-overlapping outcomes.

PIE CHART:
a circle containing wedge-shaped regions (similar to slices of a pie) that represent various traits of the population.

POPULATION:
the entire group of objects or individuals being studied.

POPULATION COMPOSITION:
the percentage (or fraction) of a population that has a particular trait.

PROBABILITY:
a number between 0 and 1 that assesses the likelihood of a chance outcome. The closer the probability is to 1, the more likely the outcome.

RANDOMIZED-RESPONSE TECHNIQUE:
a technique for surveys in which the respondents use a randomizing device to select the question or statement for their response.

RATIONAL FUNCTION:
the quotient of two polynomial expressions.

REPRESENTATIVE SAMPLE:
a sample that has roughly the same composition as the population from which it is drawn.

ROW TOTALS:
in a two-way table, the sum of the entries in each row.

SAMPLE:
a portion of the population.

TEST EFFECTIVENESS:
the percentage (or fraction) of the positive tests for a particular trait that are true positives.

TREE DIAGRAM:
a graphic representation of a step-by-step process, used when there is more than one possibility at each stage of the process.

TRUE NEGATIVE:
the test results for a particular trait are negative, and the person tested does not have the trait.

TRUE POSITIVE:
the test results for a particular trait are positive, and the person tested has the trait.

TWO-WAY TABLE:
a table in which two quantities vary, one across the columns, the other down the rows.

UNIT

8

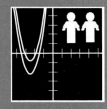

Testing 1, 2, 3

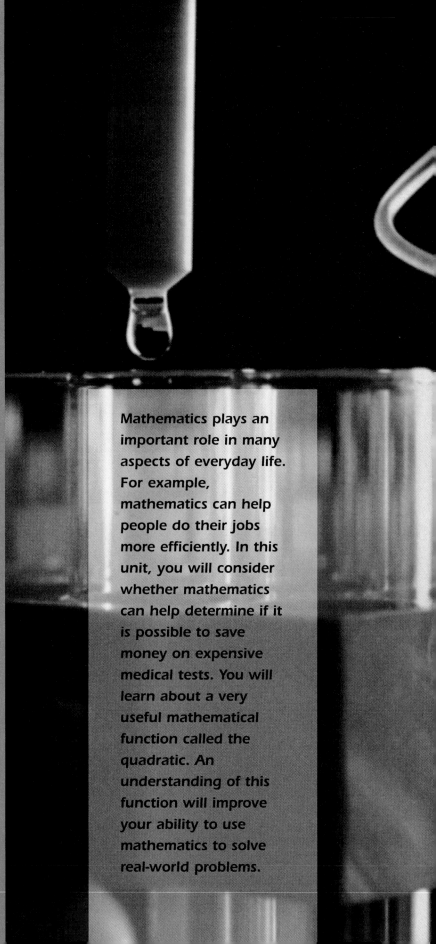

Mathematics plays an important role in many aspects of everyday life. For example, mathematics can help people do their jobs more efficiently. In this unit, you will consider whether mathematics can help determine if it is possible to save money on expensive medical tests. You will learn about a very useful mathematical function called the quadratic. An understanding of this function will improve your ability to use mathematics to solve real-world problems.

TESTING 1, 2, 3

Medical testing is common today. Sometimes medical tests are done because a patient is ill, and sometimes they are done for other reasons. In Unit 7, *Imperfect Testing*, you saw that tests used to screen populations for common ailments are seldom perfect. When these tests are done on large groups of people, those who test positive often do not have the condition the test was designed to detect.

Tests that are used to screen large populations should be inexpensive. Quite often, however, a cheap test is less accurate than an expensive one. Thus there is sometimes a tradeoff between accuracy and cost.

If a very accurate but expensive test is used, are there reasonable ways to save money? A common and fairly expensive test is used to detect steroids. Many people, from Olympic athletes to high school students, undergo steroid testing. In this unit, you will use mathematics to look for more efficient ways to test for steroids.

LESSON ONE
Steroid Testing

KEY CONCEPTS

Steroid testing

Sample pooling

Mathematical model

Probability

Simulation

Expected value

The Image Bank

PREPARATION READING

Save Money by Pooling Samples

Testing for substance abuse is not limited to athletes. Often people must pass a test for substance abuse in order to get a job. The people who require such tests, whether they are school-board members, employers, or Olympic committees, are interested in doing them economically.

One strategy commonly used to save money on tests for substance abuse is **sample pooling**. With this strategy, two or more samples are mixed together. A single test is done, and if the results are negative, all of the samples are considered negative.

The first step in using mathematics to model a real-world situation is to identify a question to answer or a problem to solve. As you know, it is often best to start simply. The simplest sample-pooling situation involves only two samples. Your first question is "When can you expect to save money by pooling two steroid test samples?"

Once you have identified a problem that mathematics might help solve, you must spend some time understanding the situation and searching for the important factors that affect the problem. In this case, you must understand the factor or factors that have the greatest effect on the cost of testing when samples are pooled. What factors are most important in determining the cost of testing two pooled samples?

In this lesson, you will consider the general issue of steroid testing and factors that affect the cost of testing pooled samples.

POOLING SAMPLES

First examine the strategy of pooling two samples to be sure that you understand it.

The logic of the sample-pooling strategy is simple: Put two or more samples together. Run one test. If the test is negative, you do not have to test the individual samples.

1. Suppose you pool two samples. You might get by with only one test, in which case you saved half the cost of two individual tests. If the pooled test is positive, however, you have to make more than one test. What is the total number of tests you might make if the pooled test is positive? Make a list of the various numbers of tests you might have to run. Write a short description of how each number could occur.

2. Imagine that you are a tester trying to save money by pooling two samples. You keep records of the number of tests you did on each pair of samples in order to help you decide whether the sample-pooling strategy is saving money. How might you organize your data in a concise way?

   Think about the various kinds of representations you have used in previous units (tables, graphs, and various kinds of diagrams) and select at least one to try here. You may not have enough information at this time to complete your representation, so make note of any missing information that would be helpful.

   Does your representation shed any light on the question of when you can expect to save money by pooling two samples?

3. Suppose a tester gives steroid tests to a large number of people, using the strategy of pooling two samples. She finds that the number of sample pairs that require one test is the same as the number of pairs that require two tests and is also the same as the number that require three tests. Each test costs $80. Does the sample-pooling strategy save the tester money?

INDIVIDUAL WORK 1

What Do You Know about Steroid Testing?

To solve a real-world problem, a mathematician must first learn about the situation. Thus your task is to learn as much as you can about steroid testing. Read one or more recent articles about the topic. Take as many notes as you can and be prepared to share them with others in your class. Write down anything you think is important. Here are some examples of questions you might be able to answer.

1. Why do people abuse steroids?

2. Why do other people try to prevent steroid abuse?

3. What does a steroid test cost?

4. What are some strategies used to save money on steroid testing?

5. What are some factors that might determine whether it is reasonable to expect to save money by pooling test samples? If you cannot find any information on this question, write down one or two conjectures. Give an explanation for each.

CONSIDER:

1. Suppose you are testing for steroid use in each of the following scenarios. You are using the strategy of pooling two samples. How many tests will you do in each school?

   a) All of the students are using steroids, and you do not know it.

   b) None of the students are using steroids, and you do not know it.

   c) Almost all of the students are using steroids, and you do not know it.

   d) Very few of the students are using steroids, and you do not know it.

2. In which schools do you expect to save money by using the strategy of pooling two samples?

INDIVIDUAL WORK 2

Saving Money on Tests

1. The bar graph in **Figure 8.1** shows the results of steroid tests given to athletes at a large high school. The strategy of pooling two samples was used.

   a) How many students were tested?

   b) Did the school save money by using the sample-pooling strategy? Explain your answer. If your research on steroid testing produced the cost of a steroid test, use the cost you found to estimate the amount of money the school saved or wasted by using the sample-pooling strategy.

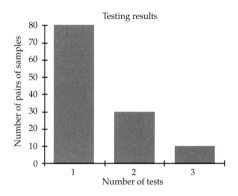

Figure 8.1

   c) What is the average number of tests required to test a pooled sample?

   d) How can the average number of tests per pooled sample be used to tell whether the school saved money?

   e) Estimate the percentage of students who tested positive for steroids. Explain your answer.

   f) Your answer in part (d) describes how the average number of tests can be used to determine if money is saved. How can you use the total numbers of tests to tell if money is saved?

2. **Figure 8.2** is a bar graph that describes steroid testing results in another school.

   a) Did this school save money by using the sample-pooling strategy? Explain your answer.

   b) Estimate the percentage of students who tested positive for steroids. Explain your answer.

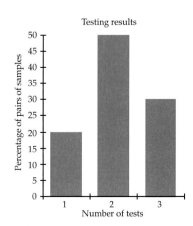

Figure 8.2.

3. As you have seen, relating the number of tests to the percentage of people who use steroids is tricky. However, it is fairly easy if all the people in a group use steroids or if none of the people do. Although these extreme cases are not likely to occur, they can be useful in creating a mathematical model for the relationship between the percentage of people using steroids and the number of tests you can expect to perform.

   a) Three tests are necessary if 100% of the people use steroids, and one test is necessary if 0% of the people use steroids. Use these facts to create a mathematical model for the relationship between the average number of tests and the percentage of people using steroids. Use at least one of the representations you learned in previous units (tables, graphs, and various kinds of diagrams) to explain your model.

   b) When can you expect to save money by pooling two steroid test samples? Explain how you used your model to obtain your answer.

4. When you create a mathematical model for a real-world situation, it is important to be aware of any assumptions you made about the situation. Later, if your model produces inaccurate results, you can question your assumptions. The following questions will help you understand the assumptions you made in creating your model.

   a) If the test used to detect steroids is not perfect, can you say that three tests are always required when 100% of the people are using steroids? Explain.

   b) If the test used to detect steroids is not perfect, can you say that one test is always required when none of the people are using steroids? Explain.

   c) List any modeling assumption you made when you created the model in Item 3.

THE TESTING GAME

A mathematical model isn't always accurate, so it is important to test your model. Your first mathematical model for the relationship between the percentage of people using steroids and the number of tests you can expect to perform may have led you to conclude that an average of two tests per pooled sample can be expected if half the people use steroids.

Your task in this activity is to design and perform a simulation to determine if an average of two tests can be expected when 50% of the people use steroids.

1. The simulation requires three people; two people play the roles of the people being tested, and one plays the role of the tester. Use a random event like flipping a coin or drawing a chip from a container to determine whether each tested person is positive. Also, use a random event to determine which person the tester tests first when individual testing is necessary.

   Once you have designed your simulation, repeat it until your group has done ten trials. When you have finished, share the results with the other groups in your class and combine all the results in a table showing the number of times one, two, and three tests occurred.

   a) Find the average number of tests needed per trial.

   b) Compare your results to the solution you obtained from your model in Individual Work 2. Does this simulation confirm the solution?

ACTIVITY

2

THE TESTING GAME

2. Use a calculator or computer program to simulate steroid testing when 50% of the people use steroids and two samples are pooled. Do a much larger number of samples than your class did in Item 1.

a) Does this simulation confirm the solution you obtained from your model?

b) Are you more confident in the result you obtained from a larger number of samples?

Note: Most computer and calculator programs work with the portion of people using steroids rather than the percentage. You will need to enter 0.5 rather than 50. Also, the program probably uses the letter p to stand for the portion using steroids. Portions will be used frequently in the remainder of this unit.

3. Summarize what you have learned from these simulations. If the results do not confirm the conclusions you obtained from your model, then return to the modeling assumptions you listed in Item 4 of Individual Work 2 and discuss which assumption(s) you think is faulty.

INDIVIDUAL WORK 3

Probability and Expected Value

Whether you can expect to save money by pooling two test samples depends on the average number of tests per pooled sample. Now you will take a closer look at the average number of tests, which is sometimes called **expected value**. You will also examine simulations like the ones in Activity 2.

1. A group of three students played the testing game. When they were finished, they organized their results into the table in **Figure 8.3**. What can you conclude about these results? Explain.

Trial number	1	2	3	4	5	6	7	8	9	10
Number of tests required	1	3	3	1	1	1	3	1	1	3

Figure 8.3.
Testing game results.

2. When you simulate a situation in which 50% of the people are using steroids, you can flip a coin. If you want to simulate a situation in which the percentage is not 50, however, coin flipping won't work. Drawing chips out of a container is a better option. For example, suppose that you want to simulate a situation in which 30% of the people are using steroids. You decide to use ten chips, some red and some blue. You decide that red represents a steroid user.

a) How many red chips and how many blue chips should be in the container?

b) In the testing game, the first person put the chip back before the second person drew a chip. Why was this necessary?

c) Suppose you use 100 chips instead of 10. Would putting the first chip back before drawing the second be as important?

3. Calculator and computer programs simulate steroid testing by generating a random decimal number. For example, suppose 30% of the people being tested use steroids. If the random decimal number is below 0.3, the program counts the person as testing positive; if the number is above 0.3, the program counts the person as testing negative. You can think of generating a random number as throwing a dart at a one-dimensional dart board (see **Figure 8.4**).

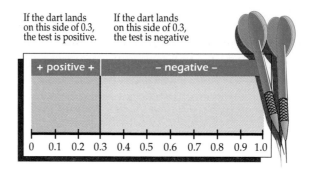

Figure 8.4.
A one-dimensional dart board.

In each of the following, tell whether the program counts the test positive or negative.

a) The percentage using steroids is 20, and the random number is 0.7873.

b) The percentage using steroids is 30, and the random number is 0.2955.

c) The percentage using steroids is 14, and the random number is 0.1433.

4. **Figure 8.5** shows two tables that display the results of the sample-pooling strategy in two schools. In each case, find the average number of tests per pooled sample, and use it to tell whether the school saved money by pooling pairs of samples.

Figure 8.5.
Testing results
in two schools.

SCHOOL 1

Number of tests	1	2	3
Number of pairs of samples	84	14	6

SCHOOL 2

Number of tests	1	2	3
Percentage of pairs of samples	28	31	41

5. The average number of tests per pooled sample is also known as the expected value. The term "expected value" can be misleading; it does not tell you the number of tests on any sample. For example, it isn't correct to say that you expect to do 2.3 tests on a given sample. In fact, you never do 2.3 tests on a given sample—you do either 1, 2, or 3 tests. The expected value is the average result over many trials.

Here is another example of expected value. Suppose that you are one of 10,000 people who pay $1 each for a raffle ticket. There is one grand prize of $1000, two second prizes of $300 each, and 50 consolation prizes of $20 each. The amounts of money you might win (or lose) and the probability you will win each amount can be organized in a table (see **Figure 8.6**).

Amount won or lost	$-1	$19	$299	$999
Probability				

Figure 8.6.
Amount won or lost in a raffle.

a) Explain the numbers in the "Amount won or lost" row.

b) Supply the missing probabilities.

c) Find the expected value. (Since probabilities are like decimal portions, you should use the same procedure you use to find the average number of tests.)

d) What does this expected value mean to the people who bought tickets?

6. Some carnivals feature a dice game in which you bet a dollar on number 1, 2, 3, 4, 5, or 6. Three dice are rolled, and you win $1 if your number comes up once, $2 if it comes up twice, or $3 if it comes up three times. If you win, you also get to keep the dollar you bet. Of course, if your number doesn't come up, you lose your dollar.

   a) Organize in a table the amounts you might win or lose. When three dice are rolled, the probabilities of a number coming up once, twice, or three times are approximately .347, .0694, and .00463.

   b) Find the expected value.

   c) What does the expected value mean to the players?

7. If a school decides to pool three tests when testing students for steroids, what result should they expect if they want to save money?

8. A tester is using a strategy of pooling three samples. If the pooled sample tests positive, the tester selects one of the individual samples to test and continues testing individual samples until no more tests are necessary. The results are shown in **Figure 8.7**.

Figure 8.7.
Test results when three samples are pooled.

Number of tests	1	2	3	4
Number of pooled samples	58	0	18	12

   a) Determine whether the tester saves money with this strategy.

   b) Explain why two tests are never needed.

   c) Is there a way the tester could change the strategy so that two tests would sometimes be needed?

LESSON TWO

Testing Models

KEY CONCEPTS

Simulation

Linear model

Residuals

Exponential model

Law of large numbers

Quadratic model

The Image Bank

PREPARATION READING

Modeling the Economics of Sample Pooling

Whether a tester can save money by pooling samples depends mostly on how common steroid use is in the group being tested. In other words, the economy of the sample-pooling strategy depends on the incidence of steroid use in the population being tested. To simplify the process of creating a mathematical model that will help determine when the sample-pooling strategy saves money, you will ignore other factors, such as the reliability of the test.

Imagine a situation in which all of the people being tested use steroids, and the tester doesn't know it. Every pooled sample results in three tests because every test, whether it is of a pooled sample or an individual sample, is positive. In contrast, imagine a situation in which no one is using steroids. A pooled sample never requires more than one test because the pooled sample always tests negative.

You can create a linear model to describe the relationship between average number of tests (expected value) and incidence by using only points (1, 3) and (0, 1), where the first member of the pair is the incidence of steroid use and the second member is the average number of tests. Whether you represent your model with a graph or with an equation, you can use the representation to conjecture that the average number of tests will be less than two if the incidence is below 0.5.

The linear model is based on only two points, so it must be tested carefully. A simulation shows that when the incidence of steroid use is 0.5, the average number of tests is more than two. Something appears to be wrong with the model. But what? Is it incorrect to assume that the relationship between average number of tests and incidence is linear?

In this lesson, you will take a closer look at linear models. To do so, you should know how to make scatter plots and perform linear regression on your calculator or computer. You should also recall how to use residual plots to evaluate a linear model. Go back to Unit 4, *Prediction*, and review these skills now, if necessary.

SIMULATIONS & LINEAR MODELS

3

In Lesson 1, you used simulations to test a linear model. Simulations can also produce data that can be used to create a model, which is the goal of this activity.

1. Use a simulation to determine the average number of tests (expected value) if the incidence of steroid use is 0.1, 0.2, 0.3, 0.4, 0.5, 0.6, 0.7, 0.8, and 0.9. Use 100 trials for each.

2. Organize the results in a table. Include incidence 0 and 1. You should have a total of 11 pairs when you are finished.

3. Prepare a scatter plot of the results. Explain why the scatter plot should show average number of tests versus incidence rather than incidence versus average number of tests. (Note: If you do the scatter plot on a graphing calculator and you do not use a statistical zoom feature, set the window a little larger than your data so that you do not have points on the very edge of the screen.)

4. Perform linear regression on your table and add the line to your scatter plot. Also add the line you used in Lesson 1—the one that connects the extreme points.

5. Discuss the scatter plot and the two linear models. Explain why you feel that each line is or isn't a good model.

INDIVIDUAL WORK 4

Evaluating Models and the Law of Large Numbers

S imulations provide mathematicians with data they can use to create and evaluate models. In this individual work, you will evaluate several models. You will consider why simulations that use a large number of trials are more reliable than simulations that use a relatively small number of trials.

1. The linear models you examined in Activity 3 are not satisfactory. An important tool in evaluating a model is a table or graph of residuals.

   a) If you have not made a table of residuals and a residual plot for both of the lines you considered in Activity 3, do so now.

   b) Write a short summary of what you learn from the residuals.

2. In Unit 6, *Wildlife*, you learned about the exponential function.

   a) Use a calculator to perform an exponential regression on your data. Prepare a scatter plot showing the data and the exponential model.

   b) What does the scatter plot you made in part (a) tell you about the exponential model?

   c) Use residuals to evaluate your exponential model.

3. The results of simulations like the ones you have done in this unit are more reliable if you use a large number of trials. The mathematical property that says you can be more confident of the result obtained with a large number of trials is the **law of large numbers**. For example, **Figure 8.8** shows the results of four simulations of a population in which half the people are using steroids ($p = 0.5$).

Figure 8.8.
Four simulations of the sample-pooling strategy with $p = 0.5$.

Number of trials	10	100	1000	10,000
Average number of tests	1.8	2.34	2.26	2.25

   a) Describe the change in the average number of tests as the number of trials increases.

   b) Based on these results, what do you think is a reasonable number of trials to use in future simulations of the testing game?

c) Based on these results, what is the best guess as to the average number of tests needed when $p = 0.5$?

d) What can you say about the average number of tests if $p < 0.5$?

4. Simulations in which $p = 0.5$ are sometimes done by flipping a coin. The law of large numbers says that the percentage of flips that show heads is more likely to be close to the theoretical 50% if the number of flips is large. In each of the following, tell which you think is more likely to happen.

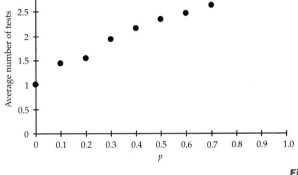

Figure 8.9.

a) Getting between 48% and 52% heads when you do 100 flips or when you do 1000.

b) Getting more than 60% heads when you do 10 flips or when you do 100.

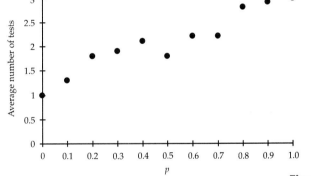

Figure 8.10.

5. The graphs of three different simulations like the one you did in Activity 3 are shown in **Figures 8.9–8.11**. The only difference in the simulations is in the number of trials done for each value of p.

Which do you think had the fewest trials? Which do you think had the most trials?

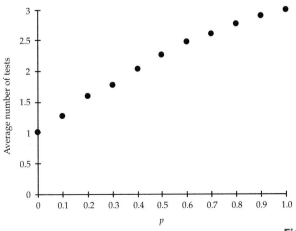

Figure 8.11.

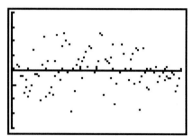

Figure 8.12.
[0,1] x [–0.8,0.8]

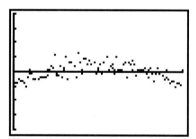

Figure 8.13.
[0,1] x [0,3.5]

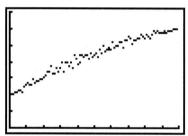

Figure 8.14.
[0,1] x [–0.8,0.8]

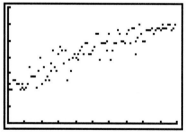

Figure 8.15.
[0,1] x [0,3.5]

6. A group of students prepared a project on linear models for the relationship between the average number of tests and steroid use. They revised *Dracula* so that it automatically performed a simulation for every value of *p* from 0.01 to 0.99 in steps of 0.01. First they ran the program with ten trials for each value of *p*. Then they set the program to do 100 trials for each value of *p* and went to lunch while it ran. When the simulations were finished, they prepared scatter plots, found linear regression models, and prepared residual plots. They transferred all the graphs from their calculator to their computer so that they could use the graphs in a report they were preparing for class. Unfortunately, they forgot to label which graph was which. Can you help them out?

   Tell whether each graph in **Figures 8.12–8.15** is a scatter plot or a residual plot for 10 trials or 100 trials.

7. Another group of students ran *Dracula* for 100 trials. The results are shown in **Figure 8.16.**

   a) Find the expected value.

   b) What is the meaning of expected value in sample-pooling situations?

   c) What value of *p* do you think the students used in their simulation? Explain your answer?

Number of tests	1	2	3
Number of pairs of samples	20	24	56

Figure 8.16.

ACTIVITY

A NEW MODEL

4

The linear and exponential functions you studied in previous units do not adequately model the relationship between the average number of tests and the incidence of steroid use. You need to learn about other kinds of models in order to answer the question "When can you expect to save money by pooling two steroid test samples?" In this activity, you will examine the quadratic function.

A **quadratic function** is characterized by a squared variable. The simplest quadratic function is $f(x) = x^2$. Other examples of quadratics are $f(x) = 3x^2 - 5$ and $f(x) = 5x^2 + 3x - 4$.

1. Most graphing calculators do a quadratic regression that finds the best quadratic model for a set of data. Use quadratic regression to find a quadratic model for your data in Activity 3. Evaluate the model and summarize your findings.

2. As you know, the law of large numbers says you can be more confident of results obtained from a large number of trials than you can of results obtained from a small number of trials. **Figure 8.17** is a table of simulation results from *Dracula* using 10,000 trials for each value of p.

p	0	0.1	0.2	0.3	0.4	0.5	0.6	0.7	0.8	0.9	1
Average number of tests	1	1.3	1.56	1.83	2.02	2.25	2.44	2.61	2.77	2.88	3

Figure 8.17.

Decide whether a linear, quadratic, or exponential model best fits these data. Explain your answer, and state the model you think is best.

3. a) Graph by itself the quadratic model you found in Item 2. Try various viewing windows until you have a good idea of its shape.

   b) The shape is called a **parabola**. Describe this parabola.

   c) Only a portion of the graph you made in part (a) makes sense in the context of steroid testing. Mark the part of your graph that makes sense.

4. Use a graphing calculator to experiment with graphs of other quadratics. How are the shapes different or similar?

Time (hrs)	Rate (liters/hr.)
0.5	32
1.5	27
2.5	24
3.5	22

Figure 8.18.

Years	E
12	12
17	25
20	39
29	92
36	109
45	222
55	547
60	755
65	1055
70	1531
80	2286
87	2455

Figure 8.19.

Number of drops	Volume (ml)
20	0.92
30	1.38
40	1.85
20	0.88
7	0.30
14	0.60
21	0.90
5	0.20
35	1.52
55	2.48

Figure 8.20.

INDIVIDUAL WORK 5

Deciding Which Model Fits Best

Linear, exponential, and quadratic functions provide the mathematician with a powerful tool kit with which to tackle a variety of real-world problems. The following questions reflect just a few of the many situations that can be modeled with one of these three functions.

In each of these questions, apply your knowledge of regression models, residual plots, and linear, exponential, and quadratic functions to find the best model. Explain your decisions.

1. **Figure 8.18** describes oil leaking out of the bottom of a container.

2. **Figure 8.19** describes electricity consumption in the United States since 1900. Electricity (E) is measured in billions of kilowatt hours.

3. **Figure 8.20** describes the volume of water (in drops) from a Beral™ pipette, which is used in chemistry experiments and is similar to a medicine dropper.

4. **Figure 8.21** describes the distance a dropped stone has fallen over time.

5. As you know from your study of linear functions in the previous units of this textbook, linear functions have a constant rate of change (or slope). Exponential functions and quadratic functions have a nonconstant rate of change, which causes the curvature in their graphs. Go back to Items 1–4, and explain why you think the rate of change in each situation is or is not constant.

Time (sec.)	Distance (m)
0	0
1	4.7
2	19.8
3	45.1
4	80.3
5	124.9

Figure 8.21.

6. In this activity and in other units, you have worked with three important kinds of mathematical functions: linear, exponential, and quadratic. Each type has a characteristic equation. These equations are summarized in **Figure 8.22**.

Type of Function	Equation
Linear	$y = ax + b$
Exponential	$y = a(b^x)$
Quadratic	$y = ax^2 + bx + c$

Figure 8.22.

Each function also has a characteristic graph. For **Figures 8.23–8.28**, state the type of function that you think produced the graph and write as much as you can about its equation.

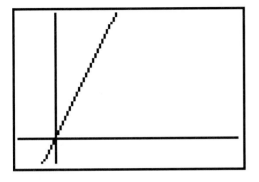

Figure 8.23.

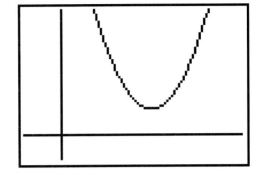

Figure 8.24.

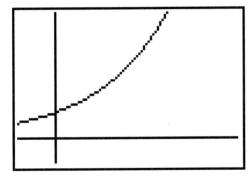

Figure 8.25.

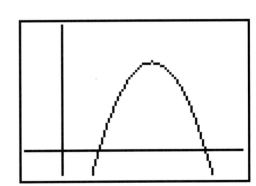

Figure 8.26.

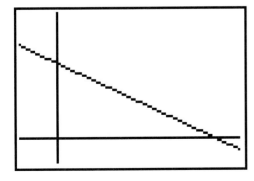

Figure 8.27.

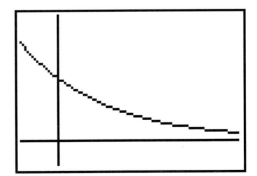

Figure 8.28.

LESSON THREE

Confirming the Model

KEY CONCEPTS

Expected value

Probability

Independent event

Area model for
expected value

Equivalent
expressions

PREPARATION READING

Confirm the Model

To determine whether you can expect to save money by pooling two steroid test samples, you must first determine the relationship between the expected number of tests and the incidence of steroid use. You know that only one test is needed if steroid use is nonexistent and three tests are needed if everyone tested is using steroids. However, these two pieces of information are not enough to determine the relationship between the expected number of tests and the incidence of steroid use.

When you do not have enough information to create a reliable mathematical model, a simulation can be helpful. Calculators or computers are useful simulators; they can perform the large number of trials needed for reliability. A simulation of the sample-pooling strategy shows that the relationship between the expected number of tests and the incidence of steroid use is not linear.

Two possible models for data that have a curved scatter plot are exponential and quadratic functions. However, an exponential regression model's curve differs from the curve of the simulation data that you have. Moreover, the exponential model's residual plot displays a clear pattern. The quadratic model's curve matches the curve of the data, and its residual plot displays no obvious pattern. The quadratic function seems to be a better model than either the exponential or the linear.

Before using a model to decide when it is reasonable to pool samples, you need to be as confident as possible of your model. Is it possible to confirm the quadratic model that you obtained from simulation data? Sometimes a model can be confirmed by examining the situation from a different viewpoint. In this lesson, you will use your knowledge of probability from Unit 7, *Imperfect Testing*, to examine the sample-pooling strategy and the quadratic model.

VISUALIZING EXPECTED VALUE

As you have seen in other units, alternative ways of representing a situation can be very useful. This activity uses area and probability to shed some light on the relationship between the expected number of steroid tests and the incidence of steroid use.

Often the tester has a rough idea of the incidence of steroid use from records of previous tests in similar situations. For example, the tester might know that approximately 17% of the people being tested are likely to use steroids. Although the tester knows that about 17% use steroids, the tester doesn't know if a person about to be tested will test positive or negative. To the tester, the probability this person will test positive is about .17.

Suppose the tester is fairly confident that approximately 20% of the people are steroid users.

1. What is the probability a person selected at random will test positive?

2. What is the probability a person selected at random will test negative?

   Again suppose that approximately 20% of the people are steroid users. In Unit 7, *Imperfect Testing,* you used multiplication to calculate the probability that both of two events occur. Suppose that the tester tests two people in a row.

3. What is the probability that the first person tests positive and the second person tests positive?

4. What is the probability that the first person tests negative and the second person tests negative?

5. What is the probability that the first person tests positive and the second person tests negative?

6. Besides the outcomes described in Items 3–5, what else can happen? Give probabilities for all answers.

7. What outcome is most likely in this case?

ACTIVITY

5

VISUALIZING EXPECTED VALUE

8. In Unit 7, you also learned that you can multiply probabilities if the two events are independent. Why might the tests of two people not be independent?

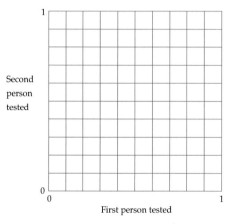

Figure 8.29.
A 10 x 10 square with scales.

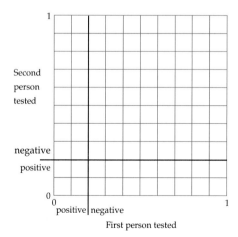

Figure 8.30.
An area model for test pooling when 20% use steroids.

The probabilities you calculated in Items 3–6 are theoretical. To show their relationship to the simulations you did in Lesson 2, make an area model for them.

To make an area model, start with a piece of graph paper. On the paper draw a 10 x 10 square.

Starting at the lower left corner of the square, scale the left and bottom edges from 0 to 1 so that each square represents a tenth. Label the bottom edge to represent the first person tested when a pooled sample is positive. Label the left edge to represent the second person tested. (See **Figure 8.29.**) (Note: There is nothing special about starting at the lower-left corner. Some people prefer to start at the upper-left corner.)

Now suppose that 20% of the people are using steroids and that the tester is pooling two samples. To show the split between positive and negative tests, draw a horizontal line at .2 and a vertical line at .2. (See **Figure 8.30.**)

VISUALIZING EXPECTED VALUE

9. How do the four rectangles created by the heavy lines represent the probabilities you found in Items 3–6?

10. Make an area model for sample pooling when 30% of the people are using steroids. Find the probabilities and interpretations associated with each region in the model.

11. Suppose that samples from two people are pooled and the pooled sample is tested. For each of the four regions of the area model you made in Item 10, tell whether one, two, or three tests are required.

12. Use your results from Items 10 and 11 to build a table showing the probability you will do one, two, or three tests when 30% of the people are using steroids. Use your table to find the expected value of the number of tests required.

13. Do your results indicate that the tester can expect to save money by pooling samples when steroid use is at 30%? Explain your answer.

CONSIDER:

1. How should the expected value obtained from an area model compare with the expected value obtained from a simulation?

2. How can an area model be used to determine whether the relationship between the average number of tests and the incidence of steroid use is quadratic?

INDIVIDUAL WORK 6

Connecting Simulations and Probability Models

Simulations and area models provide two ways of thinking about the sample-pooling strategy. It is important to be able to interpret and compare the results obtained from both, which you will do in this individual work.

1. **Figure 8.31** shows the results of a simulation done with *Dracula*.

Figure 8.31.
The results of a *Dracula* simulation.

> **What is *p*? .25**
>
> **How many trials? 1000**
>
> **576 182 242**
>
> **The average is 1.666**

a) Use an area model to represent the situation for which this simulation was done. Be sure to include the probabilities associated with each region and to interpret each of them.

b) Find the expected value and compare it to the simulation's average number of tests.

2. Based on your work thus far in this unit, make a guess at the incidence of steroid use for which the tester breaks even. Construct an area model and use it to test your guess.

3. **Figure 8.32** is an area model for the sample-pooling strategy. Estimate the incidence of steroid use that it represents and find the expected value.

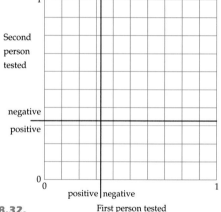

Figure 8.32.

4.a) Could **Figure 8.33** be an area model for a sample-pooling situation? Explain your answer.

b) If your answer to part (a) is no, can you think of a testing situation that Figure 8.33 could represent?

c) Use Figure 8.33 to find the expected value of the number of tests. Does the expected value indicate that the tester can expect to save money?

Figure 8.33.

5. In this unit, one of your modeling assumptions is that the tests are always correct. In Unit 7, *Imperfect Testing*, you studied situations in which tests are sometimes wrong. Suppose that a test given to a single person is correct 90% of the time and that the probability a person uses steroids is .2.

   a) Adapt the area model to this situation.

      Hint: One side of the square can represent incidence; an adjacent side can represent test reliability.

   b) What does your model tell you about the probability that someone who tests positive is not a steroid user?

   c) How is an area model different when two events are not independent?

6. A game at a county fair has a baseball diamond made from a board that is 5 feet by 5 feet. (See **Figure 8.34**.) You throw a dart at the board. If your dart lands in either the left-field bleachers or the right-field bleachers, you win a quarter. If your dart lands in both the left-field bleachers and right-field bleachers—in other words, in the center-field bleachers—you win 50¢. If your dart lands anywhere else, you lose a quarter.

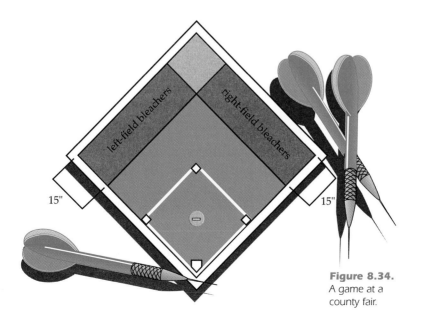

Figure 8.34.
A game at a county fair.

As a modeling assumption, consider the skill of the player irrelevant. The player throws a dart at the board.

a) What is the probability the dart lands in the left-field bleachers?

b) What is the probability the dart lands in the left-field bleachers and the right-field bleachers?

c) What is the probability the dart lands in the left-field bleachers only?

d) What is the probability the dart lands in either the left-field or the right-field bleachers (including the center-field bleachers)?

e) Find the expected value for the amount the player wins.

f) What does the expected value mean to the player?

7. The incidence of steroid use and the probability that an individual steroid test is positive are not the same thing. Explain why they are considered the same in this unit.

8. a) An area model can be used to find the product of two sums like $(4 + 5)$ and $(2 + 3)$ by doing four separate multiplications. Show how this can be done.

b) Use your answer to part (a) to show how, in general, to multiply two sums.

   Hint: Use $a + b$ and $c + d$ for the two sums.

9. It may have occurred to you that area models are used for a purpose very similar to the purpose served by tree diagrams in Unit 7, *Imperfect Testing*. If so, you are right! Area models are a more visual way to draw the same conclusions. Go back to Item 2 and use a tree diagram to analyze the guess you made.

ACTIVITY

GENERALIZING THE AREA MODEL

6

An area model is good for determining the average number of tests without requiring time-consuming simulations. If you experiment with various values for the incidence of steroid use, you can get a fairly good idea of when money can be saved by pooling two samples.

In this activity you will confirm the quadratic function as a model for the relationship between the expected number of tests and the incidence of steroid use. To accomplish this goal, you will generalize your area model.

The method of generalization, which you used in Unit 7, *Imperfect Testing,* is the key to confirming the quadratic model. Because the method of generalization requires one or more variables, the symbolic expressions that result are sometimes messy. Concentrate on writing the expressions and don't worry too much about how to manipulate them—that will come later.

Apply the method of generalization to the area model.

a) Use a variable to represent the probability that a person tests positive for steroid use.

b) Use the same variable to write an expression for the probability that a person tests negative.

c) Write algebraic expressions for the areas of each of the four regions in the area model.

d) Interpret the areas of each region and write the number of tests associated with each region.

ACTIVITY

6

GENERALIZING THE AREA MODEL

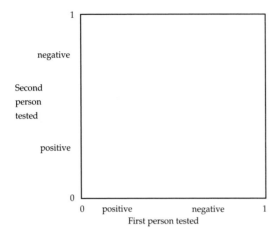

Figure 8.35.

1. Apply the first three steps to **Figure 8.35**. Since the probability of testing positive is now a variable, the division lines can be drawn anywhere. Pick a place that you think is reasonable.

2. The final step is to identify each region with a number of tests. Place each of the four expressions for a region's area in the proper column of the table in **Figure 8.36**. When two expressions belong in the same column, add them.

Number of tests	1	2	3
Probability			

Figure 8.36.

3. Use the table to write an algebraic expression for the expected value of the number of tests.

4. Is the expression you just wrote quadratic? If you're not sure, what do you need to know to be sure?

5. Go back to Lesson 2's Activity 4 and look up the quadratic model you found for the 10,000-trial data in Item 2. Compare it to the expression you wrote in Item 3 of this activity.

6. If you are still not sure that the relationship between the average number of tests and the incidence of steroid use is quadratic, what do you need to know to be sure?

INDIVIDUAL WORK 7

Comparing Models

To confirm that the relationship between the average number of tests and the incidence of steroid use is quadratic, you must compare two algebraic expressions that look very different. In the questions that follow, you will learn several ways to compare two algebraic expressions and apply these methods to sample pooling.

Here are three ways to compare two expressions. You should try them all.

TEST VALUES:

> Pick a few reasonable values of the variable and evaluate the expressions to see if they produce the same results.

GRAPHS/TABLES:

> Use a computer or calculator to create two tables or graphs and to compare them.

SYMBOLIC METHODS:

> Use symbol manipulation to simplify the expressions and compare the results.

To see how these methods work, try them with two relatively simple expressions, $x(x + 2)$ and $x^2 + 2x$.

1. If $x(x + 2)$ and $x^2 + 2x$ are equivalent expressions, any value you use for x gives identical results in both expressions.

   a) Find the value of each expression when $x = 3$.

   b) If one value of x produces the same value for both expressions, that is a good indication that the expressions are equivalent. But you should try more than one value for x; sometimes you get the same result from two expressions that are not equivalent. For example, x^2 and $2x$ are not equivalent. If you use 3 for x, one of them is 9, the other is 6. But if you pick just the right number for x, you get the same result. What number?

   c) Try a few numbers other than 3 in the expressions $x(x + 2)$ and $x^2 + 2x$ and record the results.

2. Graphs and tables let you see many values simultaneously.

   a) Use a calculator or computer to graph $y = x(x + 2)$ and $y = x^2 + 2x$ at the same time.

b) Use a calculator or computer to make a table for $y = x(x + 2)$ and $y = x^2 + 2x$.

c) How do these methods show that the two are equivalent? Which method do you prefer?

3. Some people are not comfortable concluding that two graphs are the same if only one graph can be seen. They worry they might be seeing only one graph because the other is completely off the screen and that the two expressions are not identical. Here are two easy ways to remove this uncertainty.

Make a new function by subtracting the two you are comparing. For example, if you have the two functions stored as $Y1$ and $Y2$ in your calculator, you can make a new function $Y3 = Y1 - Y2$.

Make a new function by dividing the two you are comparing.

a) Try both methods on the graphs of $y = x(x + 2)$ and $y = x^2 + 2x$. What do you notice?

b) Try both methods by adding new columns to a table containing $y = x(x + 2)$ and $y = x^2 + 2x$. What do you notice?

Symbolic methods have an advantage over testing values and graphs/tables. They guarantee that expressions produce the same result for all values of x (or other variables). The disadvantage to symbolic methods is that the symbol manipulation required to establish the equivalence of two expressions can be complicated.

4. The distributive property, which is a symbolic method, guarantees that $x(x + 2)$ and $x^2 + 2x$ are equivalent. People sometimes apply the distributive property incorrectly and conclude that $x(x + 2) = x^2 + 2$. Write a convincing argument to show that this is incorrect.

When you use symbol manipulation to simplify and compare expressions, there are two points worth remembering.

If there are more than one set of parentheses, start from the innermost set. For example, to simplify $3(x(x + 2))$, first multiply x and $x + 2$ to get $3(x^2 + 2x)$, then distribute the 3 to get $3x^2 + 6x$.

To multiply $(x + 2)$ and $(x + 3)$, multiply each part of the first by each part of the second. That is, multiply x and $x + 3$, multiply 2 and $x + 3$, and then add the results: $x(x + 3) + 2(x + 3) = x^2 + 3x + 2x + 6$ $= x^2 + 5x + 6$.

If you find it difficult to keep track of all the parts of this multiplication, you might find an area model helpful (**Figure 8.37**).

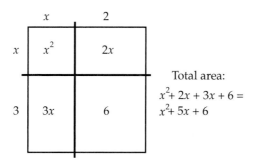

Total area:
$x^2 + 2x + 3x + 6 =$
$x^2 + 5x + 6$

Figure 8.37.
An area model for multiplying $x + 2$ and $x + 3$.

5. Try simplifying each of these expressions.

    a) $x(3(2x + 3))$

    b) $(x + 5)(2x + 1)$

    c) $(x - 1)(x + 3)$

    d) $2(x + 2)(x + 3)$

    e) $3(x(x+2) + x^2)$

6. Apply the techniques you have learned for comparing models to the expressions in Items 3 and 5 of Activity 6. Tell whether you think the expressions are equivalent and explain your reasoning. (Keep in mind that the expression derived from simulation data may not perfectly match the expression obtained from the area model.)

LESSON FOUR

Solving the Model: Tables and Graphs

KEY CONCEPTS

Solving quadratic equations

Parabola

Vertex of a parabola

Transforming graphs

Domain of a function

General form of a quadratic

Vertex form of a quadratic

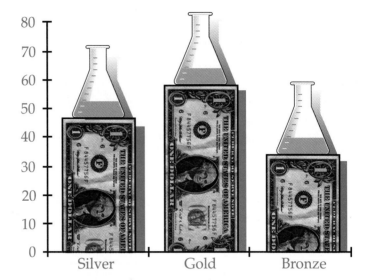

PREPARATION READING

Methods for Solving the Model

This unit begins with the question "When can you expect to save money by pooling two steroid test samples?" To answer this, you must understand the relationship between the average number of tests the tester can expect to perform and the incidence of steroid use. Data from simulations indicate that a quadratic model describes this relationship well, and a probability analysis confirms the quadratic model.

Once you have a mathematical model for a situation, you can use the model to answer questions about that situation. In this lesson, you will use the quadratic model to determine when a tester can expect to save money by pooling two steroid test samples.

Using mathematical models requires caution because models are not identical to the situations they represent. Thus a mathematical model sometimes produces answers that do not translate into reasonable statements about the situation it models. In this lesson, you will take a closer look at quadratic models and the information that they can give.

WHEN POOLING SAMPLES WILL SAVE MONEY

Simulations and area models agree. If p is the incidence of steroid use, then the model for the average number of tests you can expect to perform when you pool two samples is $1 + 3p - p^2$. Since you save money if you average fewer than two tests per pooled sample, you need to know when $1 + 3p - p^2$ is 2. In other words, you need to solve the equation $1 + 3p - p^2 = 2$.

In this activity, you will consider several ways to solve this quadratic equation.

1. One way to solve this equation is to graph $y = 1 + 3x - x^2$ on a calculator and trace until y is 2. If you can't trace to the exact point you want, you can use the zoom features of your calculator to obtain a reasonably accurate answer.

   a) Use the graph and zoom features of your calculator to solve the equation $1 + 3x - x^2 = 2$ for an answer that you feel is reasonable and accurate to the nearest 0.01. Begin by graphing $y = 1 + 3x - x^2$ and tracing as close as you can to the point with y-coordinate 2.

   b) Some people feel that the zoom-and-trace approach is a little easier if they add a horizontal line at 2. Add a second function that is equal to 2 to your graph and repeat what you did in part (a). Is this method more helpful?

   c) Some people prefer to adapt the procedure in part (b) slightly. They make a third function that is equal to the first minus the second. On many calculators, you can make $Y3 = Y1 - Y2$. Try this method. How is the new graph related to the original? Where is the point that represents the solution to the equation $1 + 3x - x^2 = 2$? Explain.

   d) People who prefer this method feel that they have an easier time setting a graphing window. Why do they?

2. The graph of a quadratic function is a parabola. Every parabola has either a high point or a low point, which is called its **vertex**. When you made your graphs in Item 1, you probably could see the vertex

ACTIVITY

WHEN POOLING SAMPLES WILL SAVE MONEY

7

in some windows, but not in others. That's because a window that gives you good precision when you solve an equation often doesn't let you see much of the parabola.

a) Show the graph of $y = 1 + 3x - x^2$ in two different windows. One should let you see the vertex. The other should show only those parts of the parabola that make sense in the sample-pooling situation. Use graphs you made in Item 1 if you think they are acceptable.

b) Recall that the domain of a mathematical function is the collection of all permissible replacements for x. In a quadratic function like $f(x) = 1 + 3x - x^2$, all numbers are in the mathematical domain because squaring and the other operations used in the function are valid for all numbers. In the situations that functions model, however, the domain is usually much more limited because fewer numbers make sense. What is the domain in the sample-pooling situation? Explain.

3. You can find the solution to the equation $1 + 3x - x^2 = 2$ by using a spreadsheet table in much the same way you use a graph. You can zoom on the table by picking a narrower range of values. Make a table with three columns: one for x, a second for $1 + 3x - x^2$, and third in which you have subtracted 2 from the second column.

4. a) Write a summary of what your quadratic model told you about saving money when pooling two steroid test samples.

b) The answer you just gave can be written with an inequality symbol if you use a variable for incidence. Use p for incidence and rewrite your answer to part (a).

CONSIDER:

1. Are solutions obtained from a mathematical model always reasonable in the situation you are modeling?

2. How is the sample-pooling model different from the sample-pooling situation?

3. How can you change graphs and tables to solve equations more easily?

THE UPS AND DOWNS OF GRAPHICAL METHODS

As you have seen, using graphs to obtain information from mathematical models is a useful skill. It is sometimes helpful to move a graph down (or up) to make it easier to locate important points. Other ways to change a graph are useful in applying mathematics to real-world problems. Some changes to a graph are called transformations. You studied several types of transformations in previous units.

In this activity, you will consider the three most important types of transformations for quadratic functions and the connection between their symbolic and graphical forms. As is usually the case, it is easiest to start simple. The simplest quadratic of all is $y = x^2$.

Here are the symbolic forms of the three types of transformations.

a) Add (or subtract) a constant to the function. For example, add 2 to x^2 in $y = x^2$ to get $y = x^2 + 2$.

b) Add (or subtract) a constant to the variable x. For example, add 2 to x in $y = x^2$ to get $y = (x + 2)^2$. Another way to describe this transformation is to say that you are replacing x with $x + 2$ or that you are substituting $x + 2$ for x.

c) Multiply the function by a constant. For example, multiply x^2 by 2 to get $y = 2x^2$. Another way to describe this transformation is to say that you are changing the coefficient of x^2.

Keep in mind that the constant in each case can be positive or negative and that it does not have to be a whole number.

ACTIVITY

THE UPS AND DOWNS OF GRAPHICAL METHODS

8

1. Experiment with transformation (a). Try different values of the constant to see how the constant controls the graph. Then write a description; be as quantitative as possible.

2. Experiment with transformation (b). Try different values of the constant to see how the constant controls the graph. Then write a description; be quantitative.

3. Experiment with transformation (c). Try different values of the constant to see how the constant controls the graph. Then write a description; be quantitative.

4. It is possible to combine two or three types of transformations. Here are a few examples. In each case, try to predict the effect on the graph of $y = x^2$. Then graph $y = x^2$ and the new function to see if your prediction is correct.

   a) $y = 2x^2 - 3$

   b) $y = (x - 2)^2 + 1$

   c) $y = 0.5(x + 3)^2$

   d) $y = 0.7(x - 1)^2 + 2$

5. Summarize what you have learned about the characteristics of the graph that are controlled by the numbers in the three locations shown in **Figure 8.38**.

Figure 8.38.
Important control numbers in a quadratic function.

$$f(x) = \boxed{}\,(x + \boxed{}\,)^2 + \boxed{}$$

INDIVIDUAL WORK 8

Information from Models

*A*s you have seen, creating mathematical models and obtaining information from them are important skills. You will use these skills answering the questions below.

1. The table in **Figure 8.39** contains information about automobile stopping distance for various speeds. (Stopping distance is the distance a car travels when braking from the given speed to 0 miles per hour.)

   a) Should a scatter plot of these data show speed versus stopping distance or stopping distance versus speed? Why?

   b) Create a mathematical model for the relationship between speed and stopping distance. Explain why you think your model is good.

   c) A police officer at an accident scene estimates that a car required 300 feet to stop. To determine how fast the car was traveling, you need to solve a quadratic equation. Write the equation and use one of the methods you learned in this lesson to determine the car's speed. Explain the method you used.

   d) Compare the domain of your model to the domain of the situation it represents.

Speed (mph)	Stopping distance (ft.)
20	42
30	74
40	116
50	173
60	248
70	343

Figure 8.39.
Automobile speed and stopping distance
(Source: U. S. Bureau of Roads).

2. In Item 1(b), you might have chosen a graphical approach that uses a parabola. The most basic of all parabolas is found by graphing $y = x^2$, which can be transformed in many ways.

   a) Explain the transformations of $y = x^2$ that produce the graph of $y = 0.6(x - 3)^2 - 4$. Do not graph the parabolas yet. Base your answers only on your knowledge of the symbolic form of the transformations.

   b) Graph both parabolas and confirm that your answer to part (a) is correct.

   c) Every parabola has a vertex, which in this case is a low point. Trace and, if necessary, zoom to find the vertex of the transformed parabola. Explain how the coordinates of the vertex are related to the equation.

3. The vertex of a parabola is sometimes important in modeling real-world situations. Here is a situation that will help you understand why. Manufacturers know that the demand for a product varies with its price. They are interested in charging a price that will produce the highest revenues for their business.

Figure 8.40 shows the records of price charged for a product and the resulting weekly revenue (rounded to the nearest $100) from one manufacturer.

Priced charged ($)	Revenue ($)
1.70	11,200
2.00	11,900
2.30	12,400
2.60	12,500
3.00	12,100
3.20	11,500

Figure 8.40.
Product price and weekly revenue.

a) Should a scatter plot of these data show price versus revenue or revenue versus price? Why?

b) Explain why a quadratic is a good model for the relationship between price charged and revenue.

c) The manufacturer wants to know the price that will produce maximum revenue. Trace and zoom to recommend a price to the manufacturer and to predict the revenue the price will produce.

d) How is the domain of your model different from the domain of the situation it represents?

4. You have seen that the equations of quadratic models come in at least two forms. One form is the kind produced by your calculator when you do quadratic regression. An example is $y = 0.5x^2 + 2x - 12$. This is called the **general form**. Another form is based on transformations of the simplest quadratic, $y = x^2$. An example is $y = 2(x + 3)^2 + 4$. This is often called the **vertex form**.

a) Why do you think the second form is called the vertex form?

b) You can use a graph to change the general form to the vertex form. Graph $y = 0.5x^2 + 2x - 12$ and use your graph to explain how to write the vertex form. (Be sure to check your answer by graphing it.)

You can use the distributive property to change the vertex form to the general form. As an example, consider $y = 2(x + 3)^2 + 4$.

First, you must multiply $(x + 3)$ by itself:

$(x + 3)(x + 3) = x(x + 3) + 3(x + 3) = x^2 + 3x + 3x + 9 = x^2 + 6x + 9.$

$x^2 + 6x + 9$ can be used in place of $(x + 3)^2$ in the original expression:

$2(x + 3)^2 + 4 = 2(x^2 + 6x + 9) + 4 = 2x^2 + 12x + 18 + 4 = 2x^2 + 12x + 22.$

c) Try it yourself. Take your answer from part (b) and change it to general form. Then compare it to $y = 0.5x^2 + 2x - 12$.

To square an expression like $(x + 3)$, apply the steps in Item 4. You could also use an area model to help you understand the steps. Draw a square and divide each side into two parts to represent $(x + 3)$. Extend the division lines across the square, and write expressions for the area of each of the four regions. The expression for the total area is the square of $(x + 3)$ (**Figure 8.41**).

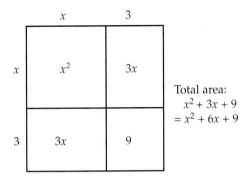

Figure 8.41.
An area model for $(x + 3)^2$.

5. Here are several squares for you to try. Use either the symbolic procedure from Item 4 or the area model. The choice is yours.

   a) $(x + 2)^2$

   b) $(x - 2)^2$

   c) $(x + 5)^2$

   d) $(2x + 1)^2$

6. In Unit 6, *Wildlife*, you used parametric equations to shift graphs vertically and horizontally.

   a) Explain how $y = x^2$ can be graphed with parametric equations.

   b) Use parametric equations to move the graph 2 units right and 3 units down.

   c) What is the vertex form of the equation of the transformed graph?

   d) Use substitution to explain how the vertex form can be obtained from the parametric form.

TARGETS

An interesting way to improve your skills with transformations of graphs of parabolas is with a calculator graphing game called *Target*, or one of its computer counterparts. Two computer graphing games are *Green Globs* and *Algebra Arcade*.

Play one of these games with quadratic functions only. Write all your functions in vertex form. Use a sheet of graph paper to keep a record of the location of the targets, the functions you used, and the targets you hit with each function. Record your score for each game. How high a score can you achieve?

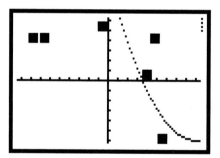

Target challenges you to hit as many targets as you can with the graph of a mathematical function. You receive a score based on the number of targets you hit.

Guess the Curve

T he calculator program *Quad* draws a parabola on the screen and asks you to guess its vertex and the number that multiplies the squared term. *Quad* uses *A* to represent the multiplier and (*H*, *K*) to represent the vertex. The program graphs your guess and the original parabola and tells you if you guessed correctly.

Run *Quad* ten times. Each time, record the approximate location of the graph and each of your guesses.

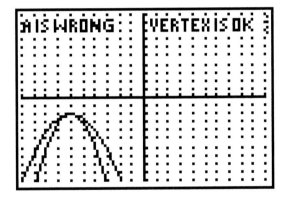

Quad challenges you to write the equation of a parabola drawn by the calculator. When you enter your guess, the calculator draws your parabola and tells you which part of your equation is correct.

LESSON FIVE

Solving the Model: Symbolic Methods

KEY CONCEPTS

Square root

Completing the square

Quadratic formula

The Image Bank

PREPARATION READING

Solving the Model by Symbolic Methods

Determining whether it is possible to save money by pooling two steroid test samples requires solving the quadratic equation $1 + 3p - p^2 = 2$. When you use tables or graphs to solve this equation, the solution appears to be about .38. That is, when the incidence of steroid use is under .38 ($p < .38$), pooling two steroid test samples is a money-saving strategy.

As you have seen, the domain of a quadratic model is usually larger than the domain of the situation it represents. For example, because the variable p represents the incidence of steroid use, the only values of p that make sense in the steroid-testing context are those between 0 and 1 (including 0 and 1). The quadratic equation $1 + 3p - p^2 = 2$ has two solutions, but only one of them is between 0 and 1. Thus only one of the solutions of $1 + 3p - p^2 = 2$ makes sense in the sample-pooling situation.

Although tables and graphs work well for solving quadratic equations, you could also solve such equations symbolically. In this lesson, you will reconsider the equation $1 + 3p - p^2 = 2$ and other quadratic equations in order to understand how they can be solved by symbolic means. You will use a number of things you've already learned, including arrow diagrams, area models, and the vertex form of the quadratic function.

ACTIVITY

SOLVING QUADRATICS THAT ARE IN VERTEX FORM

10

The easiest type of quadratic equation to solve involves the simplest quadratic expression, x^2. After that, the easiest to solve are in vertex form. The process is based on **square roots** and on the arrow diagrams you used in Unit 2, *Secret Codes and the Power of Algebra*. In this activity, you will see how to solve quadratics in vertex form.

Finding the square root of a number is the opposite of squaring. For example, since $5^2 = 25$, the square root of 25 is 5. But it is also true that $(-5)^2 = 25$, so 25 really has two square roots, 5 and –5. The main or principal square root is 5, because positive numbers are used more frequently than negative numbers in real-world situations. In fact, a few centuries ago negative numbers were considered useless.

The fact that both 5 and –5 have the same square is easily seen in the graph of $y = x^2$ (**Figure 8.42**). The horizontal line at 25 crosses the parabola at points with x-coordinates –5 and 5.

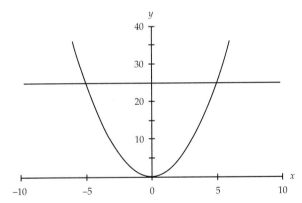

Figure 8.42.
The graphs of
$y = x^2$ and $y = 25$.

Your calculator has a square-root key, labeled √. It gives only 5 when you type √25. That's because √ means "principal square root," which is important to remember when using a calculator to find a square root.

You can use an arrow diagram to represent a quadratic equation like $x^2 = 4$ (**Figure 8.43**).

Thus the solutions of the quadratic equation $x^2 = 4$ are 2 and –2.

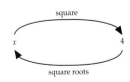

Figure 8.43.
An arrow diagram for $x^2 = 4$.

1. Try it yourself. Draw arrow diagrams and find the solutions for each quadratic equation.

   a) $x^2 = 9$

   b) $x^2 = 25$

ACTIVITY

SOLVING QUADRATICS THAT ARE IN VERTEX FORM

10

2. It is important to realize that a quadratic equation like $x^2 = -4$ has no solution because no number squared is negative. Use a graph to explain that $x^2 = -4$ has no solution.

The arrow diagram can be used to solve any equation involving a quadratic in vertex form. Suppose you want to solve $2(x + 1)^2 - 2 = 5$. The arrow diagram is shown in **Figure 8.44**.

Figure 8.44.
An arrow diagram for
$2(x + 1)^2 - 2 = 5$.

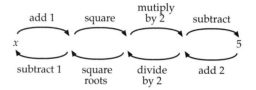

To solve $2(x + 1)^2 - 2 = 5$, work from 5 to the left.

First add 2: $5 + 2 = 7$.
Divide by 2: $7/2 = 3.5$.
Write the square roots of 3.5: $\sqrt{3.5}$ and $-\sqrt{3.5}$
Then subtract 1 from each square root: $\sqrt{3.5} - 1$, $-\sqrt{3.5} - 1$.

If necessary, you can find a decimal approximation for $\sqrt{3.5} - 1$ and $-\sqrt{3.5} - 1$ on your calculator. You can check these answers by comparing the decimal approximations to the solutions you obtain from a graph. For example, $\sqrt{3.5} - 1$ is about 0.87. The graph in **Figure 8.45** shows that $2(x + 1)^2 - 2 = 5$ has a solution near 0.87, which you can estimate more accurately by zooming.

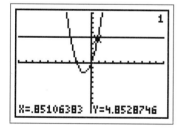

[–10, 10] x [–10, 10]

Figure 8.45.
A solution of the equation
$2(x + 1)^2 - 2 = 5$.

3. Try it yourself. Use an arrow diagram to solve each equation. Find decimal approximations for each of your solutions and check them by graphing.

a) $(x + 3)^2 = 9$

b) $2(x - 3)^2 + 1 = 7$

c) Will one of the solutions of a quadratic equation always be negative? Explain.

INDIVIDUAL WORK 10

Vertex Form and the Testing Problem

*I*n the items that follow, you will practice solving quadratic equations in vertex form and then apply your knowledge to sample pooling.

1. Symbolic methods require practice, so here are a few quadratic equations to solve. Use an arrow diagram to solve them. Find decimal approximations, and use a graph to check each solution.

   a) $x^2 = 64$

   b) $x^2 = -7$

   c) $(x + 4)^2 = 16$

   d) $(x - 5)^2 = 7$

   e) $(x + 3)^2 - 5 = 1$

   f) $(x - 1)^2 + 3 = 2$

   g) $-3(x - 1)^2 + 7 = 1$

2. In Item 4 of Individual Work 8 you used a graph to put a quadratic in vertex form. As you have seen in this lesson, you can solve equations of quadratics in vertex form.

   a) The quadratic that represents the number of tests you can expect to perform when you pool two steroid test samples is $1 + 3x - x^2$, where x is the incidence of steroid use. Use a graph to put this quadratic in vertex form.

   b) Use the vertex form to determine when the average number of tests is two. Use an arrow diagram to solve the equation.

   c) How do your solutions compare to previous solutions to the sample-pooling problem?

3. **Figure 8.46** shows the table you made in Activity 6. **Figure 8.47** is a graph of hypothetical test results from Item 1 of Lesson 1's Individual Work 2. According to the graph, 80 out of 120, or 67%, of the paired samples required one test. According to the table, the portion of paired samples that require one test is $(1 - p)^2$.

Figure 8.46.
Results from the
generalized area model.

Number of tests	1	2	3
Probability	$(1 - p)^2$	$p(1 - p)$	$p^2 + p(1 - p)$

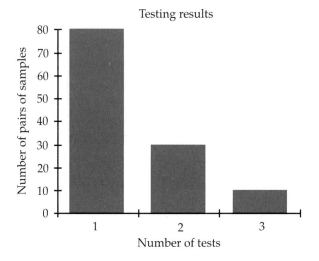

Figure 8.47.
Hypothetical testing results.

a) Use an arrow diagram to solve the quadratic equation
$(1 - p)^2 = 0.67$.

b) Do both solutions make sense in the testing situation?

c) Go back to your table and use your solution for p to evaluate the
expressions for the probability of two tests and of three tests.

d) Are the probabilities you found for two tests and three tests
consistent with the bar graph? If not, what might account for the
difference?

e) Do you think it is possible for the number of paired samples that
require one test, the number of paired samples that require two
tests, and the number of paired samples that require three tests to
be the same? Explain.

4. In general, a quadratic equation can be written in vertex form as
$a(x - h)^2 + k = 0$. What are the solutions of this equation?

ACTIVITY

FINDING VERTEX FORM

11

In this activity, you will find the vertex form symbolically. To do so, you need to think about symbolic expressions that are perfect squares.

1. Vertex form always has a part that is a perfect square. For example, in $2(x + 3)^2 - 5$, $(x + 3)^2$ is the perfect square. Consider a few quadratic functions that are perfect squares. Look at each graph and describe what they have in common.

 a) $y = x^2$

 b) $y = (x + 2)^2$

 c) $y = (x - 3.5)^2$

2. Look at the graphs of the functions below. Explain how far they need to move in order to be like the graphs in Item 1. (Tracing and zooming can help you make an accurate estimate.)

 a) $y = x^2 + 2x$

 b) $y = x^2 + 6x$

 c) $y = x^2 - 5x$

3. It's possible to get the answer to Item 2 from the equation without making a graph. Can you find the connection?

4. Another way to look at perfect-square expressions is area models. For example, **Figure 8.48** is an area model for the perfect-square expression $(x + 3)^2$.

	x	1	1	1
x	x^2	x	x	x
1	x	1	1	1
1	x	1	1	1
1	x	1	1	1

Figure 8.48.
An area model for $(x + 3)^2$.

ACTIVITY

11

FINDING VERTEX FORM

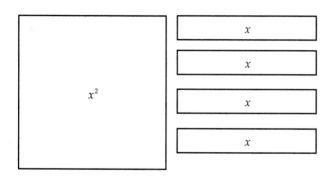

Figure 8.49.
Area pieces for $x^2 + 4x$.

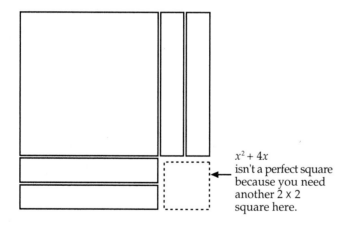

$x^2 + 4x$ isn't a perfect square because you need another 2 x 2 square here.

Figure 8.50.
Making a perfect square out of $x^2 + 4x$.

The expression $x^2 + 4x$ is not a perfect square. Use an area model to determine what this expression needs to make it a perfect square. You need a square to represent x^2 and four 1-by-x rectangles (**Figure 8.49**).

You need to make a square out of them, without using any additional x strips (**Figure 8.50**).

Thus $x^2 + 4x$ is a perfect square if you add 2 x 2, or 4, to it: $x^2 + 4x + 4$. Since each side of the square is $x + 2$, $x^2 + 4x + 4$ is the same as $(x + 2)^2$.

Use an area model to explain how to make each of these a perfect square.

a) $x^2 + 2x$

b) $x^2 + 6x$

c) $x^2 - 5x$

5. If you had trouble answering Item 3, reconsider it. How can you tell what will make each expression in Item 4 a perfect square without using a graph or an area model?

6. Summarize what you have learned about making expressions like $x^2 + 3x$ into perfect squares.

INDIVIDUAL WORK 11

Solving Quadratic Equations by Making Perfect Squares

$\mathbf{Q}$uadratic equations are solved symbolically by a process called **completing the square,** so named because the key step involves making an expression into a perfect square.

To use completing the square to solve quadratic equations, you need to know a rule and a trick that mathematicians use. The rule is that you must always do the same thing to both sides of an equation. The trick is to move any parts of an expression that are in the way to the other side of the equation.

1. Here's an example of how to use completing the square to solve the quadratic equation $x^2 + 6x - 2 = 5$.

   Since 2 doesn't make $x^2 + 6x$ a perfect square, it is in the way. Move it to the other side: $x^2 + 6x = 7$.

   Add 9 to both sides to make the left side a perfect square: $x^2 + 6x + 9 = 16$.

   Write the left side as a perfect square: $(x + 3)^2 = 16$. You can use an arrow diagram to solve this equation.

   Find the solutions. Check them by using a graph to solve the original equation $x^2 + 6x - 2 = 5$.

2. The process is slightly more complicated for equations like $2x^2 + 8x - 3 = 11$. In this case, you have to do two things before you can complete the square.

   Add 3 to both sides: $2x^2 + 8x = 14$.

   You want x^2 instead of $2x^2$, so divide by 2 to get $x^2 + 4x = 7$.

   You should be able to finish completing the square for $x^2 + 4x = 7$. Do so.

3. Solve the following quadratic equations by completing the square.

   a) $x^2 + 4x = 12$

   b) $x^2 - 6x = 0$

   c) $-24 = 2x - x^2$

   d) $x^2 + 8x - 11 = 0$

   e) $5x^2 - 10x - 30 = 0$

4. Solve the equation $1 + 3p - p^2 = 2$ by completing the square. Compare your solutions to earlier answers you got for when you can expect to save money by pooling two steroid test samples.

Some people prefer to use a formula rather than go through the process of completing the square. The **quadratic formula** works only if the quadratic is in the general form $ax^2 + bx + c = 0$. The formula says that the solutions are

$$\frac{-b + \sqrt{b^2 - 4ac}}{2a} \text{ and } \frac{-b - \sqrt{b^2 - 4ac}}{2a}.$$

Rather than write the formula twice, people usually write it once with the symbol $\pm$ to indicate that there are two solutions, one obtained by adding, the other by subtracting.

Remember, you must put the quadratic equation in general form before you can use the formula. For example, if you want to solve $2x^2 - 3x = x + 3$, move x and 3 to the left side by subtracting them, to get the equation $2x^2 - 4x - 3 = 0$.

Next put 2, –4, and –3 into the formula.

$$\frac{-(-4) \pm \sqrt{(-4)^2 - 4(2)(-3)}}{2(2)} = \frac{4 \pm \sqrt{16 + 24}}{4} = \frac{4 \pm \sqrt{40}}{4}$$

The solutions are

$$\frac{4 + \sqrt{40}}{4} \text{ and } \frac{4 - \sqrt{40}}{4}.$$

5. Use the quadratic formula to solve these quadratic equations.

a) $3x^2 + 10x + 7 = 0$

b) $2x^2 + 5x - 9 = 0$

c) $2x^2 + 3x = 7 - x^2$

Whether you solve your quadratic equations by completing the square or by the quadratic formula, you should use a graph to check your answers. Both methods have several steps, so it is easy to make mistakes.

6. You have seen that some quadratic equations do not have a solution. What happens when you apply the quadratic formula to a quadratic equation that has no solution?

7. Show how to derive the quadratic formula by applying completing the square to the general quadratic equation, $ax^2 + bx + c = 0$.

Wrapping Up Unit Eight

© 1973   King Features Syndicate

1. **Figure 8.51** is a photograph of a golf ball thrown upward. The photo was taken in a dark room by holding a camera lens open and illuminating the ball with a strobe light. The table in **Figure 8.52** shows the height of the center of the ball at the first flash of the strobe light, the second flash, and so forth.

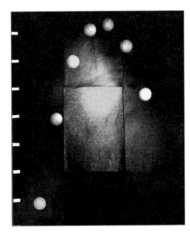

Flash number	Height (cm)
1	19.0
2	48.7
3	69.5
4	80.7
5	82.4
6	74.8
7	57.5

Figure 8.51.
A golf ball thrown upward.

Figure 8.52.
The height of the ball at each flash of the strobe light.

UNIT SUMMARY

a) Explain why a quadratic function is a good model for the relationship between the height of the ball and the time it has traveled.

b) You can find the time at which the ball is at height 0 by solving a quadratic equation. Write the correct equation and solve it.

c) Use your answer to part (b) and your quadratic model to determine the maximum height of the golf ball.

2. Use an area model to explain how to square $2x + 1$.

3. The proprietor of a pizza shop keeps track of the cost of the ingredients needed to make different sizes of deluxe pizzas, as shown in the table in **Figure 8.53**.

Diameter (in.)	Cost ($)
8	1.10
10	1.70
12	2.50
14	3.40

Figure 8.53.
The cost of pizza ingredients.

   a) Show that a quadratic is a good model for the relationship between cost of ingredients and pizza diameter.

   b) You can solve a quadratic equation to determine the approximate size of a pizza that has $2.00 worth of ingredients. Show how to do this.

   c) What does your model tell you about the cost of ingredients on a 15-inch pizza?

4. Solve each quadratic equation by any method.

   a) $x^2 + 3x = 1$

   b) $2x^2 - 4x + 7 = 1$

   c) $x^2 - 6x - 4 = 0$

   d) $3x^2 - 7 = 0$

5. Pierre and Shana are members of a team playing the graphing calculator game *Target*. They have three targets left, as shown in **Figure 8.54**. They are trying to find a quadratic that has a good chance of hitting all three. Help them out.

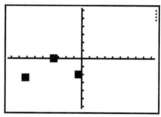

Figure 8.54.

6. In late November 1996, a major retail chain advertised the prices in **Figure 8.55** for various sizes of one brand of color television. Note: The size of a television set is measured by the length of the diagonal of the picture screen.

Size of Set (in.)	Price ($)
13	120.00
19	150.00
25	250.00
27	290.00

Figure 8.55.
Television prices.

   a) Find a mathematical model for the relationship between the price and the size of the set.

   b) If $200 is the most you can spend on a television, what is the largest set of this brand you could expect to buy? Explain how you got your answer.

   c) What does your model say the price of a 5-inch television set should be? Do you think this is accurate?

7. A steroid tester has collected approximately the same number of samples from two groups of people. He expects the incidence of steroid use to be 8% in one group and 18% in the other. He intends to pool two samples when conducting the test. He is considering two options. One is to pool pairs of samples from each group separately. The other is to pool one sample from one group with one sample from the other group. Which is the best strategy?

8. Exposure to ultraviolet radiation kills yeast. The table in **Figure 8.56** gives results of experiments done at Kansas State University.

   Is a quadratic a good model for the relationship between the average number of yeast cells remaining and the exposure time?

Exposure time (sec.)	Average number of yeast colonies
5	117
10	80
15	32.5
20	13.3
25	4.7

Figure 8.56. Ultraviolet radiation exposure of yeast.

9. In Unit 2, *Secret Codes and the Power of Algebra*, you used symbolic expressions to model number tricks. Here is another number trick.

   • Pick a number between 1 and 10.

   • Double it.

   • Subtract the result from 20.

   • Multiply by your original number.

   a) Use a symbolic expression to model this number trick.

   b) The magician does the trick for someone who gets a result of 42. Help the magician find the number with which this person started.

   c) This time the magician does the trick for someone who gets a result of 60. Help the magician find the number with which this person started.

   d) Is this a good number trick for the magician to use? Explain.

Mathematical Summary

Medical tests such as steroid tests are expensive. Testers who test large groups of people must find ways to control the costs. One way that they attempt to save money is by pooling two or more samples.

The simplest sample-pooling strategy involves pooling two samples. The central question that guides this unit is "When can you expect to save money by pooling two steroid test samples?"

Pooling two samples can save a lot of money if very few people use steroids. In fact, if there are no steroid users in the group tested, each pooled sample requires only one test. If there are a lot of steroid users in the group tested, however, a pooled sample requires three tests, which is one more than needed if everyone is tested individually.

Since two individual tests costs twice as much as a single pooled test, pooling two samples saves money if the number of tests performed on a sample is less than two. Thus, to save money in the long run, the average number of tests per pooled sample must be under two.

In considering when you can expect to save money by pooling two steroid test samples, you must first identify the factors to vary and observe. The incidence of steroid use and the average number of tests performed are variables, and the relationship between these two variables must be understood. For example, how does the average number of tests change when the incidence of steroid use changes?

Understanding the relationship between the number of tests you expect for each pooled sample and the incidence of steroid use requires data. Without much data, information about the number of tests needed when everyone uses steroids and when no one uses steroids can be used to create a linear model for the relationship. The linear model predicts that you can expect to save money if the incidence of steroid use is below 50%.

Simulations by calculator or computer can provide plenty of data. An analysis of these data shows that the relationship between the number of tests you expect for each pooled sample and the incidence of steroid use is not linear. Comparison of linear, exponential, and quadratic models shows that the quadratic is the best model.

To determine the number of tests you can expect to perform, you calculate the average number of tests per pooled sample in the simulation data. This average is the expected value. If a large number of trials are simulated, the resulting average number of trials should match the theoretical expected value fairly well.

A probability analysis of the sample-pooling strategy produces the quadratic model $1 + 3p - p^2$ for the average number of tests, where p represents the incidence of steroid use. This model matches the quadratic model obtained from simulation data very well. The probability analysis that produces the theoretical quadratic model uses area diagrams to show how symbolic expressions are multiplied.

To answer the question "When can you expect to save money by pooling two steroid test samples?" you must solve the quadratic equation $1 + 3p - p^2 = 2$. There are several ways to solve quadratic equations, including tabular and graphical methods that allow you to zoom on the table or graph to increase precision. Two symbolic methods for solving quadratic equations are completing the square and the quadratic formula.

If you use graphing calculators or computer graphing utilities to solve quadratic equations, you can get precise, although approximate, answers. One way to use a graphing utility to solve the quadratic equation $1 + 3p - p^2 = 2$ is to graph $y = 1 + 3x - x^2$ and $y = 2$ and trace to the point of intersection. An alternative strategy is subtracting these two functions and graphing the resulting function, $y = 1 + 3x - x^2 - 2$. The solution is on the x-axis and therefore easier to locate on a graphing utility screen.

Subtracting 2 from a quadratic lowers its graph 2 units. Moving a graph vertically is an example of a transformation of the graph. Other transformations include horizontal movement and flattening.

Transforming a quadratic by manipulating its equation is easiest when the equation is in vertex form. For example, $y = 2(x - 3)^2 + 4$ is in vertex form. The vertex of the quadratic is at $(3, 4)$, which can be read from the equation. The constants h and k in the general vertex form $y = a(x - h)^2 + k$ control the location of the vertex. The constant a controls the "flatness" of the parabola. The sign (+ or –) of a controls whether the parabola opens upward or downward.

For a specific value of y, you find exact solutions by using the appropriate quadratic formula. There are formulas for the vertex form and the general form. ($y = 1 + 3x - x^2$ is an example of the general form of the quadratic.) The solutions can be checked by converting them to approximations and comparing the approximations to answers you get from a graphing calculator or a spreadsheet.

When the quadratic equation $1 + 3p - p^2 = 2$ is solved, the conclusion is that pooling two samples saves money if the incidence of steroid use is below 38%. In other words, if p is the incidence of steroid use, you can expect to save money by pooling two steroid test samples if $p < 0.38$.

Glossary

COMPLETING THE SQUARE:
The process of changing the general form of a quadratic to vertex form. Completing the square is used to solve quadratic equations.

EXPECTED VALUE:
The average result over many trials. Expected value is calculated by multiplying the numerical value associated with each outcome by the portion of the time it occurs, then adding the results of all the multiplications. The meaning of expected value varies. In the case of sample pooling, it is the average number of tests per pooled sample.

GENERAL FORM OF A QUADRATIC:
$ax^2 + bx + c$.

LAW OF LARGE NUMBERS:
The average result is more likely to be close to the theoretical result if the number of trials is large than if the number of trials is small. For example, theoretically, a flipped coin should show heads 50% of the time. If you flip a coin many times, the percentage of heads is more likely to be closer to 50% if you do 100 flips than if you do only 10 flips.

PARABOLA:
The shape of a graph of a quadratic function.

QUADRATIC FORMULA:
A formula used to solve quadratic equations of the form $ax^2 + bx + c = 0$. The quadratic formula is

$$\frac{-b + \sqrt{b^2 - 4ac}}{2a}$$

$$\text{and } \frac{-b - \sqrt{b^2 - 4ac}}{2a},$$

$$\text{or } \frac{-b \pm \sqrt{b^2 - 4ac}}{2a}.$$

For quadratics of the form $a(x - h)^2 + k = 0$, another type of quadratic formula gives the solutions $h \pm \sqrt{\frac{-k}{a}}$.

QUADRATIC FUNCTION:
A function of the form $y = ax^2 + bx + c$. The highest power of x in a quadratic is 2.

SAMPLE POOLING:
Combining two or more test samples and testing the combined sample. If a positive test results, additional tests must be performed on one or more of the individual samples.

SQUARE ROOT:
A factor of a number that, when squared, gives the original number. For example, 5 and –5 are square roots of 25 because $(-5)^2 = 25$ and $5^2 = 25$.

√ (SQUARE ROOT SYMBOL):
The principal square root of a number. The principal square root of a positive number is the positive square root.

VERTEX FORM OF A QUADRATIC:
$a(x - h)^2 + k$. The point (h, k) is the vertex of the graph of the quadratic.

VERTEX OF A PARABOLA:
The point at which a parabola changes from sloping upward to sloping downward, or vice versa.

Index

Glossary terms and the pages on which they are defined in the text appear in boldface.

Acknowledgements

PROJECT LEADERSHIP:

Solomon Garfunkel
COMAP, INC., LEXINGTON, MA

Landy Godbold,
THE WESTMINSTER SCHOOLS, ATLANTA, GA

Henry Pollak
TEACHERS COLLEGE, COLUMBIA UNIVERSITY, NY

EDITOR:

Landy Godbold

AUTHORS:

Allan Bellman
WATKINS MILL HIGH SCHOOL, GAITHERSBURG, MD

John Burnette
KINKAID SCHOOL, HOUSTON, TX

Horace Butler
GREENVILLE HIGH SCHOOL, GREENVILLE, SC

Claudia Carter
MISSISSIPPI SCHOOL FOR MATH AND SCIENCE,
COLUMBUS, MS

Nancy Crisler
PATTONVILLE SCHOOL DISTRICT, ST. ANN, MO

Marsha Davis
EASTERN CONNECTICUT STATE UNIVERSITY,
WILLIMANTIC, CT

Gary Froelich
SECONDARY SCHOOL PROJECTS MANAGER, COMAP, INC.,
LEXINGTON, MA

Landy Godbold
THE WESTMINSTER SCHOOLS, ATLANTA, GA

Bruce Grip
ETIWANDA HIGH SCHOOL, ETIWANDA, CA

Rick Jennings
EISENHOWER HIGH SCHOOL, YAKIMA, WA

Paul Kehle
INDIANA UNIVERSITY, BLOOMINGTON, IN

Darien Lauten
OYSTER RIVER HIGH SCHOOL, DURHAM, NH

Sheila McGrail
CHARLOTTE COUNTRY DAY SCHOOL, CHARLOTTE, NC

Geraldine Oliveto
THOMAS JEFFERSON HIGH SCHOOL FOR
SCIENCE AND TECHNOLOGY, ALEXANDRIA, VA

Henry Pollak
TEACHERS COLLEGE, COLUMBIA UNIVERSITY, NY

J.J. Price
PURDUE UNIVERSITY, WEST LAFAYETTE, IN

Joan Reinthaler
SIDWELL FRIENDS SCHOOL, WASHINGTON, D.C.

James Swift
ALBERNI SCHOOL DISTRICT, BRITISH COLUMBIA, CANADA

Brandon Thacker
BOUNTIFUL HIGH SCHOOL, BOUNTIFUL, UT

Paul Thomas
MINDQ, FORMERLY OF THOMAS JEFFERSON HIGH SCHOOL
FOR SCIENCE AND TECHNOLOGY, ALEXANDRIA, VA

REVIEWERS:

Dédé de Haan, Jan de Lange,
Henk van der Kooij
FREUDENTHAL INSTITUTE, THE NETHERLANDS

David Moore
PURDUE UNIVERSITY, WEST LAFAYETTE, IN

Henry Pollak
TEACHERS COLLEGE, COLUMBIA UNIVERSITY, NY

ASSESSMENT:

Dédé de Haan, Jan de Lange,
Kees Lagerwaard, Anton Roodhardt,
Henk van der Kooij
THE FREUDENTHAL INSTITUTE, THE NETHERLANDS

REVISION TEAM

Marsha Davis, Gary Froelich,
Landy Godbold, Bruce Grip

EVALUATION:

Barbara Flagg
MULTIMEDIA RESEARCH, BELLPORT, NY

TEACHER TRAINING:

Allan Bellman, Claudia Carter,
Nancy Crisler, Beatriz D'Ambrosio,
Rick Jennings, Paul Kehle,
Geraldine Oliveto, Paul Thomas

FIELD TEST SCHOOLS AND TEACHERS:

Clear Brook High School,
Friendswood, TX
JEAN FRANKIE, TOM HYLE, LEE YEAGER

Clear Creek Middle School, Gresham, OR
DAVID DROM, JOHN MCPARTIN, NICOLE RIGELMAN

Damascus Middle School, Boring, OR
MARIAH MCCARTY, CLAUDIA MURRAY

Dexter McCarty Middle School,
Gresham, OR
CONNIE RICE

Dr. James Hogan Senior High School,
Vallejo, CA
GEORGIA APPLEGATE, PAM HUTCHISON, JERRY LEGE,
TOM LEWIS

Foxborough High School,
Foxborough, MA
BERT ANDERSON, SUE CARLE, MAUREEN DOLAN,
JOHN MARINO, MARY PARKER, DAVE WALKINS,
LEN YUTKINS

Frontier Regional High School,
South Deerfield, MA
LINDA DIDGE, DON GORDEN, PATRICIA TAYLOR

Gordon Russell Middle School,
Gresham, OR
MARGARET HEYDEN, TIFFANI JEFFERIS, KEITH KEARSLEY

Gresham Union High School,
Gresham, OR
DAVE DUBOIS, KAY FRANCIS, ERIN HALL,
THERESA HUBBARD, RICK JIMISON, GAYLE MEIER,
CRAIG OLSEN

Jefferson High School, Portland, OR
STEVE BECK, DAVE DAMCKE, LYNN INGRAHAM,
MARTHA LANSDOWNE, JOHN OPPEDISANO, LISA WILSON

Lincoln School, Providence, RI
JOAN COUNTRYMAN

Mills E. Godwin High School,
Richmond, VA
KEVIN O'BRYANT, ANN W. SEBRELL

New School of Northern Virginia,
Fairfax, VA
JOHN BUZZARD, VICKIE HAVELAND,
BARBARA HERR, LISA TEDORA

Northside High School, Fort Wayne, IN
ROBERT LOVELL, EUGENE MERKLE

Ossining High School, Ossining, NY
JOSEPH DICARLUCCI

Pattonville High School,
Maryland Heights, MO
SUZANNE GITTEMEIER, ANN PERRY

Price Laboratory School, Cedar Falls, IA
DENNIS KETTNER, JIM MALTAS

Rex Putnam High School,
Milwaukie, OR
JEREMY SHIBLEY, KATHY WALSH

Sam Barlow High School, Gresham, OR
BRAD GARRETT, KATHY GRAVES, COY ZIMMERMAN

Simon Gratz High School,
Philadelphia, PA
LINDA ANDERSON, ANNE BOURGEOIS, WILLIAM ELLERBEE

Ursuline Academy, Dallas, TX
SUSAN BAUER, FRANCINE FLAUTT, DEBBIE JOHNSTON,
MARGARET KIDD, ELAINE MEYER, MARGARET NOULLET,
MARY PAWLOWICZ, SHARON PIGHETTI, PATTY WALLACE,
KATHY WARD

West Orient Middle School, Gresham, OR
DAN MCELHANEY

COMAP STAFF

Solomon Garfunkel, Laurie Aragon,
Sheila Sconiers, Gary Froelich, Roland
Cheyney, Roger Slade, George Ward,
Michele Barry, Emily Sacca, Daiva Kiliulis,
David Barber, Gail Wessell,
Gary Feldman, Annette Moccia, Clarice
Callahan, Brenda McDonald,
George Jones, Rafael Aragon,
Peter Bousquet

PHOTOGRAPHY/ART

The Image Bank
BOSTON, MA

Corbis
BELLEVUE, WA

Space Imaging EOSAT - Intermountain
Digital Imaging
SALT LAKE CITY, UT

King Features Syndicate
NEW YORK, NY
TUMBLEWEEDS REPRINTED WITH PERMISSION

E.C. Publications, Inc.
NEW YORK, NY
ALFRED E. NEUMAN AND MAD ARE TRADEMARKS OF E.C.
PUBLICATIONS, INC.©1960. ALL RIGHTS RESERVED. USED
WITH PERMISSION.

COVER ART

The Image Bank
BOSTON, MA

INDEX EDITOR

Seth Maislin
FOCUS PUBLISHING SERVICES, WATERTOWN, MA

References

UNIT 1:

Webb, N. M. and S. H. Farivar, "Are Your Students Prepared for Group Work?" *Middle School Journal* 25 (1994): 3, 51–54. Columbus, OH: National Middle School Association. Used by permission of the publisher.

UNIT 2:

Adams, Susan, "The Code Breakers," *Forbes* (February 26, 1996): 100.

Froelich, Gary, and Joseph Malkevitch. 1993. *Loads of Codes* (HiMAP Module 22). Lexington, MA: The Consortium for Mathematics and Its Applications (COMAP), Inc.

Janeczko, Paul B. 1984. *Loads of Codes and Secret Ciphers*. New York: Macmillan.

Friedman, William F. and Lambros D. Callimahos. 1956. *Military Cryptanalytics*. Washington, D. C.: National Security Agency.

UNIT 3:

Crabb, Charlene. January, 1993. "Frankincense," *Discover* 14 (1), 56. Charlene Crabb/© 1993. Reprinted with permission of *Discover* Magazine.

Monastersky, Richard. July 22, 1989. "Spotting erosion from space," *SCIENCE NEWS* 136 (4), 61. Reprinted with permission from *SCIENCE NEWS*, the weekly newsmagazine of science, copyright 1989 by Science Service.

————— March 17, 1990. "New picture of California plate puzzle," *SCIENCE NEWS* 137 (11), 175. Reprinted with permission from *SCIENCE NEWS*, the weekly newsmagazine of science, copyright 1990 by Science Service.

UNIT 4:

Anscombe, F. J., "Graphs in Statistical Analysis," *The American Statistician* 27 (1973): 17–21.

Behrman, Richard E. ed. 1992. *Textbook of Pediatrics*. Philadelphia, PA: W. B. Saunders Company.

Conroy, Glenn, et. al., 1992. "Obituary: Mildred Trotter, Ph.D. (Feb. 2, 1899–Aug. 23, 1991)," *American Journal of Physical Anthropology* 87:373–374.

Stanford, Craig, "Chimpanzee Hunting Behavior and Human Evolution," *American Scientist* (May/June 1995).

Trotter, Mildred, and Goldine Gleser. "Estimation of Stature from Long Bones of American Whites and Negroes," (1952) *American Journal of Physical Anthropology*, n.s., 10:463–514a.

UNIT 5:

Anselmo, Joseph, "Satellite Data Plays Key Role in Bosnia Peace Treaty," *Aviation Week and Space Technology* (December 11, 1995): 29.

Scott, Joan E. 1982. *Introduction to Interactive Computer Graphics*. New York: John Wiley and Sons, 126–128.

Solomon, Charles and Ron Stark. 1983. *The Complete Kodak Animation Book*, New York: Eastman Kodak Company.

Vince, John. 1992. *3-D Computer Animation*. Reading, MA: Addison-Wesley Publishing Company, Inc., vii–11.

Watt, Alan and Mark Watt. 1992. *Advanced Animation and Rendering Techniques, Theory and Practice*. Reading, MA: Addison-Wesley Publishing Company, Inc.

UNIT 6:

Biddle, Nina Archer and Spencer Reiss. "The Strep-A Scare," *Newsweek* 123:25 (June 20, 1994): 32–3.

Houston, J. "Rise of Mount Trashmore; Virginia Beach," *Parks and Recreation* 8 (January, 1973): 28–30.

Lemonick, Michael D. "Streptomania Hits Home," *Time* 143 (June 20, 1994): 54.

Nordland, Rod. "Deadly Lessons," *Newsweek* 119:10 (March 9, 1992): 25–8.

Rathje ,W. and C. Murphy. 1992. *Rubbish! The Archeology of Garbage*. New York: Harper-Collins Publishers, 34–35.

Stephens, Ronald D. "Gangs, Guns, and School Violence," *USA Today* 122 (January, 1994): 29–34.

Satchell, Michael. "A discouraging word for the buffalo," *U.S.News and World Report* 121:13 (September 30, 1996): 61–3.

UNIT 7:

Anderson, William A., Richard R. Albrecht, and Douglas B. McKeag. *A Second Replication of the National Study of the Substance Use and Abuse Habits of College Student-Athletes—Final Report*: Presented to National Collegiate Association, Overland Park, Kansas. (July 30, 1993).

Sloand, Elaine M., "HIV Testing: State of the Art," *Journal of the American Medical Association* 266 (1991): 2861–2866.

UNIT 8:

Insel, Arnold J. 1992. *Cassette Tapes: Predicting Recording Time* (UMAP Module 641). Lexington, MA: The Consortium for Mathematics and Its Applications (COMAP), Inc.

Wright, J. E., and V. 1990. Cowart. *Anabolic Steroids, Altered States*. Carmel, IN: Benchmark Press.